SIMULATING COPULAS

Stochastic Models,
Sampling Algorithms, and Applications

Series in Quantitative Finance

ISSN: 1756-1604

Published

Vol. 1 An Introduction to Computational Finance
by Ömür Uğur

Vol. 2 Advanced Asset Pricing Theory
by Chenghu Ma

Vol. 3 Option Pricing in Incomplete Markets:
Modeling Based on Geometric Lévy Processes and Minimal Entropy Martingale Measures
by Yoshio Miyahara

Vol. 4 Simulating Copulas:
Stochastic Models, Sampling Algorithms, and Applications
by Jan-Frederik Mai and Matthias Scherer

Series in Quantitative Finance – Vol. 4

SIMULATING COPULAS

Stochastic Models, Sampling Algorithms, and Applications

Jan-Frederik Mai

Assenagon Credit Management GmbH, Germany

Matthias Scherer

Technische Universität München, Germany

ICP Imperial College Press

Published by

Imperial College Press
57 Shelton Street
Covent Garden
London WC2H 9HE

Distributed by

World Scientific Publishing Co. Pte. Ltd.
5 Toh Tuck Link, Singapore 596224
USA office: 27 Warren Street, Suite 401-402, Hackensack, NJ 07601
UK office: 57 Shelton Street, Covent Garden, London WC2H 9HE

British Library Cataloguing-in-Publication Data
A catalogue record for this book is available from the British Library.

Series in Quantitative Finance — Vol. 4
SIMULATING COPULAS
Stochastic Models, Sampling Algorithms, and Applications

ISBN-13 978-1-84816-874-9
ISBN-10 1-84816-874-8

Printed in Singapore by World Scientific Printers.

To our amazing wives, Jasna and Leni.
And to Paul Jonah, who has just seen the light of day.

Preface

The joint treatment of $d \geq 2$ random variables requires vector-valued stochastic models. In the financial industry, multivariate models are applied to, e.g., asset allocation problems (portfolio optimization), the pricing of basket options, risk management, and the modeling of credit portfolios. In particular, the development during the past years highlighted that the financial industry is in urgent need of realistic and viable models in large dimensions. Other fields of application for multivariate stochastic models include geostatistics, hydrology, insurance mathematics, medicine, and reliability theory.

Besides specifying the univariate marginals, for multivariate distributions it is additionally required to appropriately define the dependence structure among the modeled objects. In most applications, a portfolio perspective is significantly more demanding compared to modeling univariate marginals. One consequence is that analytical solutions for the aforementioned applications can typically be derived under restrictive assumptions only. An increasingly popular alternative to accepting unrealistic simplifications is to solve the model in question by Monte Carlo simulation. This allows for very general models but requires efficient simulation schemes for multivariate distributions. This book aims at providing a toolbox for the simulation of random vectors with a considerable spectrum of dependence structures.

Why Sampling Copulas?

This book focuses on sampling copulas, i.e. distribution functions on $[0,1]^d$ with uniform univariate marginals. On the first view, this standardization to univariate margins seems to be a rather artificial assumption. The justification for considering copulas instead of more general multivariate distribution functions is provided by Sklar's seminal decomposition (see Sklar (1959) and Section 1.1.2). Heuristically speaking, Sklar's theorem allows us to decompose any d-dimensional multivariate distribution function F into its univariate margins $F_1, \ldots, F_d$ and the dependence structure among them. The latter is described by the copula behind the model, denoted C. More precisely, we have $F(x_1, \ldots, x_d) = C(F_1(x_1), \ldots, F_d(x_d))$ for $(x_1, \ldots, x_d) \in \mathbb{R}^d$. The converse implication also holds, i.e. coupling univariate margins with some copula yields a multivariate distribution. This observation is especially convenient for the specification of a multivariate model, since a separate treatment of the dependence structure and univariate margins is usually easier compared to specifying the multivariate distribution in one step.

Sklar's decomposition also applies to sampling applications. Assume that we want to simulate from a multivariate distribution function F with univariate marginal distribution functions $F_1, \ldots, F_d$ and copula C. Given a sampling scheme for the copula C, the following algorithm generates a sample from the distribution F by applying the generalized inverses $F_1^{-1}, \ldots, F_d^{-1}$ (see Lemma 1.4) to the sample of the copula.

Algorithm 0.1 (Sampling Multivariate Distributions)
Let $F(x_1, \ldots, x_d) = C(F_1(x_1), \ldots, F_d(x_d))$ be a d-dimensional distribution function. Let sample_C() be a function that returns a sample from C. Sampling F is then possible via the following scheme:

FUNCTION sample_F()

Set $(U_1, \ldots, U_d) :=$ *sample_C()*

RETURN $\left(F_1^{-1}(U_1), \ldots, F_d^{-1}(U_d)\right)$

Why Another Book on Copulas?

Our main motivation for writing this book was to summarize the fast-growing literature on simulation algorithms for copulas. Several results on new sampling techniques for classical copulas, e.g. the Archimedean and Marshall–Olkin families, have lately been published. Moreover, new families and construction principles have been discovered; an example is the pair-copula construction. At the same time, the financial industry has become aware that copula models (beyond a Gaussian dependence structure) are required to realistically model various aspects of quantitative finance. This book takes account of this fact by providing a comprehensive toolbox for financial engineering, and, of course, for other applications as well. All algorithms are described in pseudo-code. Thus, they can easily be implemented in the user's preferred programming language. Moreover, we aim at being comprehensive with respect to sampling schemes for univariate random variables as well as with respect to the use of Monte Carlo sampling engines in general. We purposely included sampling schemes for very basic copulas, even though this might not be required for an expert in the field. Another intention is to provide an elementary introduction to copulas from the perspective of probabilistic representations. Hence, an experienced researcher might skip some parts of the book. But someone who is new to the field of copulas can use the book as a stand-alone textbook. The book, however, does not treat statistical estimation of dependence models.

Especially for sampling applications, the dimension of the copula plays a crucial role. To give an example, the original probabilistic model behind the so-called d-dimensional Marshall–Olkin copula is based on $2^d - 1$ random variables, i.e. the dimension d enters exponentially. Hence, this book explicitly focuses on the d-dimensional case and discusses the efficiency of the provided algorithms with respect to their dimension. Especially in the field of portfolio credit risk modeling, there are some applications requiring high-dimensional models with $d = 125$ or even more.

Copulas can be investigated from two (not necessarily disjoint) perspectives: (1) analytically, i.e. viewing them as d-dimensional functions, and (2) probabilistically, i.e. viewing them as the dependence structure behind some random vector. Both perspectives have their distinct advantages.

(1) The analytical perspective aims at deriving statements about copulas from their functional form. This is especially successful for analytically tractable families. In this case, it is often possible

to deduce very general dependence properties from the parametric form of the respective copula. For high-dimensional sampling applications, however, this approach is of minor use, since the functional form does not, in general, provide a convenient rule for the construction of a sampling algorithm.

(2) Investigating copulas from a probabilistic perspective is based on stochastic representations of the dependence structure. This means that there is an explicit (and preferably simple) probability space, on which the random vector associated with a copula is defined. The immediate advantage is that such probabilistic constructions provide a recipe for a sampling algorithm. Besides sampling, it is also possible to investigate the copula from the respective representation.

This book pursues a mostly probabilistic treatment. This is especially suitable for sampling applications.

Acknowledgments

First of all, we would like to thank C. Czado, E. Korn, R. Korn, and J. Stöber for providing the chapters on pair copulas, univariate sampling algorithms, and Monte Carlo techniques. We also would like to thank several friends and colleagues for patiently answering questions, reading previous versions of the manuscript, and pointing us at several improvements. These are K. Bannör, G. Bernhart, F. Durante, C. Hering, P. Hieber, M. Hofert, H. Joe, D. Krause, A. Min, A. Reuß, D. Neykova, W. Richter, S. Schenk, D. Selch, and N. Shenkman. Finally, we would like to thank our editor, Tasha D'Cruz, for her extremely valuable feedback and professional handling of the manuscript.

Jan-Frederik Mai and Matthias Scherer

Contents

Chapter 1

Introduction

Before we start, let us clarify some notations.

General comment: The dimension of a random vector is typically denoted by $d \geq 2$.

Important sets: $\mathbb{N}$ denotes the set of natural numbers $\{1, 2, \ldots\}$, and $\mathbb{N}_0 := \{0\} \cup \mathbb{N}$. $\mathbb{R}$ denotes the set of real numbers. Moreover, for $d \in \mathbb{N}$, $\mathbb{R}^d$ denotes the set of all d-dimensional row vectors with entries in $\mathbb{R}$. For $\boldsymbol{v} := (v_1, \ldots, v_d) \in \mathbb{R}^d$, we denote by $\boldsymbol{v}'$ its transpose. For some set A, we denote by $\mathcal{B}(A)$ the corresponding Borel σ-algebra, which is generated by all open subsets of A. The cardinality of a set A is denoted by $|A|$. Subsets and proper subsets are denoted by $A \subset B$ and $A \subsetneq B$, respectively.

Probability spaces: A probability space is denoted by $(\Omega, \mathcal{F}, \mathbb{P})$, with σ-algebra $\mathcal{F}$ and probability measure $\mathbb{P}$. The corresponding expectation operator is denoted by $\mathbb{E}$. The variance, covariance, and correlation operators are written as Var, Cov, Corr, respectively. Random variables (or vectors) are mostly denoted by the letter X (respectively $\mathbf{X} := (X_1, \ldots, X_d)$). As an exception, we write $\mathbf{U} := (U_1, \ldots, U_d)$ for a d-dimensional random vector with a copula as joint distribution function.[1] If two random variables X_1, X_2 are equal in distribution, we write $X_1 \stackrel{d}{=} X_2$. Similarly, $\stackrel{d}{\to}$ denotes convergence in distribution. Elements of the space Ω, usually denoted by ω, are almost always omitted as arguments of random variables, i.e. instead of writing $X(\omega)$, we simply write X. Finally, the acronym i.i.d. stands for "**i**ndependent and **i**dentically **d**istributed".

Functions: Univariate as well as d-dimensional distribution functions are denoted by capital letters, mostly F or G. Their corresponding survival functions are denoted $\bar{F}, \bar{G}$. As an exception, a copula is denoted by the letter C; its arguments are denoted $(u_1, \ldots, u_d) \in [0, 1]^d$. The characteristic

[1] The letter U indicates that $U_1, \ldots, U_d$ are uniformly distributed on the unit interval.

function of a random variable X is denoted by $\phi_X(x) := \mathbb{E}[\exp(i\,x\,X)]$. The Laplace transform of a non-negative random variable X is denoted by $\varphi_X(x) := \mathbb{E}[\exp(-x\,X)]$. Moreover, the nth derivative of a real-valued function f is abbreviated as $f^{(n)}$; for the first derivative we also write f'. The natural logarithm is denoted log.

Stochastic processes: A stochastic process $X : \Omega \times [0, \infty) \to \mathbb{R}$ on a probability space $(\Omega, \mathcal{F}, \mathbb{P})$ is denoted by $X = \{X_t\}_{t \geq 0}$, i.e. we omit the argument $\omega \in \Omega$. The time argument t is written as a subindex, i.e. X_t instead of $X(t)$. This is in order to avoid confusion with deterministic functions f, whose arguments are written in brackets, i.e. $f(x)$.

Important univariate distributions: Some frequently used probability distributions are introduced here. Sampling univariate random variables is discussed in Chapter 6.

(1) $U[a, b]$ denotes the uniform distribution on $[a, b]$ for $-\infty < a < b < \infty$. Its density is given by $f(x) = \mathbb{1}_{\{x \in [a,b]\}}\,(b - a)^{-1}$ for $x \in \mathbb{R}$.

(2) $Exp(\lambda)$ denotes the exponential distribution with parameter $\lambda > 0$, i.e. with density $f(x) = \lambda\,\exp(-\lambda\,x)\,\mathbb{1}_{\{x>0\}}$ for $x \in \mathbb{R}$.

(3) $\mathcal{N}(\mu, \sigma^2)$ denotes the normal distribution with mean $\mu \in \mathbb{R}$ and variance $\sigma^2 > 0$. Its density is given by

$$f(x) = \frac{1}{\sqrt{2\pi\sigma^2}} e^{-\frac{1}{2}\frac{(x-\mu)^2}{\sigma^2}}, \quad x \in \mathbb{R}.$$

(4) $\mathcal{LN}(\mu, \sigma^2)$ denotes the lognormal distribution. Its density is given by

$$f(x) = \frac{1}{x\sqrt{2\pi\sigma^2}} e^{-\frac{1}{2}\frac{(\log(x)-\mu)^2}{\sigma^2}}\,\mathbb{1}_{\{x>0\}}, \quad x \in \mathbb{R}.$$

(5) $\Gamma(\beta, \eta)$ denotes the Gamma distribution with parameters $\beta, \eta > 0$, i.e. with density

$$f(x) = \frac{\eta^\beta}{\Gamma(\beta)} e^{-\eta\,x}\,x^{\beta-1}\,\mathbb{1}_{\{x>0\}}, \quad x \in \mathbb{R}.$$

Note in particular that the exponential law $Exp(\lambda) = \Gamma(1, \lambda)$ is a special case of the Gamma distribution.

(6) $\chi^2(\nu)$ denotes the χ^2-distribution (Chi-square) with $\nu \in \mathbb{N}$ degrees of freedom. The density satisfies

$$f(x) = \frac{1}{2^{\nu/2}\Gamma(\nu/2)} x^{\nu/2-1} e^{-x/2}\,\mathbb{1}_{\{x>0\}}, \quad x \in \mathbb{R}.$$

From the density we can easily see that $\chi^2(\nu) = \Gamma(\nu/2, 1/2)$ is a second important special case of the Gamma distribution. If Z has a $\chi^2(\nu)$-distribution and $X_1, \ldots, X_\nu$ are i.i.d. standard normally distributed random variables, then $Z \stackrel{d}{=} X_1^2 + \ldots + X_\nu^2$.

(7) $IG(\beta, \eta)$ denotes the inverse Gaussian distribution with parameters $\eta, \beta > 0$, i.e. with density

$$f(x) = \frac{\beta}{x^{\frac{3}{2}}\sqrt{2\pi}} e^{\eta\beta - \frac{1}{2}(\beta^2/x + \eta^2 x)} \mathbb{1}_{\{x>0\}}, \quad x \in \mathbb{R}.$$

(8) $\mathcal{S}(\alpha, h)$ denotes the (exponentially tilted) α-stable distribution with $0 < \alpha < 1$ and $h \geq 0$, as discussed for instance in Barndorff-Nielsen and Shephard (2001). This distribution is characterized via its Laplace transform. If X has a $\mathcal{S}(\alpha, h)$-distribution, then

$$\mathbb{E}\left[e^{-xX}\right] = e^{-((x+h)^\alpha - h^\alpha)}, \quad x > 0.$$

We write $\mathcal{S}(\alpha, 0) =: \mathcal{S}(\alpha)$. Note: $\mathcal{S}(1/2, h) = IG(2^{-1/2}, \sqrt{2}\,h)$.

(9) $Bin(n, p)$ denotes the binomial distribution with n trials and success probability $p \in (0, 1)$. The (discrete) density of X with $Bin(n, p)$-distribution is

$$\mathbb{P}(X = k) = \binom{n}{k} p^k (1-p)^{n-k}, \quad k \in \{0, \ldots, n\}.$$

(10) $Poi(\lambda)$ denotes the Poisson distribution with mean $\lambda > 0$. The (discrete) density of X with $Poi(\lambda)$-distribution is $\mathbb{P}(X = k) = \lambda^k \exp(-\lambda)/k!$ for $k \in \mathbb{N}_0$.

(11) $t(\nu)$ denotes the Student's t-distribution with $\nu \in \mathbb{N}$ degrees of freedom. The density[2] is given by

$$f(x) = \frac{\Gamma\left(\frac{\nu+1}{2}\right)}{\sqrt{\pi\nu}\,\Gamma\left(\frac{\nu}{2}\right)} \left(1 + \frac{x^2}{\nu}\right)^{-\left(\frac{\nu+1}{2}\right)}, \quad x \in \mathbb{R}.$$

(12) $t(\mu, \nu)$ denotes the (non-central) Student's t-distribution with non-centrality parameter $\mu \in \mathbb{R}$ and $\nu \in \mathbb{N}$ degrees of freedom. This distribution is composed of Z, a normally distributed random variable with unit variance and zero mean, and V, a Chi-square distributed random variable with ν degrees of freedom (independent of Z), via $(Z + \mu)/\sqrt{V/\nu}$.

(13) $Pareto(\alpha, x_0)$ denotes the Pareto distribution with parameters $\alpha, x_0 > 0$, i.e. with survival function

$$\bar{F}(x) = 1 - F(x) = \left(\frac{x_0}{x}\right)^\alpha \mathbb{1}_{\{x \geq x_0\}} + \mathbb{1}_{\{x < x_0\}}, \quad x \in \mathbb{R}.$$

[2] The motivation for this distribution is the composition of Z, a normally distributed random variable with unit variance and zero mean, and V, a Chi-square distributed random variable with ν degrees of freedom (independent of Z): $Z/\sqrt{V/\nu}$ has a $t(\nu)$-distribution.

(14) *Geo*(p) denotes the geometric distribution with success probability $p \in (0,1]$. The (discrete) density of X with *Geo*(p)-distribution is $\mathbb{P}(X = k) = (1-p)^{k-1}\,p$ for $k \in \mathbb{N}$.

(15) *Beta*(a,b) denotes the Beta distribution with parameters $a, b > 0$. Its density is given by

$$f(x) = \mathbb{1}_{\{x \in (0,1)\}} \frac{\Gamma(a+b)}{\Gamma(a)\,\Gamma(b)}\, x^{a-1}\,(1-x)^{b-1}.$$

Abbreviations for other distributions are introduced when they first appear. The symbol $\sim$ means "distributed according to", e.g. $E \sim Exp(1)$ means that E is an exponential random variable with unit mean.

1.1 Copulas

The law of a d-dimensional random vector $\boldsymbol{X} := (X_1, \ldots, X_d)$, defined on a probability space $(\Omega, \mathcal{F}, \mathbb{P})$, is usually studied from its *distribution function*

$$F(x_1, \ldots, x_d) := \mathbb{P}(X_1 \leq x_1, \ldots, X_d \leq x_d), \quad x_1, \ldots, x_d \in \mathbb{R}.$$

For $i = 1, \ldots, d$ the distribution function F_i of X_i is called the (univariate) *marginal law* or *margin* and can be retrieved from F via

$$F_i(x_i) := \mathbb{P}(X_i \leq x_i) = F(\infty, \ldots, \infty, x_i, \infty, \ldots, \infty), \quad x_i \in \mathbb{R}.$$

Taking ∞ as an argument of F is used as a shorthand notation for taking the limits as the arguments $x_1, \ldots, x_{i-1}, x_{i+1}, \ldots, x_d$ tend to infinity. It is important to mention that it is not enough to know the margins $F_1, \ldots, F_d$ in order to determine F. Additionally it is required to know how the marginal laws are coupled. This is achieved by means of a *copula* of $(X_1, \ldots, X_d)$. Generally speaking, knowing the margins and a copula is equivalent to knowing the distribution. It is now appropriate to give the definition of a copula.

Definition 1.1 (Copula)

(1) A function $C : [0,1]^d \to [0,1]$ is called a (d-dimensional) copula, *if there is a probability space $(\Omega, \mathcal{F}, \mathbb{P})$ supporting a random vector $(U_1, \ldots, U_d)$ such that $U_k \sim U[0,1]$ for all $k = 1, \ldots, d$ and*

$$C(u_1, \ldots, u_d) = \mathbb{P}(U_1 \leq u_1, \ldots, U_d \leq u_d), \quad u_1, \ldots, u_d \in [0,1].$$

(2) On a probability space $(\Omega, \mathcal{F}, \mathbb{P})$ let $(U_1, \ldots, U_d)$ be a random vector on $[0,1]^d$ whose joint distribution function (restricted to $[0,1]^d$) is a copula $C : [0,1]^d \to [0,1]$. For $i = 2, \ldots, d$ and indices $1 \le j_1 < \ldots < j_i \le d$ the notation $C_{j_1,\ldots,j_i} : [0,1]^i \to [0,1]$ is introduced for the joint distribution function of the random subvector $(U_{j_1}, \ldots, U_{j_i})$. It is itself a copula and called an i-margin *of C.*

For a random vector $(U_1, \ldots, U_d) \in [0,1]^d$ on the d-dimensional unit cube the values of its distribution function on $\mathbb{R}^d \setminus [0,1]^d$ are completely determined by its values on $[0,1]^d$. Thus, copulas are defined on $[0,1]^d$ only. A d-dimensional copula C induces a probability measure dC on the unit cube $[0,1]^d$. More clearly, if a random vector $(U_1, \ldots, U_d)$ on $[0,1]^d$ is defined on the probability space $(\Omega, \mathcal{F}, \mathbb{P})$ and has distribution function C, then

$$dC(B) := \mathbb{P}\big((U_1, \ldots, U_d) \in B\big), \quad B \in \mathcal{B}\big([0,1]^d\big).$$

The measure dC is called the *probability measure associated with the copula* C. It is uniquely determined by C. The three simplest examples of copulas are defined in the following examples.

Example 1.1 (Independence Copula)
The function $\Pi : [0,1]^d \to [0,1]$, given by

$$\Pi(u_1, \ldots, u_d) := \prod_{i=1}^{d} u_i, \quad u_1, \ldots, u_d \in [0,1],$$

is called the independence copula. *To see that Π actually is a copula, consider a probability space $(\Omega, \mathcal{F}, \mathbb{P})$ supporting i.i.d. random variables $U_1, \ldots, U_d$ with $U_1 \sim U[0,1]$. The random vector $(U_1, \ldots, U_d)$ then has $U[0,1]$-distributed margins and joint distribution function*

$$\begin{aligned}\mathbb{P}(U_1 \le u_1, \ldots, U_d \le u_d) &= \prod_{i=1}^{d} \mathbb{P}(U_i \le u_i) = \prod_{i=1}^{d} u_i \\ &= \Pi(u_1, \ldots, u_d), \quad u_1, \ldots, u_d \in [0,1].\end{aligned}$$

The independence of the components of Π explains the nomenclature.

Example 1.2 (Comonotonicity Copula)
Considering a probability space $(\Omega, \mathcal{F}, \mathbb{P})$ supporting a single random variable $U \sim U[0,1]$, the random vector $(U_1, \ldots, U_d) := (U, \ldots, U) \in [0,1]^d$ has $U[0,1]$-distributed margins and joint distribution function

$$\begin{aligned}\mathbb{P}(U_1 \le u_1, \ldots, U_d \le u_d) &= \mathbb{P}(U \le \min\{u_1, \ldots, u_d\}) \\ &= \min\{u_1, \ldots, u_d\}, \quad u_1, \ldots, u_d \in [0,1].\end{aligned}$$

Consequently, the function $M : [0,1]^d \to [0,1]$, *defined by*

$$M(u_1,\dots,u_d) := \min\{u_1,\dots,u_d\}, \quad u_1,\dots,u_d \in [0,1],$$

is a copula called the copula of complete comonotonicity *(also called the* upper Fréchet–Hoeffding bound, *see Lemma 1.3).*

Example 1.3 (Countermonotonicity Copula)
Considering a probability space $(\Omega, \mathcal{F}, \mathbb{P})$ *supporting a single random variable* $U \sim U[0,1]$, *the bivariate random vector* $(U_1, U_2) := (U, 1-U) \in [0,1]^2$ *has perfectly negatively associated components (i.e. if* U_1 *is large, then* U_2 *is small, and vice versa). This random vector has* $U[0,1]$*-distributed margins and joint distribution function*

$$\begin{aligned}\mathbb{P}(U_1 \le u_1, U_2 \le u_2) &= \mathbb{P}(1 - u_2 \le U \le u_1) \\ &= (u_1 + u_2 - 1)\,\mathbb{1}_{\{1-u_2 \le u_1\}}, \quad u_1, u_2 \in [0,1].\end{aligned}$$

Consequently, the function $W : [0,1]^2 \to [0,1]$, *defined by*

$$W(u_1,u_2) := (u_1 + u_2 - 1)\,\mathbb{1}_{\{u_1+u_2 \ge 1\}}, \quad u_1, u_2 \in [0,1],$$

is a bivariate copula called the copula of complete countermonotonicity *(also called the* lower Fréchet–Hoeffding bound, *see Lemma 1.3).*

When dealing with copulas, especially in larger dimensions, many computations exhibit a combinatorial character. This is mainly due to the fact that the dimension $d \ge 2$ is a "discrete" number. The following basic fact from probability calculus will sometimes be useful in this regard and is stated here for later reference. It shows how to compute the probability of a union of events in terms of probabilities of intersections.

Lemma 1.1 (Principle of Inclusion and Exclusion)
Let $(\Omega, \mathcal{F}, \mathbb{P})$ *be a probability space. With* $A_1, \dots, A_n \in \mathcal{F}$ *one has*[3]

$$\begin{aligned}\mathbb{P}\Big(\bigcup_{i=1}^n A_i\Big) &= \sum_{\emptyset \ne I \subset \{1,\dots,n\}} (-1)^{|I|+1}\,\mathbb{P}\Big(\bigcap_{j\in I} A_j\Big) \\ &= \sum_{k=1}^n (-1)^{k+1} \sum_{1 \le i_1 < \dots < i_k \le n} \mathbb{P}\Big(\bigcap_{j=1}^k A_{i_j}\Big).\end{aligned}$$

Proof. See, e.g., Billingsley (1995, p. 24). □

[3] For $n = 2$: $\mathbb{P}(A_1 \cup A_2) = \mathbb{P}(A_1) + \mathbb{P}(A_2) - \mathbb{P}(A_1 \cap A_2)$.

1.1.1 *Analytical Properties*

Univariate distribution functions can be characterized by analytical properties. It is well known that a function $F : \mathbb{R} \to [0,1]$ is the distribution function of some (real-valued) random variable X if and only if it is right-continuous, non-decreasing, $\lim_{x\to-\infty} F(x) = 0$, and $\lim_{x\to\infty} F(x) = 1$ (see, e.g., Billingsley (1995, Theorem 12.4, p. 176)). There is an analogous characterization of multivariate distribution functions via analytical properties, using the notions of *groundedness* and *d-increasingness.*

Definition 1.2 (Groundedness and d-Increasingness)
Let $C : [0,1]^d \to [0,1]$ be an arbitrary function.

(1) C is called grounded, *if $C(u_1,\ldots,u_d) = 0$ whenever $u_i = 0$ for at least one component $i \in \{1,\ldots,d\}$.*
(2) C is called d-increasing, *if for all*

$$u := (u_1,\ldots,u_d), \quad v := (v_1,\ldots,v_d) \in [0,1]^d$$

with $u_i < v_i$ for all $i = 1,\ldots,d$ one has[4]

$$\sum_{(w_1,\ldots,w_d)\in\times_{i=1}^d\{u_i,v_i\}} (-1)^{|\{i\,:\,w_i=u_i\}|}\, C(w_1,\ldots,w_d) \geq 0.$$

Obviously, a copula $C : [0,1]^d \to [0,1]$ is a grounded function by Definition 1.1. To see that a d-dimensional copula is also d-increasing, consider a probability space $(\Omega, \mathcal{F}, \mathbb{P})$ supporting a random vector $(U_1,\ldots,U_d) \sim C$ and let $\boldsymbol{u} := (u_1,\ldots,u_d), \boldsymbol{v} := (v_1,\ldots,v_d) \in [0,1]^d$ with $u_i < v_i$ for all $i = 1,\ldots,d$. For each subset $\emptyset \neq I \subset \{1,\ldots,d\}$, define the vector $\boldsymbol{w}^{(I)} := (w_1^{(I)},\ldots,w_d^{(I)})$ by

$$w_k^{(I)} := \begin{cases} u_k \;, k \in I \\ v_k \;, k \notin I \end{cases}, \quad k = 1,\ldots,d.$$

[4]For $d=2$: $dC\big((u_1,u_2)\times(v_1,v_2)\big) = C(v_1,v_2) - C(v_1,u_2) - C(u_1,v_2) + C(u_1,u_2) \geq 0.$

Using the principle of inclusion and exclusion (see Lemma 1.1) in the following third equality, one checks that[5]

$$
\begin{aligned}
0 \leq \mathbb{P}\Big(\bigcap_{i=1}^{d}\{u_i < U_i < v_i\}\Big) &= \mathbb{P}\Big(\bigcap_{i=1}^{d}\{U_i < v_i\} \setminus \bigcup_{i=1}^{d}\{U_i < u_i\}\Big) \\
&= \mathbb{P}\Big(\bigcap_{i=1}^{d}\{U_i < v_i\}\Big) - \mathbb{P}\Big(\bigcup_{i=1}^{d}\big(\{U_i < u_i\} \cap \{U_1 < v_1, \ldots, U_d < v_d\}\big)\Big) \\
&= C(v_1, \ldots, v_d) + \sum_{\emptyset \neq I \subset \{1,\ldots,d\}} (-1)^{|I|}\, C(w_1^{(I)}, \ldots, w_d^{(I)}) \\
&= \sum_{(w_1,\ldots,w_d) \in \times_{i=1}^{d}\{u_i, v_i\}} (-1)^{|\{i\,:\, w_i = u_i\}|}\, C(w_1, \ldots, w_d).
\end{aligned}
$$

Hence, the complicated condition of d-increasingness in the case of a copula C means that the associated probability measure dC assigns non-negative mass to any non-trivial d-box $[u_1, v_1] \times \ldots \times [u_d, v_d]$, and is therefore well defined. Conversely, one can show that groundedness and d-increasingness are sufficient to define a proper distribution function. This result is based on a canonical construction of a probability space from a given (distribution) function. However, this construction is useless for sampling applications and only serves as an existence proof. Therefore it is omitted in this introduction and the interested reader is referred to Ressel (2011). Summing up, the following characterization of copulas via analytical properties is obtained. Theorem 1.1 is used to define a copula in many textbooks (see, e.g., Nelsen (2006, p. 45)).

Theorem 1.1 (Characterization of Copulas via Analytical Properties)
A function $C : [0,1]^d \to [0,1]$ is a copula if and only if it is grounded, d-increasing, and satisfies $C(1, \ldots, 1, u_i, 1, \ldots, 1) = u_i$ for all components $i = 1, \ldots, d$ and $u_i \in [0,1]$.

Proof. This statement can be retrieved, e.g., from a result in Ressel (2011), which characterizes arbitrary multivariate distribution functions. Note in particular that the last condition of the claimed statement corresponds precisely to the fact that the marginal laws associated with a copula have to be uniform on $[0,1]$. □

[5] Note that the absolute continuity of the $U[0,1]$-law implies the almost sure equality of events such as $\{U_i < v_i\}$ and $\{U_i \leq v_i\}$, which is used frequently here.

1.1.1.1 *Testing if C is a Copula*

Checking whether a given function $C : [0,1]^d \to [0,1]$ is a copula might be a tedious exercise. To solve it, one can either

(1) try to construct a probability space $(\Omega, \mathcal{F}, \mathbb{P})$ supporting a random vector $(U_1, \ldots, U_d)$ whose distribution function is C, or
(2) try to verify the three analytic conditions of Theorem 1.1.

If approach (1) is taken and the probability space $(\Omega, \mathcal{F}, \mathbb{P})$ is constructed "conveniently", a simulation algorithm for the respective copula C is immediately available: one simply has to mimic the construction of $(U_1, \ldots, U_d) \sim C$. However, finding the right probability space requires one to already have a good understanding of the given copula C. Approach (2) is purely analytic and does not provide a simulation algorithm. But checking the characterizing properties of a copula analytically might be a difficult task. In particular, it is often difficult, not to say impossible, to check whether a given function is d-increasing or not. Only in very special cases, and mostly only in dimension $d = 2$, can this condition be checked explicitly. An example is provided below. For another example, namely bivariate Archimedean copulas, see Nelsen (2006, Theorem 4.1.4, p. 111).

Example 1.4 (Bivariate Cuadras–Augé Copula)
With a parameter $\alpha \in [0,1]$ we define the function $C_\alpha : [0,1]^2 \to [0,1]$ by

$$C_\alpha(u_1, u_2) := \min\{u_1, u_2\} \max\{u_1, u_2\}^{1-\alpha}, \quad u_1, u_2 \in [0,1].$$

Theorem 1.1 is used in the sequel to prove analytically that C_α is a proper copula. First observe that groundedness and $C_\alpha(u, 1) = C_\alpha(1, u) = u$, $u \in [0,1]$, are obvious properties. Hence, the only non-obvious property is two-increasingness. To this end, let $0 \leq u_1 < v_1 \leq 1$ and $0 \leq u_2 < v_2 \leq 1$. We have to show that

$$C_\alpha(u_1, u_2) + C_\alpha(v_1, v_2) - C_\alpha(u_1, v_2) - C_\alpha(v_1, u_2) \geq 0. \tag{1.1}$$

We distinguish all possible cases:

(1) $u_1 < v_1 \leq u_2 < v_2$: In this case, (1.1) boils down to

$$\begin{aligned} u_1\, u_2^{1-\alpha} + v_1\, v_2^{1-\alpha} - u_1\, v_2^{1-\alpha} - v_1\, u_2^{1-\alpha} &\geq 0 \\ \Leftrightarrow (v_1 - u_1)\, (v_2^{1-\alpha} - u_2^{1-\alpha}) &\geq 0. \end{aligned}$$

Clearly, this is satisfied.

(2) $u_1 \leq u_2 \leq v_1 \leq v_2$: *In this case, we have that* $u_2^{\alpha} \leq v_1^{\alpha}$, *implying* $u_2\, v_1^{1-\alpha} \leq u_2^{1-\alpha}\, v_1$. *Therefore,*

$$\begin{aligned} &C_{\alpha}(u_1,u_2) + C_{\alpha}(v_1,v_2) - C_{\alpha}(u_1,v_2) - C_{\alpha}(v_1,u_2) \\ &\quad = u_1\, u_2^{1-\alpha} + v_1\, v_2^{1-\alpha} - u_1\, v_2^{1-\alpha} - u_2\, v_1^{1-\alpha} \\ &\quad \geq u_1\, u_2^{1-\alpha} + v_1\, v_2^{1-\alpha} - u_1\, v_2^{1-\alpha} - u_2^{1-\alpha}\, v_1 \\ &\quad = (v_2^{1-\alpha} - u_2^{1-\alpha})\,(v_1 - u_1) \geq 0. \end{aligned}$$

Hence, (1.1) is established.

(3) $u_2 \leq u_1 < v_1 \leq v_2$: *On the one hand, since* $0 \leq \alpha \leq 1$ *and* $0 \leq v_2 \leq 1$, *it follows that* $v_2^{1-\alpha} \geq v_2$, *implying* $v_2^{1-\alpha}/u_2 \geq v_2/u_2 \geq 1$.[6] *On the other hand, the function* $g : [0,1] \to \mathbb{R}$, *defined by* $g(x) := (v_1^x - u_1^x)/(v_1 - u_1)$, *can easily be checked to be continuous and increasing from* $g(0) = 0$ *to* $g(1) = 1$. *Both facts together imply*

$$\frac{v_1^{1-\alpha} - u_1^{1-\alpha}}{v_1 - u_1} = g(1-\alpha) \leq 1 \leq \frac{v_2^{1-\alpha}}{u_2},$$

which, by rearranging the terms, is precisely equivalent to (1.1) in this case.

(4) $u_2 < v_2 \leq u_1 < v_1$: *This case is symmetric to case (1).*

(5) $u_2 \leq u_1 \leq v_2 \leq v_1$: *This case is symmetric to case (2).*

(6) $u_1 \leq u_2 < v_2 \leq v_1$: *This case is symmetric to case (3).*

Hence, we have shown that C_{α} *is indeed a copula, using the analytic characterization of Theorem 1.1. But the reader will agree that the proof is quite tedious and doesn't provide insight into the kind of dependence structure induced by the copula* C_{α}. *Therefore, this proof is of limited practical value for applications. A much simpler and more elegant proof is provided in Example 1.8. Additionally, it immediately provides a sampling algorithm.*

A point of discontinuity x of the distribution function F of a (univariate) random variable X on a probability space $(\Omega, \mathcal{F}, \mathbb{P})$ corresponds to an atom of X, i.e. $\lim_{y \uparrow x} F(y) < F(x)$ implies $\mathbb{P}(X = x) > 0$. In contrast, a copula cannot have an atom, since the margins are standardized to $U[0,1]$-distributions, which have a continuous (univariate) distribution function. This implies that a copula is always a continuous function. Moreover, the partial derivatives of a copula with respect to each argument exist (Lebesgue) almost everywhere.

[6]We assume without loss of generality (w.l.o.g.) that $u_2 > 0$, since if $u_2 = 0$, the statement is trivial.

Lemma 1.2 (Smoothness of a Copula)
Let $C : [0,1]^d \to [0,1]$ be a copula.

(1) *For every $(u_1, \ldots, u_d), (v_1, \ldots, v_d) \in [0,1]^d$ it holds that*

$$\left|C(u_1, \ldots, u_d) - C(v_1, \ldots, v_d)\right| \leq \sum_{i=1}^{d} |u_i - v_i|.$$

In particular, C is Lipschitz continuous with the Lipschitz constant equal to 1.

(2) *For $k = 1, \ldots, d$ and fixed $(u_1, \ldots, u_{k-1}, u_{k+1}, \ldots, u_d) \in [0,1]^{d-1}$, the partial derivative $u_k \mapsto \frac{\partial}{\partial u_k} C(u_1, \ldots, u_d)$ exists (Lebesgue) almost everywhere on $[0,1]$ and takes values in $[0,1]$.*

Proof. See Schweizer and Sklar (1983, Lemma 6.1.9, p. 82) for a proof of part (1). See also Darsow et al. (1992) for a proof in the bivariate case. Part (2) is an immediate consequence of the monotonicity and continuity of the functions $u_k \mapsto C(u_1, \ldots, u_d)$ for $k = 1, \ldots, d$. The boundedness of the partial derivatives by $[0,1]$ follows from the Lipschitz constant 1. □

Each copula C is pointwise bounded from above by the comonotonicity copula M. This is intuitive, since the comonotonicity copula implies the strongest positive association possible between components. In dimension $d = 2$, the countermonotonicity copula W is a pointwise lower bound. In contrast, in dimensions $d \geq 3$ there is no "smallest" copula. The explanation for this fact is the following: for $d = 2$ it is immediate from Example 1.3 that the two components in a random vector $(U_1, U_2) \sim W$, defined on $(\Omega, \mathcal{F}, \mathbb{P})$, are perfectly negatively associated. More clearly, it holds almost surely (a.s.) that $U_1 = 1 - U_2$, i.e. the minus sign implies that if one variable moves in one direction, the other moves precisely in the opposite direction. However, for $d \geq 3$ there are at least three directions and it is not clear how to define three directions to be perfect opposites of each other. Nevertheless, the following result provides a sharp lower bound for arbitrary copulas.

Lemma 1.3 (Fréchet–Hoeffding Bounds)
Let $C : [0,1]^d \to [0,1]$ be a copula and M the upper Fréchet-Hoeffding bound. Then it holds for every $(u_1, \ldots, u_d) \in [0,1]^d$ that

$$W(u_1, \ldots, u_d) := \max\left\{\left(\sum_{i=1}^{d} u_i\right) - (d-1), 0\right\}$$
$$\leq C(u_1, \ldots, u_d) \leq M(u_1, \ldots, u_d).$$

Moreover, the lower bound is sharp in the sense that for a given $(u_1, \ldots, u_d) \in [0,1]^d$ *one can find a copula* C *such that* $W(u_1, \ldots, u_d) = C(u_1, \ldots, u_d)$. *However, for* $d \geq 3$ *the function* W *is not a copula.*

Proof. Originally due to Fréchet (1957); see also Mikusinski et al. (1992). For the claimed sharpness statement, see Nelsen (2006, p. 48). Given a random vector $(U_1, \ldots, U_d) \sim C$ on a probability space $(\Omega, \mathcal{F}, \mathbb{P})$, the upper bound follows from the fact that for each $k = 1, \ldots, d$ one has

$$\bigcap_{i=1}^{d} \{U_i \leq u_i\} \subset \{U_k \leq u_k\},$$

implying

$$C(u_1, \ldots, u_d) = \mathbb{P}\Big(\bigcap_{i=1}^{d} \{U_i \leq u_i\}\Big) \leq \mathbb{P}(U_k \leq u_k) = u_k.$$

Taking the minimum over $k = 1, \ldots, d$ on both sides of the last equation implies the claimed upper bound. The lower bound is obtained as follows:

$$\begin{aligned} C(u_1, \ldots, u_d) &= \mathbb{P}\Big(\bigcap_{i=1}^{d} \{U_i \leq u_i\}\Big) = 1 - \mathbb{P}\Big(\bigcup_{i=1}^{d} \{U_i > u_i\}\Big) \\ &\geq 1 - \sum_{i=1}^{d} \mathbb{P}(U_i > u_i) = 1 - \sum_{i=1}^{d} (1 - u_i) = 1 - d + \sum_{i=1}^{d} u_i. \end{aligned}$$

Since $C(u_1, \ldots, u_d) \geq 0$, the claimed lower bound is established. □

If the probability measure dC associated with a copula C is absolutely continuous with respect to the Lebesgue measure on $[0,1]^d$, then by the theorem of Radon–Nikodym there is a (Lebesgue) almost everywhere unique function $c : [0,1]^d \to [0, \infty)$ such that

$$C(u_1, \ldots, u_d) = \int_0^{u_1} \ldots \int_0^{u_d} c(v_1, \ldots, v_d)\, dv_d \ldots dv_1, \quad u_1, \ldots, u_d \in [0,1].$$

In this case, the copula C is called *absolutely continuous* with *copula density* c. Analytically, a d-dimensional copula C is absolutely continuous, if it is d times differentiable, and in this case it is almost everywhere (see, e.g., McNeil et al. (2005, p. 197)):

$$c(u_1, \ldots, u_d) = \frac{\partial}{\partial u_1} \cdots \frac{\partial}{\partial u_d} C(u_1, \ldots, u_d), \quad u_1, \ldots, u_d \in (0,1).$$

Example 1.5 (Density of the Independence Copula)
The independence copula Π *from Example 1.1 is absolutely continuous, since for all* $u_1, \dots, u_d \in [0,1]$ *one has*

$$\Pi(u_1, \dots, u_d) = \prod_{k=1}^{d} u_k = \int_0^{u_1} \dots \int_0^{u_d} 1 \, dv_d \dots dv_1.$$

Hence, the copula density of Π *is equal to* 1 *on* $[0,1]^d$.

This book also discusses copulas that are not absolutely continuous. For some copulas C, their associated measure dC can be decomposed into $dC = dC^{abs} + dC^{sing}$, where dC^{abs} is absolutely continuous with respect to the Lebesgue measure, but dC^{sing} is not. In this case, C is said to have a *singular component*. A singular component often causes analytical difficulties when working with the respective copula. For example, if no density exists, standard maximum likelihood techniques are not directly applicable for parametric estimation purposes. Nevertheless, such distributions can still have interesting and useful properties for applications. Analytically, a singular component can often be detected by finding points $(u_1, \dots, u_d) \in [0,1]^d$, where some (existing) partial derivative of the copula has a point of discontinuity. An example is provided by the comonotonicity copula.

Example 1.6 (Copula with Singular Component)
Consider the copula

$$M(u_1, \dots, u_d) = \min\{u_1, \dots, u_d\}$$

from Example 1.2. We have already seen that the components of a random vector $(U_1, \dots, U_d) \sim M$ *are almost surely all identical. In other words, the associated probability measure* dM *assigns all mass to the diagonal of the unit d-cube* $[0,1]^d$*, a set with zero d-dimensional Lebesgue measure. Hence,* M *is not absolutely continuous. Indeed, we can check that the partial derivatives*

$$\frac{\partial}{\partial u_k} M(u_1, \dots, u_d) = \begin{cases} 1 \, , \, u_k < \min\{u_1, \dots, u_{k-1}, u_{k+1}, \dots, u_d\} \\ 0 \, , \, u_k > \min\{u_1, \dots, u_{k-1}, u_{k+1}, \dots, u_d\} \end{cases},$$

$k = 1, \dots, d$*, exhibit a jump at* $\min\{u_1, \dots, u_{k-1}, u_{k+1}, \dots, u_d\}$. *Note in particular that the partial derivative w.r.t. the kth component is not defined at* $\min\{u_1, \dots, u_{k-1}, u_{k+1}, \dots, u_d\}$*, which is a set with (one-dimensional) Lebesgue measure zero.*

1.1.2 *Sklar's Theorem and Survival Copulas*

1.1.2.1 *Sklar's Theorem*

At the heart of copula theory stands the seminal theorem of Sklar. In a nutshell, it states that any multivariate probability distribution can be split into its univariate margins and a copula. Conversely, combining some given margins with a given copula, one can build a multivariate distribution. This result allows us to divide the treatment of a multivariate distribution into two often easier subtreatments: (1) investigation of the univariate marginal laws, and (2) investigation of a copula, i.e. a standardized and therefore often more convenient multivariate distribution function. To prepare the proof of Sklar's theorem, the following lemma is a useful intermediate step. Given a random variable $X \sim F$, it shows how to transform it such that one obtains a uniform distribution. It is called the *distributional transform* of F. A short overview of its applications, as well as the following lemma, can be found in Rüschendorf (2009).

Lemma 1.4 (Standardizing a Probability Distribution)
Consider a probability space $(\Omega, \mathcal{F}, \mathbb{P})$ supporting two independent random variables V and X. It is assumed that $V \sim U[0,1]$ and $X \sim F$ for an arbitrary distribution function F. The following notation is introduced for $x \in \mathbb{R}$ and $u, v \in (0,1)$:

$$F(x-) := \lim_{y \uparrow x} F(y), \quad \Delta F(x) := F(x) - F(x-),$$
$$F^{-1}(u) := \inf\{y \in \mathbb{R} : F(y) \geq u\}, \quad F^{(v)}(x) := F(x-) + v\,\Delta F(x).$$

The function F^{-1} is called the generalized inverse *of F. The following statements are valid.*

(1) $F^{(V)}(X) \sim U[0,1]$.
(2) $F^{-1}\big(F^{(V)}(X)\big) = X$ *almost surely (a.s.).*
(3) $F^{-1}(V) \sim F$.

Proof. For part (1), let $U := F^{(V)}(X)$, fix $u \in (0,1)$, and define $q_u := \sup\{y \in \mathbb{R} : F(y) < u\}$. We have to prove that $\mathbb{P}(U \leq u) = u$.

First we prove that for $(x, v) \in (q_u, \infty) \times (0,1)$ we have

$$F(x-) + v\,\Delta F(x) \leq u \Leftrightarrow u = F(q_u) = F(x). \tag{1.2}$$

To see "$\Leftarrow$", notice that $x > q_u$ and $F(q_u) = F(x) = u$ imply $F(x) = F(x-) = u$, yielding $F(x-) + v\,\Delta F(x) = u$. To see "$\Rightarrow$", notice that

$$u \geq F(x-) + v\,\Delta F(x) = v \underbrace{F(x)}_{\geq F(x-)} + (1-v)\,F(x-)$$
$$\geq F(x-) \geq F(q_u) \geq u.$$

The last inequality is due to the right-continuity of F. This implies $u = F(q_u) = F(x-) + v\,\Delta F(x) = F(x-)$. But this implies that $\Delta F(x) = 0$, or put differently, $F(x) = F(x-) = u$, hence (1.2) is established.

Moreover, the definition of q_u and the fact that $F(x-) + v\,\Delta F(x) \leq F(x)$ imply

$$(-\infty, q_u) \times (0,1) \subset \{(x,v) \in \mathbb{R} \times (0,1) \,:\, F(x-) + v\,\Delta F(x) < u\}. \quad (1.3)$$

Therefore, we obtain from (1.2) and (1.3) that

$$\begin{aligned} &\{F(X-) + V\,\Delta F(X) \leq u\} \\ &\quad = \{F(X-) + V\,\Delta F(X) \leq u,\ X < q_u\} \\ &\qquad \cup \{F(X-) + V\,\Delta F(X) \leq u,\ X = q_u\} \\ &\qquad \cup \{F(X-) + V\,\Delta F(X) \leq u,\ X > q_u\} \\ &\quad = \{X < q_u\} \cup \{F(q_u-) + V\,\Delta F(q_u) \leq u,\ X = q_u\} \\ &\qquad \cup \{F(X) = F(q_u) = u,\ X > q_u\}. \end{aligned} \quad (1.4)$$

Now (1.2) further implies that

$$\begin{aligned} &\mathbb{P}(F(X) = F(q_u) = u,\ X > q_u) \\ &\qquad = \mathbb{P}\big(X \in (q_u, \sup\{x \geq q_u \,:\, F(x) = u\})\big) = 0, \end{aligned} \quad (1.5)$$

where the last probability is 0, since F is constant on the set $(q_u, \sup\{x \geq q_u \,:\, F(x) = u\})$ or the interval is even empty in case $\{x \geq q_u \,:\, F(x) = u\} = \emptyset$. Combining (1.4) and (1.5), and recalling the independence of X and V, we obtain

$$\begin{aligned} \mathbb{P}(U \leq u) &= F(q_u-) + \mathbb{P}(F(q_u-) + V\,\Delta F(q_u) \leq u)\,\mathbb{P}(X = q_u) \\ &= F(q_u-) + \mathbb{P}(F(q_u-) + V\,\Delta F(q_u) \leq u)\,\Delta F(q_u). \end{aligned}$$

If $\Delta F(q_u) > 0$, then this implies the claim, since

$$\mathbb{P}(F(q_u-) + V\,\Delta F(q_u) \leq u) = \mathbb{P}\Big(V \leq \frac{u - F(q_u-)}{\Delta F(q_u)}\Big) = \frac{u - F(q_u-)}{\Delta F(q_u)}.$$

If $\Delta F(q_u) = 0$, then this also implies the claim, since in this case $F(q_u-) = F(q_u) = u$. To see the latter, notice that the definition of the supremum allows us to find a sequence $x_n \uparrow q_u$ satisfying $F(x_n) < u$. Together with the continuity of F in q_u this implies $F(q_u) = \lim_{n\to\infty} F(x_n) \leq u$. But $F(q_u) \geq u$ is trivial by the right-continuity of F. Hence, part (1) is established.

Part (2) is easier to check. First, it is clear by definition that $F(X-) \leq F^{(V)}(X) \leq F(X)$ almost surely. Second, it holds for all $x \in \mathbb{R}$ with $\Delta F(x) > 0$ that $u \in (F(x-), F(x)] \Rightarrow F^{-1}(u) = x$, also by definition. Combining both simple facts directly implies $F^{-1}\big(F^{(V)}(X)\big) = X$ almost surely, as claimed.

Part (3) follows directly from the fact that $F^{-1}(v) \leq x \Leftrightarrow F(x) \geq v$ for all $x \in \mathbb{R}$, $v \in (0,1)$. The latter is a direct consequence of the right-continuity of F. $\square$

Remark 1.1 (Continuous Case)
If the distribution function F in Lemma 1.4 is continuous, then $F^{(V)} = F$ does not depend on V. This means that one actually has $F(X) \sim U[0,1]$ and $F^{-1}\big(F(X)\big) = X$ almost surely. This means that in the continuous case, the auxiliary random variable V from Lemma 1.4 is not required to standardize the margins.

Theorem 1.2 (Sklar's Theorem)
Let F be a d-dimensional distribution function with margins $F_1, \ldots, F_d$. Then there exists a d-dimensional copula C such that for all $(x_1, \ldots, x_d) \in \mathbb{R}^d$ it holds that

$$F(x_1, \ldots, x_d) = C\big(F_1(x_1), \ldots, F_d(x_d)\big). \tag{1.6}$$

If $F_1, \ldots, F_d$ are continuous, then C is unique. Conversely, if C is a d-dimensional copula and $F_1, \ldots, F_d$ are univariate distribution functions, then the function F defined via (1.6) is a d-dimensional distribution function.

Proof. Originally due to Sklar (1959). The following proof is taken from Rüschendorf (2009, Theorem 2.2); see also Moore and Spruill (1975). Consider a probability space $(\Omega, \mathcal{F}, \mathbb{P})$ supporting a random vector $(X_1, \ldots, X_d) \sim F$ and an independent random variable $V \sim U[0,1]$. Using the notation from Lemma 1.4, define the random vector $(U_1, \ldots, U_d)$ by $U_k := F_k^{(V)}(X_k) = F_k(X_k-) + V\,\Delta F_k(X_k)$, $k = 1, \ldots, d$. Then by part (1) of Lemma 1.4 we obtain $U_k \sim U[0,1]$ for all $k = 1, \ldots, d$, hence the distribution function of $(U_1, \ldots, U_d)$ is a copula by definition. We denote this copula by C and claim that it satisfies (1.6). To see this,

$$\begin{aligned} F(x_1,\dots,x_d) &= \mathbb{P}(X_k \le x_k, \text{ for } k=1,\dots,d) \\ &= \mathbb{P}\big(F_k^{-1}(F_k^{(V)}(X_k)) \le x_k, \text{ for } k=1,\dots,d\big) \\ &= \mathbb{P}\big(F_k^{(V)}(X_k) \le F_k(x_k), \text{ for } k=1,\dots,d\big) \\ &= C\big(F_1(x_1),\dots,F_d(x_d)\big), \end{aligned}$$

where the second equality follows from part (2) in Lemma 1.4 and the third follows from the fact that

$$F_k^{-1}(u) \le x \Leftrightarrow F_k(x) \ge u \text{ for all } x \in \mathbb{R},\ u \in (0,1),\ k=1,\dots,d. \qquad (1.7)$$

The latter is a direct consequence of the right-continuity of F_k. To see the uniqueness of the copula in the case of continuous margins, one simply has to observe that for all $u_1,\dots,u_d \in (0,1)$ one can find $x_1,\dots,x_d \in \mathbb{R}$ satisfying $F_k(x_k) = u_k$, $k=1,\dots,d$, by the intermediate value theorem. Hence, assuming the existence of two copulas C_1, C_2 satisfying (1.6) one obtains

$$\begin{aligned} C_1(u_1,\dots,u_d) &= C_1\big(F_1(x_1),\dots,F_d(x_d)\big) = F(x_1,\dots,x_d) \\ &= C_2\big(F_1(x_1),\dots,F_d(x_d)\big) = C_2(u_1,\dots,u_d). \end{aligned}$$

Hence, $C_1 = C_2$.

The converse part of the statement is proved as follows. Let $(\Omega, \mathcal{F}, \mathbb{P})$ be a probability space supporting a random vector $(U_1,\dots,U_d) \sim C$. Define the random vector $(X_1,\dots,X_d)$ by $X_k := F_k^{-1}(U_k)$, $k=1,\dots,d$. Lemma 1.4(3) implies that $X_k \sim F_k$ for $k=1,\dots,d$. Furthermore,

$$\begin{aligned} \mathbb{P}(X_k \le x_k, \text{ for } k=1,\dots,d) &= \mathbb{P}\big(F_k^{-1}(U_k) \le x_k, \text{ for } k=1,\dots,d\big) \\ &= \mathbb{P}\big(U_k \le F_k(x_k), \text{ for } k=1,\dots,d\big) \\ &= C\big(F_1(x_1),\dots,F_d(x_d)\big). \end{aligned}$$

The second equality again follows from (1.7). □

If a random vector $(X_1,\dots,X_d) \sim F$ has continuous margins $F_1,\dots,F_d$, the copula in Theorem 1.2 is unique, and we refer to it as "the" copula of $(X_1,\dots,X_d)$. The proof of Lemma 1.4, and consequently of Sklar's theorem, would be less technical in the case of continuous margins. For similar technical reasons, we will often assume later on in this book that a multivariate distribution has continuous margins. For most applications, this assumption is justified anyway. In the case of discontinuous margins, one sometimes faces certain subtleties that complicate things. For a reader-friendly overview of some of these subtleties we refer the reader to the article by Genest and Nešlehová (2007).

Sklar's theorem allows us to conveniently construct multivariate distribution functions in two steps. In a first step one may choose the univariate margins, and in a second step a copula connecting them. Thus, having a repertoire of parametric models for the margins and the copula, it is possible to fit a multivariate distribution to given data by first fitting the parameters of the margins and subsequently the copula parameters (see Joe and Xu (1996)). This is the main reason for the popularity of copulas in statistical modeling. And, as already outlined, this two-step procedure can also be used for simulation.

Example 1.7 (Gaussian Copula)
On a probability space $(\Omega, \mathcal{F}, \mathbb{P})$, let $(X_1, \ldots, X_d)$ be a normally distributed random vector with joint distribution function

$$F(x_1, \ldots, x_d) := \int_{\times_{i=1}^{d}(-\infty, x_i]} (2\pi)^{-\frac{d}{2}} \det(\Sigma)^{-\frac{1}{2}} \exp\Big(-\frac{1}{2}(\boldsymbol{s}-\boldsymbol{\mu})\,\Sigma^{-1}\,(\boldsymbol{s}-\boldsymbol{\mu})'\Big)\,d\boldsymbol{s},$$

for a symmetric, positive-definite matrix Σ and a mean vector $\boldsymbol{\mu} = (\mu_1, \ldots, \mu_d) \in \mathbb{R}^d$, where $\boldsymbol{s} := (s_1, \ldots, s_d)$ and $\det(\Sigma)$ is the determinant of Σ. Denoting by $\sigma_1^2 := \Sigma_{11}, \ldots, \sigma_d^2 := \Sigma_{dd} > 0$ the diagonal entries of Σ, the marginal law F_i of X_i is a normal distribution with mean μ_i and variance σ_i^2, $i = 1, \ldots, d$. The copula C_Σ^{Gauss} of $(X_1, \ldots, X_d)$ is called the Gaussian copula *and is given by*

$$C_\Sigma^{Gauss}(u_1, \ldots, u_d) := F\big(F_1^{-1}(u_1), \ldots, F_d^{-1}(u_d)\big). \tag{1.8}$$

The copula of a multivariate distribution F with strictly increasing continuous margins $F_1, \ldots, F_d$ is always implicitly given by (1.8), but sometimes this expression can be computed explicitly. In the Gaussian case, however, this is not possible due to the fact that no closed-form antiderivatives of normal densities are known.

1.1.2.2 *Survival Copulas*

Sometimes it is more convenient to describe the distribution of a random vector $(X_1, \ldots, X_d)$ by means of its survival function instead of its distribution function. Especially when the components X_k are interpreted as lifetimes, this description is more intuitive. Letting $(X_1, \ldots, X_d)$ be defined on a probability space $(\Omega, \mathcal{F}, \mathbb{P})$, its survival function is defined as

$$\bar{F}(x_1, \ldots, x_d) := \mathbb{P}(X_1 > x_1, \ldots, X_d > x_d), \quad x_1, \ldots, x_d \in \mathbb{R}.$$

For $k = 1, \ldots, d$ the univariate marginal survival function $\bar{F}_k := 1 - F_k$ of X_k can be retrieved from $\bar{F}$ via

$$\bar{F}_k(x_k) = \mathbb{P}(X_k > x_k) = \bar{F}(-\infty, \ldots, -\infty, x_k, -\infty, \ldots, -\infty), \quad x_k \in \mathbb{R}.$$

Using $-\infty, \ldots, -\infty, x_k, -\infty, \ldots, -\infty$ as the argument of $\bar{F}$ is a shorthand notation for taking the limits as $x_1, \ldots, x_{k-1}, x_{k+1}, \ldots, x_d$ tend to $-\infty$. Analogously to Sklar's theorem (see Theorem 1.2), a d-dimensional survival function can be decomposed into a copula and its marginal survival functions.

Theorem 1.3 (Survival Analog of Sklar's Theorem)
Let $\bar{F}$ be a d-dimensional survival function with marginal survival functions $\bar{F}_1, \ldots, \bar{F}_d$. Then there exists a d-dimensional copula $\hat{C}$ such that for all $(x_1, \ldots, x_d) \in \mathbb{R}^d$ it holds that

$$\bar{F}(x_1, \ldots, x_d) = \hat{C}\big(\bar{F}_1(x_1), \ldots, \bar{F}_d(x_d)\big). \tag{1.9}$$

If $\bar{F}_1, \ldots, \bar{F}_d$ are continuous, then $\hat{C}$ is unique. Conversely, if $\hat{C}$ is a d-dimensional copula and $\bar{F}_1, \ldots, \bar{F}_d$ are univariate survival functions, then the function $\bar{F}$ defined via (1.9) is a d-dimensional survival function.

Proof. Similar to the proof of Theorem 1.2. □

Due to uniqueness in the case of continuous margins, the copula $\hat{C}$ in Theorem 1.3 is called "the" survival copula of a random vector $(X_1, \ldots, X_d)$ with survival function $\bar{F}$. It is important to stress that the survival copula $\hat{C}$ is a proper copula, i.e. a distribution function and not a survival function. For some multivariate distributions, the survival function (survival copula) is the more natural object to study, compared to its distribution function (copula). For example, the analytical expression of the survival function (survival copula) might be much more convenient to work with. Popular examples comprise the families of Archimedean copulas and Marshall–Olkin copulas, which naturally arise as survival copulas of simple stochastic models (see Chapters 2 and 3). Therefore, it is useful to know both concepts: the copula as well as the survival copula.

1.1.2.3 *The Connection between C and $\hat{C}$*

Knowing the copula of a random vector allows us to compute its survival copula, and vice versa. This computation is accomplished by the principle of inclusion and exclusion (see Lemma 1.1). Given the copula C of a random vector $(X_1, \ldots, X_d)$ on a probability space $(\Omega, \mathcal{F}, \mathbb{P})$ with continuous margins $F_1, \ldots, F_d$, its survival copula $\hat{C}$ is computed as follows: we recall from Lemma 1.4 the notation $F_i^{-1}(x) := \inf\{y \in \mathbb{R} \,|\, F_i(y) \geq x\}$, $x \in (0,1)$, for the *generalized inverse* of F_i, $i = 1, \ldots, d$. The continuity of the margins implies that $F_i \circ F_i^{-1}(x) = x$ for all $x \in (0,1)$ and $i = 1, \ldots, d$ (see Mai (2010, p. 22–23)). For $u_1, \ldots, u_d \in (0,1)$ it follows that

$$
\begin{aligned}
&\hat{C}(u_1, \ldots, u_d) \\
&= \hat{C}\Big(\bar{F}_1\big(F_1^{-1}(1-u_1)\big), \ldots, \bar{F}_d\big(F_d^{-1}(1-u_d)\big)\Big) \\
&= \mathbb{P}\big(X_1 > F_1^{-1}(1-u_1), \ldots, X_d > F_d^{-1}(1-u_d)\big) \\
&= \mathbb{P}\Big(\bigcap_{k=1}^{d} \{X_k > F_k^{-1}(1-u_k)\}\Big) \\
&= 1 - \mathbb{P}\Big(\bigcup_{k=1}^{d} \{X_k \leq F_k^{-1}(1-u_k)\}\Big) \\
&= 1 - \sum_{k=1}^{d} (-1)^{k+1} \sum_{1 \leq j_1 < \ldots < j_k \leq d} \mathbb{P}\Big(\bigcap_{i=1}^{k} \{X_{j_i} \leq F_{j_i}^{-1}(1-u_{j_i})\}\Big) \\
&= 1 + \sum_{k=1}^{d} (-1)^{k} \sum_{1 \leq j_1 < \ldots < j_k \leq d} \times \\
&\qquad \times C_{j_1,\ldots,j_k}\Big(F_{j_1}\big(F_{j_1}^{-1}(1-u_{j_1})\big), \ldots, F_{j_k}\big(F_{j_k}^{-1}(1-u_{j_k})\big)\Big) \\
&= 1 + \sum_{k=1}^{d} (-1)^{k} \sum_{1 \leq j_1 < \ldots < j_k \leq d} C_{j_1,\ldots,j_k}\big(1-u_{j_1}, \ldots, 1-u_{j_k}\big).
\end{aligned} \tag{1.10}
$$

In the above computation, the second and the sixth equalities follow from Theorems 1.3 and 1.2, respectively. Interchanging the roles of $\hat{C}$ and C yields by a similar computation

$$
C(u_1, \ldots, u_d) = 1 + \sum_{k=1}^{d} (-1)^{k} \sum_{1 \leq j_1 < \ldots < j_k \leq d} \hat{C}_{j_1,\ldots,j_k}\big(1-u_{j_1}, \ldots, 1-u_{j_k}\big).
$$

In the bivariate case $d = 2$ this simplifies to

$$
C(u_1, u_2) = \hat{C}(1-u_1, 1-u_2) + u_1 + u_2 - 1.
$$

An algorithm for computing volumes of d-dimensional survival copulas from their associated copula is provided in Cherubini and Romagnoli (2009). An alternative view on the copula C and the survival copula $\hat{C}$ of a random vector $(X_1, \ldots, X_d)$ with continuous[7] margins $F_1, \ldots, F_d$ can be extracted from the proofs of Theorems 1.2 and 1.3: C is the distribution function of $\big(F_1(X_1), \ldots, F_d(X_d)\big)$ and $\hat{C}$ is the distribution function of $\big(\bar{F}_1(X_1), \ldots, \bar{F}_d(X_d)\big)$. Sometimes it is useful to switch from a copula C to its survival copula $\hat{C}$, or vice versa. In probabilistic terms, if $(U_1, \ldots, U_d) \sim C$, then $(1 - U_1, \ldots, 1 - U_d) \sim \hat{C}$. This also explains that $\hat{\hat{C}} = C$, i.e. the survival copula of the survival copula is the original copula. The following sampling algorithm is included for later reference.

Algorithm 1.1 (Sampling the Survival Copula)
Input: A sampling algorithm for the copula C.

(1) *Sample* $(U_1, \ldots, U_d) \sim C$.
(2) *Return* $(1 - U_1, \ldots, 1 - U_d) \sim \hat{C}$.

Finally, the copula C (and hence also the survival copula $\hat{C}$ by the earlier computation) of a random vector $(X_1, \ldots, X_d)$ with continuous margins is invariant under strictly increasing transformations.

Lemma 1.5 (Invariance under Strictly $\nearrow$ Transformations)
Let $(X_1, \ldots, X_d)$ be a random vector with continuous margins and copula C. For strictly increasing functions $g_1, \ldots, g_d : \mathbb{R} \to \mathbb{R}$, the copula of $\big(g_1(X_1), \ldots, g_d(X_d)\big)$ is again C.

Proof. See Embrechts et al. (2003, Theorem 2.6). □

Lemma 1.5 is often used to transform or standardize the marginal laws without changing the copula. For example, let F_1, F_2 be two continuous and strictly increasing distribution functions and (U_1, U_2) be a random vector with the copula C as the joint distribution function. Then the random vector $(F_1^{-1}(U_1), F_2^{-1}(U_2))$ has copula C, but margins F_1, F_2.

As a corollary to Lemma 1.5, we observe that when strictly decreasing transformations are applied to the univariate marginals of some random vector with copula C, the resulting random vector has copula $\hat{C}$, i.e. the survival copula of C.

[7]In the general case one has to work with the auxiliary random variable such as in Lemma 1.4.

Corollary 1.1 (Strictly ↘ Transformations)
Let $(X_1,\dots,X_d)$ *be a random vector with continuous margins and copula* C. *For strictly decreasing functions* $g_1,\dots,g_d : \mathbb{R} \to \mathbb{R}$, *the copula of* $\big(g_1(X_1),\dots,g_d(X_d)\big)$ *is* $\hat{C}$.

Proof. From Lemma 1.5 we know that the copula of some random vector is invariant under strictly increasing transformations. Moreover, we know that $\big(\bar{F}_1(X_1),\dots,\bar{F}_d(X_d)\big)$ has distribution function $\hat{C}$. We notice that the functions $\big(g_i \circ \bar{F}_i^{-1}\big)(x) = g_i\big(\bar{F}_i^{-1}(x)\big)$, $i = 1,\dots,d$, are strictly increasing, since all g_i and $\bar{F}_i^{-1}$ are strictly decreasing. Consequently, applying $g_i \circ \bar{F}_i^{-1}$, $i = 1,\dots,d$, to the respective components of $\big(\bar{F}_1(X_1),\dots,\bar{F}_d(X_d)\big)$ does not alter its copula $\hat{C}$, and the resulting random vector (after cancellation of $\bar{F}_i^{-1} \circ \bar{F}_i$, $i = 1,\dots,d$) is simply $\big(g_1(X_1),\dots,g_d(X_d)\big)$. □

1.1.3 *General Sampling Methodology in Low Dimensions*

For a given copula $C : [0,1]^2 \to [0,1]$ there is a general methodology of how to simulate from C. This simulation algorithm uses the fact that the partial derivatives of a copula have a probabilistic meaning. To see this it is useful to consider an easy special case first. Assume that a given bivariate copula C is absolutely continuous with copula density c. Now fix a possible realization of the second component U_2, say $U_2 = u_2 \in (0,1)$. It is well known from basic probability calculus that the conditional density $f_{U_1|U_2=u_2}$ of U_1 given the event $\{U_2 = u_2\}$ equals the fraction of the joint density of (U_1,U_2) and the density of U_2 evaluated at u_2. Since the density of $U_2 \sim U[0,1]$ is constantly 1, it follows that $f_{U_1|U_2=u_2}(u_1) = c(u_1,u_2)$ for $u_1 \in [0,1]$. Thus, we obtain

$$\begin{aligned}\frac{\partial}{\partial u_2}C(u_1,u_2) &= \frac{\partial}{\partial u_2}\int_0^{u_2}\int_0^{u_1} c(v_1,v_2)dv_1\,dv_2 = \int_0^{u_1} c(v_1,u_2)dv_1 \\ &= \int_0^{u_1} f_{U_1|U_2=u_2}(v_1)\,dv_1 = \mathbb{P}(U_1 \le u_1 \,|\, U_2 = u_2).\end{aligned}$$

In other words, for fixed $u_2 \in (0,1)$ the function $u_1 \mapsto \frac{\partial}{\partial u_2}C(u_1,u_2)$ equals the (conditional) distribution function $F_{U_1|U_2=u_2}$ of the first component U_1 conditioned on the event that $\{U_2 = u_2\}$. This means that one can first simulate the second component $U_2 \sim U[0,1]$, then fix it (i.e. set $u_2 := U_2$), and subsequently use a univariate sampling technique (see Chapter 6) to simulate U_1 from the distribution function $F_{U_1|U_2=u_2}$. Recalling the standard inversion algorithm to simulate a random variable from a given distribution function, see, e.g., Lemma 1.4(3), one obtains the following generic sampling scheme (see Algorithm 1.2). It is shown in Darsow et al. (1992) that the right-continuous version of the (almost everywhere existing) partial derivative of a bivariate copula can always be interpreted as the conditional distribution function of one variable given the other, i.e. in the general, not necessarily absolutely continuous, case. Hence, Algorithm 1.2 is valid for an arbitrary bivariate copula C.

Algorithm 1.2 (Conditional Sampling for Bivariate Copulas)
The input for the algorithm is a bivariate copula $C : [0,1]^2 \to [0,1]$.

(1) Simulate $U_2 \sim U[0,1]$.
(2) Compute (the right-continuous version of) the function

$$F_{U_1|U_2}(u_1) := \frac{\partial}{\partial u_2} C(u_1, u_2)\Big|_{u_2=U_2}, \quad u_1 \in [0,1].$$

(3) Compute the generalized inverse of $F_{U_1|U_2}$, *i.e.*

$$F^{-1}_{U_1|U_2}(v) := \inf\{u_1 > 0 \,:\, F_{U_1|U_2}(u_1) \geq v\}, \quad v \in (0,1).$$

(4) Simulate $V \sim U[0,1]$, *independent of* U_2.
(5) Set $U_1 := F^{-1}_{U_1|U_2}(V)$ *and return* $(U_1, U_2) \sim C$.

This algorithm is based on a canonical probability space $(\Omega, \mathcal{F}, \mathbb{P})$ supporting only two i.i.d. random variables $U_2, V \sim U[0,1]$. Since this algorithm is by no means restricted to any specific class of copulas, it can in principle be used to sample arbitrary bivariate copulas. The only tedious step is the computation of the partial derivative and its generalized inverse, which typically requires a nice analytic form of the copula. However, even though Algorithm 1.2 is quite general, it is not always the most intuitive approach for a specific copula. If the probabilistic motivation of a certain copula is well understood, one typically finds an easier and more intuitive algorithm. The following example illustrates this fact.

Example 1.8 (Sampling a Bivariate Cuadras–Augé Copula)
Consider the copula C_α *from Example 1.4 with a parameter* $\alpha \in (0,1)$. *With fixed* $u_2 \in (0,1)$ *we compute*

$$F_{U_1|U_2=u_2}(u_1) := \frac{\partial}{\partial u_2} C_\alpha(u_1, u_2) = \begin{cases} u_1\,(1-\alpha)\,u_2^{-\alpha}, & u_1 < u_2 \\ u_1^{1-\alpha} & , u_1 > u_2 \end{cases},$$

where $u_1 \in [0,1] \setminus \{u_2\}$. *Keep in mind that the function* $F_{U_1|U_2=u_2}$ *is not defined at the point* u_2. *In order to make it a distribution function (i.e. right-continuous), we therefore have to set* $F_{U_1|U_2=u_2}(u_2) := u_2^{1-\alpha}$. *The generalized inverse of this conditional distribution function is computed to be*

$$F^{-1}_{U_1|U_2=u_2}(v) = \begin{cases} \frac{v}{1-\alpha}\, u_2^{\alpha} & , v < (1-\alpha)\, u_2^{1-\alpha} \\ u_2 & , (1-\alpha)\, u_2^{1-\alpha} \leq v < u_2^{1-\alpha} \\ v^{\frac{1}{1-\alpha}} & , v \geq u_2^{1-\alpha} \end{cases}, \quad v \in (0,1).$$

Algorithm 1.2 implies simulating i.i.d. variables $U_2, V \sim U[0,1]$ *and then returning* $\big(F^{-1}_{U_1|U_2}(V), U_2\big)$.

In this example, one can clearly see that the computation of the generalized inverse can be tedious. Moreover, the analytic expression of this generalized inverse is difficult to investigate further in order to obtain a deeper understanding of the dependence structure induced by C_α. *Therefore, this copula is a good example to demonstrate the use of an alternative probabilistic construction. To this end, let* $(\Omega, \mathcal{F}, \mathbb{P})$ *be a probability space on which three independent exponential random variables* $E_1, E_2, E_{1,2}$ *are defined. We assume* $E_1, E_2 \sim Exp\big((1-\alpha)/\alpha\big)$ *and* $E_{1,2} \sim Exp(1)$. *Define the two random variables* U_1 *and* U_2 *by*

$$U_1 := \exp\Big(-\frac{1}{\alpha}\min\big\{E_1, E_{1,2}\big\}\Big),$$
$$U_2 := \exp\Big(-\frac{1}{\alpha}\min\big\{E_2, E_{1,2}\big\}\Big). \tag{1.11}$$

Then, it follows for $0 < u_1, u_2 < 1$ *that*

$$\begin{aligned} &\mathbb{P}(U_1 \leq u_1, U_2 \leq u_2) \\ &\quad = \mathbb{P}\Big(\min\big\{E_1, E_{1,2}\big\} \geq -\alpha\log(u_1), \min\big\{E_2, E_{1,2}\big\} \geq -\alpha\log(u_2)\Big) \\ &\quad = \mathbb{P}\Big(E_1 \geq -\alpha\log(u_1), E_2 \geq -\alpha\log(u_2), \\ &\qquad\quad E_{1,2} \geq \max\big\{-\alpha\log(u_1), -\alpha\log(u_2)\big\}\Big) \\ &\quad = \mathbb{P}\big(E_1 \geq -\alpha\log(u_1)\big)\,\mathbb{P}\big(E_2 \geq -\alpha\log(u_2)\big)\times \\ &\qquad \times \mathbb{P}\big(E_{1,2} \geq -\alpha\log(\min\{u_1, u_2\})\big) \\ &\quad = u_1^{1-\alpha}\, u_2^{1-\alpha}\, \min\{u_1, u_2\}^{\alpha} = C_\alpha(u_1, u_2). \end{aligned}$$

Hence, we can simulate C_α *alternatively by simulating* $E_1, E_2, E_{1,2}$ *independently of each other and then returning* (U_1, U_2) *as defined in (1.11). The above construction additionally provides a second (this time probabilistic and simple) proof for the fact that* C_α *is a copula, which was established in a quite tedious way in Example 1.4.*

Example 1.9 (Sampling a Copula of Class BC_2)
The function $C_{a,b} : [0,1]^2 \to [0,1]$ for parameters $a, b \in (0,1)$ is shown below to define a copula, where

$$C_{a,b}(u_1, u_2) = \min\{u_1^a, u_2^b\} \min\{u_1^{1-a}, u_2^{1-b}\}, \quad u_1, u_2 \in [0,1].$$

This family of copulas is named BC_2 in Mai and Scherer (2011e), since it can be considered a class of "Building Components" for so-called extreme-value copulas. To show that $C_{a,b}$ actually defines a proper copula consider a probability space $(\Omega, \mathcal{F}, \mathbb{P})$ supporting two i.i.d. random variables $V_1, V_2 \sim U[0,1]$. Define the random vector (U_1, U_2) by

$$U_1 := \max\big\{V_1^{\frac{1}{a}}, V_2^{\frac{1}{1-a}}\big\}, \quad U_2 := \max\big\{V_1^{\frac{1}{b}}, V_2^{\frac{1}{1-b}}\big\}.$$

A small computation shows that $(U_1, U_2) \sim C_{a,b}$. This construction obviously implies a very easy sampling algorithm for $C_{a,b}$. If the general conditional method from Algorithm 1.2 is used for sampling instead, the effort is considerably larger, since the computation of the generalized inverse is quite tedious.

Example 1.10 (Sampling the Comonotonicity Copula)
Consider the copula M from Example 1.2 in the bivariate case $d = 2$. With fixed $u_2 \in (0,1)$ we know from Example 1.6 that

$$F_{U_1|U_2=u_2}(u_1) := \frac{\partial}{\partial u_2} C(u_1, u_2) = \begin{cases} 1 \,, & u_2 < u_1 \\ 0 \,, & u_2 > u_1 \end{cases}, \quad u_1 \in [0,1] \setminus \{u_2\}.$$

Keep in mind that the function $F_{U_1|U_2=u_2}$ is not defined at the point u_2. In order to make it a distribution function (i.e. right-continuous), we therefore have to set $F_{U_1|U_2=u_2}(u_2) := 1$. The generalized inverse of this conditional distribution function is clearly given by

$$F^{-1}_{U_1|U_2=u_2}(v) \equiv u_2, \quad v \in (0,1).$$

Hence, Algorithm 1.2 implies simulating U_2 and then setting $U_1 := F^{-1}_{U_1|U_2}(V) = U_2$, independent of V. A scatterplot is given in Figure 1.4.

There is a result lifting Algorithm 1.2 to arbitrary dimensions $d \geq 2$. It is based on the so-called *multivariate quantile transform*, which was introduced in O'Brien (1975), Arjas and Lehtonen (1978), and Rüschendorf (1981). Given an arbitrary multivariate distribution function F with margins $F_1, \ldots, F_d$, it provides a canonical construction of a random vector $\boldsymbol{X} := (X_1, \ldots, X_d) \sim F$ as a function of i.i.d. random variables $V_1, \ldots, V_d$

with $V_1 \sim U[0,1]$ on a probability space $(\Omega, \mathcal{F}, \mathbb{P})$. More clearly, $\boldsymbol{X}$ is defined recursively as follows:

$$X_1 := F_1^{-1}(V_1),$$
$$X_k := F_{k|1,\ldots,k-1}^{-1}(V_k \,|\, X_1, \ldots, X_{k-1}), \quad k = 2, \ldots, d,$$

where $F_{k|1,\ldots,k-1}(. \,|\, x_1, \ldots, x_{k-1})$ denotes the conditional distribution function of X_k, conditioned on the event $\{X_1 = x_1, \ldots, X_{k-1} = x_{k-1}\}$. Again, these required conditional distribution functions can be computed analytically from F via successive partial derivatives. However, these computations are often tedious or even impossible. Therefore, this general sampling strategy has its limitations for practical applications. In particular, if the dimension d is large, it requires the computation of higher-order partial derivatives.

1.1.4 *Graphical Visualization*

The most immediate and intuitive step when investigating a probability law is trying to find ways to visualize it by a graph. A bivariate copula C can be visualized by a three-dimensional function plot, i.e. for each pair of arguments (u_1, u_2) in the plane $[0,1]^2$ the value $C(u_1, u_2)$ is plotted on the height axis. As an example, Figure 1.1 illustrates function plots of the bivariate Cuadras–Augé copula C_α, as introduced in Example 1.4, for several choices of α.

Similarly, instead of depicting the whole function plot, it is possible to only plot a discrete grid of level sets

$$L_{k,n} := \{(u_1, u_2) \in [0,1]^2 \,:\, C(u_1, u_2) = k/n\}, \quad k = 0, 1, \ldots, n.$$

Due to the fact that $C(u, 1) = C(1, u) = u$, $u \in [0,1]$, each level set $L_{k,n}$ is a continuous line segment from the point $(k/n, 1)$ to $(1, k/n)$, which makes it superfluous to indicate the height k/n of the level set within the graph. Figure 1.2 illustrates such level set plots of the bivariate Cuadras–Augé copula C_α, as introduced in Example 1.4, for several choices of α.

In the case of one-dimensional probability laws, the plot of the distribution function is somehow inconvenient for judging the probabilistic properties of the underlying distribution. If the distribution is absolutely continuous, it is more illuminating to plot its density. Analogously, in the case of an absolutely continuous bivariate copula, a function plot of the copula density is more illuminating than the function plot of the copula itself. However, this is impossible in the case of copulas with a singular component, as for example for C_α from Example 1.4.

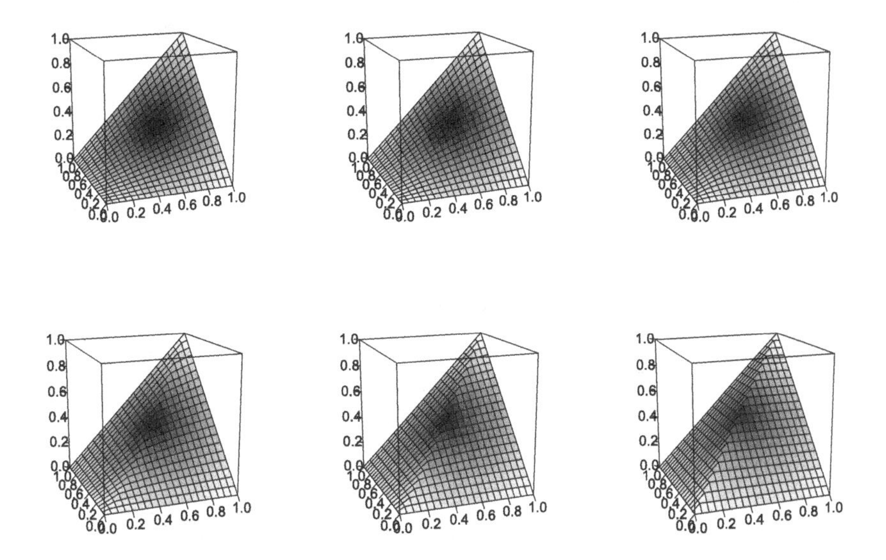

Fig. 1.1 Function plots of $C_\alpha(u_1, u_2) = \min\{u_1, u_2\} \max\{u_1, u_2\}^{1-\alpha}$ for $\alpha = 0$ (upper left, independence case), $\alpha = 0.2$, $\alpha = 0.4$, $\alpha = 0.6$, $\alpha = 0.8$, and $\alpha = 1$ (lower right, case of complete comonotonicity). One observes that the bend on the diagonal emerges with increasing parameter α.

In two (or three) dimensions it is also possible to visualize the distribution associated with a copula C by means of a two- (or three-) dimensional *scatterplot*. This requires a sampling algorithm for C. The idea is to simulate $n \in \mathbb{N}$ independent realizations of a random vector associated with C. These points are then plotted using a two- (or three-) dimensional coordinate system. This technique is somehow the analog of a histogram in the univariate case. Figure 1.3 depicts two-dimensional scatterplots of the bivariate Cuadras–Augé copula C_α, as introduced in Example 1.4, for several choices of α. Notice in particular that $C_0 = \Pi$ and $C_1 = M$. The interpretation of such scatterplots is the following:

$$\frac{\text{number of points in } B}{\text{number of all points}} \approx \mathbb{P}\big((U_1, U_2) \in B\big) = dC(B), \quad B \in \mathcal{B}(\mathbb{R}^2).$$

If $d \geq 4$, the visualization of a copula suffers from the lack of dimensions. One possible approach to still obtain a visualization is to consider all possible bivariate subpairs of components and arrange bivariate visualizations in a matrix. This is particularly useful if the dependence between different pairs is heterogeneous. In later chapters, we will encounter such

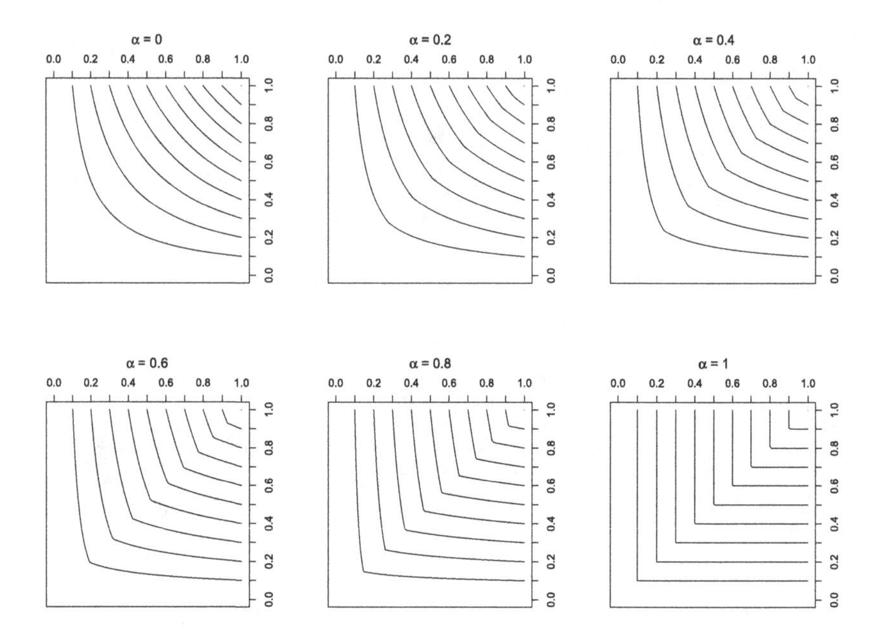

Fig. 1.2 Level set plots of $C_\alpha(u_1, u_2) = \min\{u_1, u_2\} \max\{u_1, u_2\}^{1-\alpha}$ for $\alpha = 0, 0.2, 0.4, 0.6, 0.8, 1$.

cases and provide examples of graphical visualizations of these structures (see Figure 2.10).

1.1.5 *Concordance Measures*

For many applications it is desirable to encode the strength of dependence induced by a copula in a single number. This leads to the notion of so-called *concordance measures*. One says that two pairs $(u_1, u_2), (v_1, v_2) \in [0, 1]^2$ are *concordant*, if both components u_1, u_2 are either both greater or both less than their respective components of the second pair v_1, v_2, i.e. if $(u_1 - v_1)\,(u_2 - v_2) > 0$. Otherwise, they are called *discordant*. Introduced in Scarsini (1984), concordance measures are functions from the set of copulas to the interval $[-1, 1]$ satisfying certain desirable axioms. For example, the value 1 should stand for the strongest positive dependence and the independence copula should be mapped to the value 0, so that the range $[-1, 1]$ can be interpreted as interpolating from perfect negative to perfect positive association. More details can be found in Nelsen (2006, p. 157ff).

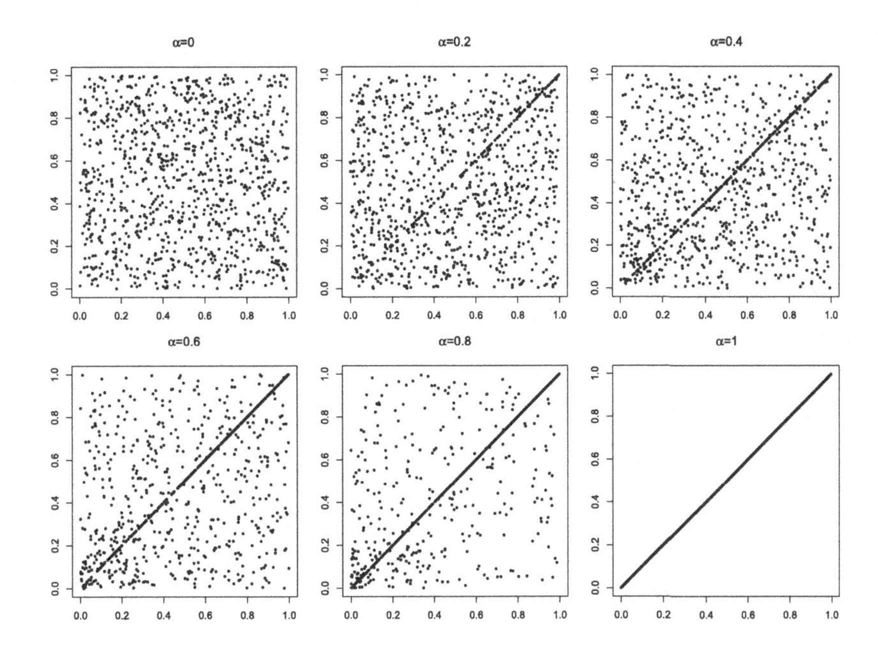

Fig. 1.3 Scatterplots of 1 000 samples from $C_\alpha(u_1, u_2) = \min\{u_1, u_2\} \max\{u_1, u_2\}^{1-\alpha}$ for $\alpha = 0, 0.2, 0.4, 0.6, 0.8, 1$.

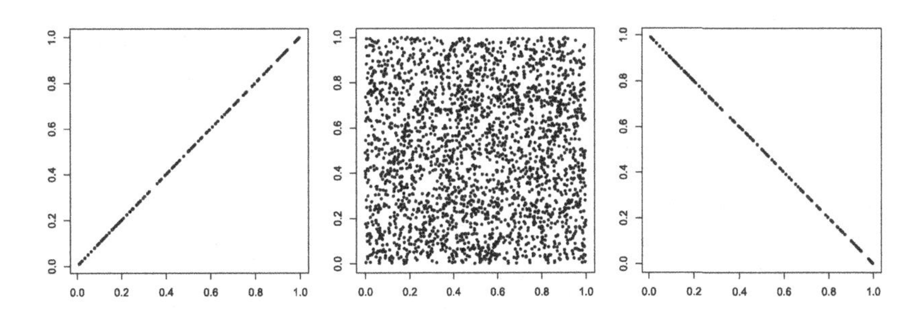

Fig. 1.4 Scatterplots of 250 samples from the comonotonicity copula (left), the countermonotonicity copula (right), and 2 500 samples from the independence copula (middle).

In the case of bivariate copulas, we introduce the two most prominent examples for later reference: *Kendall's tau* and *Spearman's rho*.

Definition 1.3 (Kendall's Tau)
Let C be a bivariate copula. Consider a probability space $(\Omega, \mathcal{F}, \mathbb{P})$ supporting $(U_1, U_2) \sim C$. The value

$$\tau_C := 4\,\mathbb{E}\big[C(U_1, U_2)\big] - 1$$

is called Kendall's tau *of the copula C.*

Some of the most important properties of Kendall's tau are listed in the following lemma.

Lemma 1.6 (Properties of Kendall's Tau)
Let $C, \tilde{C}$ be bivariate copulas.

(1) Consider a probability space $(\Omega, \mathcal{F}, \mathbb{P})$ supporting two i.i.d. random vectors $(U_1, U_2), (V_1, V_2) \sim C$. Then

$$\begin{aligned}\tau_C &= \mathbb{P}\big((U_1 - V_1)(U_2 - V_2) > 0\big) - \mathbb{P}\big((U_1 - V_1)(U_2 - V_2) < 0\big) \\ &= \mathbb{E}\big[sign\big((U_1 - V_1)(U_2 - V_2)\big)\big].\end{aligned}$$

(2) If $C \leq \tilde{C}$ pointwise, then $\tau_C \leq \tau_{\tilde{C}}$.

(3) For the independence copula, we have $\tau_\Pi = 0$. Furthermore, $\tau_C = 1 \Leftrightarrow C = M$.

(4) One has the representation

$$\tau_C = 1 - 4 \int_0^1 \int_0^1 \frac{\partial}{\partial u_1} C(u_1, u_2) \frac{\partial}{\partial u_2} C(u_1, u_2)\, du_1\, du_2.$$

Proof. See Nelsen (2006, Theorem 5.1.1, p. 159) for part (1). Parts (2) and (3) follow from Nelsen (2006, Theorem 5.1.9, p. 169). Part (4) follows from Li et al. (2002). Recall in particular from Lemma 1.2(2) that the partial derivatives of a copula exist almost everywhere. □

In particular, part (1) of Lemma 1.6 shows why it is reasonable to call Kendall's tau a *concordance measure*: it equals the difference of the probability of concordance and the probability of discordance of two pairs of i.i.d. random vectors. The formula of part (4) often helps to compute Kendall's tau. Here is an example.

Example 1.11 (Kendall's Tau of the Cuadras–Augé Copula)
Consider once again the copula C_α from Example 1.4. It is easy to compute

$$\frac{\partial}{\partial u_1} C_\alpha(u_1, u_2) = \begin{cases} u_2^{1-\alpha} & , u_1 < u_2 \\ (1-\alpha)\, u_2\, u_1^{-\alpha} & , u_1 > u_2 \end{cases},$$

$$\frac{\partial}{\partial u_2} C_\alpha(u_1, u_2) = \begin{cases} u_1^{1-\alpha} & , u_2 < u_1 \\ (1-\alpha)\, u_1\, u_2^{-\alpha} & , u_2 > u_1 \end{cases}.$$

By virtue of Lemma 1.6(4) this implies that

$$\begin{aligned}\tau_{C_\alpha} &= 1 - 4 \int_0^1 \int_0^{u_2} u_2^{1-\alpha}\,(1-\alpha)\,u_1\,u_2^{-\alpha}\,du_1 \\ &\quad + \int_{u_2}^1 u_1^{1-\alpha}\,(1-\alpha)\,u_2\,u_1^{-\alpha}\,du_1\,du_2 \\ &= 1 - 4 \int_0^1 u_2^{1-2\alpha}\,(1-\alpha)\,\frac{u_2^2}{2} + (1-\alpha)\,u_2\,\frac{1}{2-2\alpha}\,\big(1 - u_2^{2-2\alpha}\big)\,du_2 \\ &= 1 - 4\Big(\frac{1-\alpha}{2}\int_0^1 u_2^{3-2\alpha}\,du_2 + \frac{1}{2}\Big(\frac{1}{2} - \int_0^1 u_2^{3-2\alpha}\,du_2\Big)\Big) \\ &= 1 - 4\Big(\frac{1}{4} + \Big(\frac{1-\alpha}{2} - \frac{1}{2}\Big)\frac{1}{4-2\alpha}\Big) = \frac{\alpha}{2-\alpha}.\end{aligned}$$

Notice in particular that for α ranging in $[0,1]$, Kendall's tau interpolates between 0 *and* 1. *This is consistent with the fact that $C_0 = \Pi$ and $C_1 = M$.*

Another popular concordance measure is Spearman's rho.

Definition 1.4 (Spearman's Rho)
Let C be a bivariate copula. Consider a probability space $(\Omega, \mathcal{F}, \mathbb{P})$ supporting $(U_1, U_2) \sim C$. The value

$$\rho_C := Corr(U_1, U_2) = 12\,Cov(U_1, U_2)$$

is called Spearman's rho *of the copula C.*

Some of the most important properties of Spearman's rho are listed in the following lemma.

Lemma 1.7 (Properties of Spearman's Rho)
Let $C, \tilde{C}$ be bivariate copulas.

(1) Consider $(\Omega, \mathcal{F}, \mathbb{P})$ supporting three i.i.d. random vectors (U_1, U_2), (V_1, V_2), $(W_1, W_2) \sim C$. Then one has $\rho_C =$

$$3\Big(\mathbb{P}\big((U_1 - V_1)\,(U_2 - W_2) > 0\big) - \mathbb{P}\big((U_1 - V_1)\,(U_2 - W_2) < 0\big)\Big).$$

(2) If $C \le \tilde{C}$ pointwise, then $\rho_C \le \rho_{\tilde{C}}$.
(3) One has $\rho_\Pi = 0$. Furthermore, $\rho_C = 1 \Leftrightarrow C = M$.
(4) One has

$$\rho_C = 12 \int_0^1 \int_0^1 C(u_1, u_2)\,du_1\,du_2 - 3.$$

Proof. See Nelsen (2006, Theorem 5.1.6, p. 167) for part (1). Parts (2) and (3) follow from Nelsen (2006, Theorem 5.1.9, p. 169). Part (4) follows from Fubini's theorem:

$$\begin{aligned}\mathbb{E}[U_1\,U_2] &= \mathbb{E}\Big[\int_0^1\int_0^1 \mathbb{1}_{\{u_1<U_1,u_2<U_2\}}\,du_1\,du_2\Big]\\ &= \int_0^1\int_0^1 \mathbb{P}(U_1>u_1,U_2>u_2)du_1\,du_2,\end{aligned}$$

which, together with $\mathbb{P}(U_1>u_1,U_2>u_2)=1-u_1-u_2+C(u_1,u_2)$, implies the claimed formula. □

Again, the formula of part (4) sometimes helps to compute Spearman's rho. Here is an example.

Example 1.12 (Spearman's Rho of the Cuadras–Augé Copula) *Consider once again the copula C_α from Example 1.4. One computes*

$$\begin{aligned}\rho_{C_\alpha} &= 12\int_0^1\int_0^1 \min\{u_1,u_2\}\,\max\{u_1,u_2\}^{1-\alpha}\,du_1\,du_2-3\\ &= 12\int_0^1 u_2^{1-\alpha}\int_0^{u_2} u_1\,du_1+u_2\int_{u_2}^1 u_1^{1-\alpha}du_1\,du_2-3\\ &= 12\int_0^1 u_2^{1-\alpha}\frac{u_2^2}{2}+u_2\frac{1}{2-\alpha}\left(1-u_2^{2-\alpha}\right)du_2-3\\ &= 12\left(\frac{1}{2}\int_0^1 u_2^{3-\alpha}\,du_2+\frac{1}{2-\alpha}\left(\frac{1}{2}-\int_0^1 u_2^{3-\alpha}\,du_2\right)\right)-3\\ &= 12\left(\frac{1}{2\,(2-\alpha)}-\left(\frac{1}{2-\alpha}-\frac{1}{2}\right)\frac{1}{4-\alpha}\right)-3=\frac{3\,\alpha}{4-\alpha}.\end{aligned}$$

Like in the case of Kendall's tau, one verifies that $\rho_{C_0}=0$ and $\rho_{C_1}=1$, which is consistent with the facts that $C_0=\Pi$ and $C_1=M$.

Rewriting the formula of Lemma 1.7(4) for ρ_C, one checks that

$$\rho_C=\frac{\int_0^1\int_0^1 C(u_1,u_2)\,du_1\,du_2-\int_0^1\int_0^1\Pi(u_1,u_2)\,du_1\,du_2}{\int_0^1\int_0^1 M(u_1,u_2)\,du_1\,du_2-\int_0^1\int_0^1\Pi(u_1,u_2)\,du_1\,du_2}.\tag{1.12}$$

This means that ρ_C can be considered a normalized average distance between the copula C and the independence copula Π. Formula (1.12) suggests a natural extension to higher dimensions $d>2$, due to Wolff (1980).

Definition 1.5 (Multivariate Spearman's Rho)
For a given copula $C : [0,1]^d \to [0,1]$ Spearman's multivariate rho *is defined by* $\rho_C :=$

$$\frac{\int_{[0,1]^d} C(u_1,\ldots,u_d)\,d(u_1,\ldots,u_d) - \int_{[0,1]^d} \Pi(u_1,\ldots,u_d)\,d(u_1,\ldots,u_d)}{\int_{[0,1]^d} M(u_1,\ldots,u_d)\,d(u_1,\ldots,u_d) - \int_{[0,1]^d} \Pi(u_1,\ldots,u_d)\,d(u_1,\ldots,u_d)}.$$

Solving the integrals over Π and M in the definition of ρ_C, it follows that

$$\rho_C = \frac{d+1}{2^d-(d+1)}\Big(2^d \int_{[0,1]^d} C(u_1,\ldots,u_d)\,d(u_1,\ldots,u_d) - 1\Big).$$

Like in the bivariate case, ρ_C obviously satisfies $\rho_\Pi = 0$ and $\rho_M = 1$. This and other multivariate extensions of bivariate concordance measures can also be found, e.g., in Schmid and Schmidt (2006, 2007a,b).

1.1.6 *Measures of Extremal Dependence*

Especially for some applications in the context of mathematical finance and insurance risk, measures of extremal dependence have gained increasing interest in recent years. The reason for this is that various empirical studies suggest that during economic crises, the dependence between economic entities is often "extreme". A typical question would be: Given one entity suffers badly from a crisis, how much does this affect another entity? To quantify the answers to such questions, several measures of extremal dependence have been invented. Some of them are listed as follows.

Definition 1.6 (Upper- and Lower-Tail Dependence)
For a bivariate copula C the coefficients of upper- and lower-tail dependence UTD_C *and* LTD_C *are defined as*

$$UTD_C := \lim_{u\uparrow 1} \frac{C(u,u)-2\,u+1}{1-u}, \quad LTD_C := \lim_{u\downarrow 0} \frac{C(u,u)}{u},$$

if the respective limit exists (see Larsson and Nešlehová (2011) for copulas without existing lower-tail dependence).

The intuition behind Definition 1.6 is that for a random vector (U_1, U_2) on a probability space $(\Omega, \mathcal{F}, \mathbb{P})$, whose joint distribution function is the copula

C, it holds that

$$\begin{aligned} UTD_C &= \lim_{u \uparrow 1} \frac{C(u,u) - 2\,u + 1}{1-u} \\ &= \lim_{u \uparrow 1} \frac{\mathbb{P}(U_1 \le u, U_2 \le u) - \mathbb{P}(U_1 \le u) - \mathbb{P}(U_2 \le u) + 1}{\mathbb{P}(U_2 > u)} \\ &\overset{L.1.1}{=} \lim_{u \uparrow 1} \frac{\mathbb{P}(U_1 > u, U_2 > u)}{\mathbb{P}(U_2 > u)} = \lim_{u \uparrow 1} \mathbb{P}(U_1 > u \,|\, U_2 > u). \end{aligned}$$

Thus, the coefficient of upper-tail dependence equals the probability that U_1 is large given U_2 is large. Similarly, LTD_C equals the probability that U_1 is small given U_2 is small. Positive upper- or lower-tail dependence are desirable in stochastic models that support extreme scenarios. A bivariate Gaussian copula is a popular example for a distribution whose tail dependencies are both 0 (see Lemma 4.7). As a consequence, models based on normality assumptions are often criticized for their lack of extreme scenarios.

Remark 1.2 (Tail Dependence for Arbitrary Random Vectors)
The notions of upper- and lower-tail dependence are more generally defined for arbitrary bivariate random vectors (X_1, X_2) with continuous, but not necessarily uniform, marginal laws F_1, F_2. The definition in this more general case is the same as before, where the copula C is the copula of the corresponding random vector. More clearly, the lower-tail dependence coefficient of the bivariate random vector (X_1, X_2), defined on $(\Omega, \mathcal{F}, \mathbb{P})$, is given by

$$\lim_{x \downarrow 0} \mathbb{P}\big(X_1 \le F_1^{-1}(x) \,|\, X_2 \le F_2^{-1}(x)\big) = \lim_{u \downarrow 0} \frac{C(u,u)}{u} = LTD_C,$$

provided the existence of the limit. For the upper-tail dependence coefficient the corresponding similar definition of UTD_C applies. Notice in particular that in the definition of LTD_C it does not matter whether we exchange X_1 and X_2.

Remark 1.3 (Tail Dependence of the Survival Copula)
Existence provided, the upper-tail dependence coefficient of the survival copula $\hat{C}$ of a random vector equals the lower-tail dependence coefficient of the respective copula C, i.e. $UTD_{\hat{C}} = LTD_C$. This follows from the computation (1.10):

$$
\begin{aligned}
UTD_{\hat{C}} &= \lim_{u \uparrow 1} \frac{\hat{C}(u,u) - 2\,u + 1}{1-u} \\
&\stackrel{(1.10)}{=} \lim_{u \uparrow 1} \frac{1 + C(1-u, 1-u) - 2\,(1-u) - 2\,u + 1}{1-u} \\
&= \lim_{u \downarrow 0} \frac{C(u,u)}{u} = LTD_C.
\end{aligned}
$$

Similarly, $LTD_{\hat{C}} = UTD_C$, existence provided.

Example 1.13 (Tail Dependence of the Cuadras–Augé Copula)
In the case of a bivariate Cuadras–Augé copula C_α, as introduced in Example 1.4, one may check (using the rule of de l'Hospital) that $UTD_{C_\alpha} = \alpha$, since

$$
\begin{aligned}
UTD_{C_\alpha} &= \lim_{u \uparrow 1} \frac{C_\alpha(u,u) - 2\,u + 1}{1-u} \\
&= \lim_{u \uparrow 1} \frac{u^{1+1-\alpha} - 2\,u + 1}{1-u} = \lim_{u \uparrow 1} \frac{(2-\alpha)\,u^{1-\alpha} - 2}{-1} = \alpha.
\end{aligned}
$$

In contrast, C_α has zero lower-tail dependence (unless $\alpha = 1$):

$$
LTD_{C_\alpha} = \lim_{u \downarrow 0} \frac{u^{1+1-\alpha}}{u} = \lim_{u \downarrow 0} u^{1-\alpha} = \mathbb{1}_{\{\alpha=1\}}.
$$

Asymmetric tail dependence parameters, i.e. $UTD_{C_\alpha} \neq LTD_{C_\alpha}$, are sometimes desirable. For example if C_α is the survival copula of two companies' bankruptcy times, then positive upper-tail dependence of C_α – correspondingly lower-tail dependence of the bankruptcy times by Remark 1.3 – has an intuitive interpretation. It implies that the early default of one firm is likely to coincide with the early default of the other firm. In contrast, zero lower-tail dependence of C_α means that the extraordinarily long survival of one firm does not automatically induce longevity of the other.

One possible extension of the bivariate concept of tail dependence to dimensions $d > 2$ is introduced by Frahm (2006) and is presented here.

Definition 1.7 (Upper- and Lower-Extremal Dependence Coefficient)
For a d-dimensional copula $C : [0,1]^d \to [0,1]$ the upper- and lower-extremal dependence coefficients *$UEDC_C$ and $LEDC_C$ are defined by*

$$
\begin{aligned}
UEDC_C &:= \lim_{u \uparrow 1} \frac{1 + \sum_{i=1}^{d} (-1)^i \sum_{1 \leq j_1 < \ldots < j_i \leq d} C_{j_1,\ldots,j_i}(u,\ldots,u)}{1 - C(u,\ldots,u)}, \\
LEDC_C &:= \lim_{u \downarrow 0} \frac{C(u_1,\ldots,u_d)}{\sum_{i=1}^{d} (-1)^{i+1} \sum_{1 \leq j_1 < \ldots < j_i \leq d} C_{j_1,\ldots,j_i}(u,\ldots,u)},
\end{aligned}
$$

if the corresponding limit exists.

Using the principle of inclusion and exclusion (see Lemma 1.1) it is possible to understand the intuition behind this definition. Letting $(U_1, \ldots, U_d)$ be defined on $(\Omega, \mathcal{F}, \mathbb{P})$ and having as joint distribution function the copula C, it holds that

$$\begin{aligned}
UEDC_C &= \lim_{u \uparrow 1} \frac{1 + \sum_{i=1}^{d} (-1)^i \sum_{1 \le j_1 < \ldots < j_i \le d} C_{j_1, \ldots, j_i}(u, \ldots, u)}{1 - C(u, \ldots, u)} \\
&= \lim_{u \uparrow 1} \frac{1 + \sum_{i=1}^{d} (-1)^i \sum_{1 \le j_1 < \ldots < j_i \le d} \mathbb{P}(U_{j_1} \le u, \ldots, U_{j_i} \le u)}{1 - \mathbb{P}(U_1 \le u, \ldots, U_d \le u)} \\
&\overset{L.1.1}{=} \lim_{u \uparrow 1} \frac{\mathbb{P}(U_1 > u, \ldots, U_d > u)}{\mathbb{P}\big(\max\{U_1, \ldots, U_d\} > u\big)} \\
&= \lim_{u \uparrow 1} \mathbb{P}\big(\min\{U_1, \ldots, U_d\} > u \,\big|\, \max\{U_1, \ldots, U_d\} > u\big).
\end{aligned}$$

Thus, $UEDC_C$ gives the probability that all components of $(U_1, \ldots, U_d) \sim C$ are large given at least one of them is large. Similarly, $LEDC_C$ gives the probability that all components are small given at least one is small. Let us point out that in the case $d = 2$ the $UEDC$ is **not** the same as the UTD, the difference being that one conditions on the maximum of U_1, U_2 being greater than u, instead of just conditioning on U_2 being greater than u. There are other versions of upper-extremal dependence coefficients which really extend the bivariate UTD (see, e.g., Li (2009)). To conclude, the UTD as well as the $UEDC$ are limits of probabilities, hence they take values in $[0, 1]$, and the closer they are to 1, the more likely are extreme events. Similar interpretations hold for LTD and $LEDC$.

1.2 General Classifications of Copulas

This section provides a structural classification of dependence structures. Many specific parametric families of copulas encountered in later chapters can be categorized according to this classification. Some structural properties discussed here are useful to understand the probabilistic model behind some specific parametric families of copulas.

1.2.1 *Radial Symmetry*

A random variable $X \in \mathbb{R}$ is called *symmetric* about $x \in \mathbb{R}$, if $X - x$ has the same distribution as $x - X$; or equivalently, if $X \overset{d}{=} 2\,x - X$. In terms of the distribution function F and the survival function $\bar{F}$ of X, this is equivalent to the fact that $F(y) = 1 - F\big((2\,x - y) -\big)$ for all $y \in \mathbb{R}$, since

$$\begin{aligned}F(y) &= \mathbb{P}(X \leq y) = \mathbb{P}(2\,x - X \leq y) = \mathbb{P}(X \geq 2\,x - y) \\ &= \bar{F}(2\,x - y) + \Delta F(2\,x - y) \\ &= 1 - F(2\,x - y) + F(2\,x - y) - F\big((2\,x - y) -\big) \\ &= 1 - F\big((2\,x - y) -\big).\end{aligned}$$

If F is continuous, then this is equivalent to $F(y) = \bar{F}(2\,x - y)$, $y \in \mathbb{R}$. If X has finite mean, it follows that $\mathbb{E}[X] = x$. The normal distribution is a prominent example of a distribution which is symmetric about its mean. This definition of symmetry is extended to the multivariate setup in the following definition.

Definition 1.8 (Radially Symmetric)

(1) A d-dimensional random vector $(X_1, \ldots, X_d)$ is called radially symmetric *about $(x_1, \ldots, x_d) \in \mathbb{R}^d$ if*

$$(X_1, \ldots, X_d) \stackrel{d}{=} (2\,x_1 - X_1, \ldots, 2\,x_d - X_d).$$

(2) A copula C is called radially symmetric *if it is its own survival copula, i.e. if $C = \hat{C}$.*

It is important to note that radial symmetry is a stronger notion than componentwise symmetry. By definition, the components of a radially symmetric random vector are symmetric themselves. However, there are random vectors with symmetric components that are not radially symmetric.[8] Here is an example.

Example 1.14 (Componentwise Symmetry $\not\Rightarrow$ Radial Symmetry)
We consider a probability space $(\Omega, \mathcal{F}, \mathbb{P})$ supporting two i.i.d. $U[0,1]$-distributed random variables V_1, V_2, and define (U_1, U_2) by

$$(U_1, U_2) := \Big(\max\{V_1^2, V_2^2\}, \max\{V_1^3, V_2^{\frac{3}{2}}\}\Big).$$

Recall from Example 1.9 with $(a, b) = (1/2, 1/3)$ that the joint distribution function of (U_1, U_2) is the copula C given by

$$C(u_1, u_2) = \min\Big\{u_1^{\frac{1}{2}}, u_2^{\frac{1}{3}}\Big\} \min\Big\{u_1^{\frac{1}{2}}, u_2^{\frac{2}{3}}\Big\}, \quad u_1, u_2 \in [0,1].$$

---

[8]Every random vector $(U_1, \ldots, U_d)$ which has a copula as its distribution function has symmetric components, since $U_k \sim U[0,1]$ is symmetric about $1/2$.

In particular, U_1, U_2 are both $U[0,1]$-distributed, i.e. symmetric about $1/2$. However, (U_1, U_2) is not radially symmetric about $(1/2, 1/2)$. This can be seen as follows: on the one hand, we have

$$\begin{aligned}\mathbb{P}\Big(U_1 - \frac{1}{2} \leq \frac{1}{4}, U_2 - \frac{1}{2} \leq \frac{1}{4}\Big) &= \mathbb{P}\Big(U_1 \leq \frac{3}{4}, U_2 \leq \frac{3}{4}\Big)\\ &= C\Big(\frac{3}{4}, \frac{3}{4}\Big) = \Big(\frac{3}{4}\Big)^{\frac{7}{6}} \approx 0.715.\end{aligned}$$

On the other hand, we compute that

$$\begin{aligned}&\mathbb{P}\Big(\frac{1}{2} - U_1 \leq \frac{1}{4}, \frac{1}{2} - U_2 \leq \frac{1}{4}\Big)\\ &\quad = \mathbb{P}\Big(U_1 \geq \frac{1}{4}, U_2 \geq \frac{1}{4}\Big)\\ &\quad = 1 - \mathbb{P}\Big(U_1 \leq \frac{1}{4}\Big) - \mathbb{P}\Big(U_2 \leq \frac{1}{4}\Big) + \mathbb{P}\Big(U_1 \leq \frac{1}{4}, U_2 \leq \frac{1}{4}\Big)\\ &\quad = \frac{1}{2} + \Big(\frac{1}{4}\Big)^{\frac{7}{6}} \approx 0.698.\end{aligned}$$

To understand part (2) of Definition 1.8, observe that the $U[0,1]$-law is symmetric about $1/2$. If C is a copula and $(U_1, \ldots, U_d) \sim C$ on a probability space $(\Omega, \mathcal{F}, \mathbb{P})$, then $(U_1, \ldots, U_d)$ is radially symmetric (necessarily about $(1/2, \ldots, 1/2)$), if

$$(U_1, \ldots, U_d) \stackrel{d}{=} (1 - U_1, \ldots, 1 - U_d).$$

Since $(1 - U_1, \ldots, 1 - U_d) \sim \hat{C}$, this means precisely that C is its own survival copula, i.e. $C = \hat{C}$. One can further show that radially symmetric copulas are precisely the copulas of radially symmetric random vectors. To see this in the simpler case of continuous margins, consider the following argument. On a probability space $(\Omega, \mathcal{F}, \mathbb{P})$, suppose that $(X_1, \ldots, X_d)$ is radially symmetric about $(x_1, \ldots, x_d)$ and has continuous margins. Denote its margins by $F_1, \ldots, F_d$, its unique copula by C, and its unique survival copula by $\hat{C}$. Using the definition of radial symmetry in Equation $(*)$ below, it follows for all $y_1, \ldots, y_d \in \mathbb{R}$ that

$$\begin{aligned}&C\big(F_1(y_1), \ldots, F_d(y_d)\big) = \mathbb{P}(X_1 \leq y_1, \ldots, X_d \leq y_d)\\ &\quad = \mathbb{P}(X_1 - x_1 \leq y_1 - x_1, \ldots, X_d - x_d \leq y_d - x_d)\\ &\quad \stackrel{(*)}{=} \mathbb{P}(x_1 - X_1 \leq y_1 - x_1, \ldots, x_d - X_d \leq y_d - x_d)\\ &\quad = \mathbb{P}(X_1 \geq 2\,x_1 - y_1, \ldots, X_d \geq 2\,x_d - y_d)\\ &\quad = \hat{C}\big(\bar{F}_1(2\,x_1 - y_1), \ldots, \bar{F}_d(2\,x_d - y_d)\big) \stackrel{(**)}{=} \hat{C}\big(F_1(y_1), \ldots, F_d(y_d)\big).\end{aligned}$$

Equation (**) follows from the componentwise symmetry again. Hence, radially symmetric random vectors are characterized by symmetric margins and a copula which equals its own survival copula, i.e. $C = \hat{C}$.

Concerning the tail dependence properties of a bivariate radially symmetric random vector, it is immediately clear from $C = \hat{C}$ that the upper- and lower-tail dependence coefficients are identical. Moreover, radial symmetry of a bivariate copula means that a scatterplot exhibits a great level of symmetry with respect to the counterdiagonal $\{u_2 = 1 - u_1\}$.

1.2.2 *Exchangeability*

Generally speaking, the distribution function F of a random vector $(X_1, \ldots, X_d)$ can be a very complicated object. In particular for large dimensions $d \gg 2$ there are only a limited number of examples of distribution functions F that allow for a successful analytical study. If the distribution is *exchangeable*, however, the associated distribution function F can often be simplified and written in a quite compact form, which is convenient to explore further. Exchangeability intuitively means that the dependence structure between the components of a random vector is completely symmetric and does not depend on the ordering of the components. Such symmetric dependence structures sometimes arise naturally and are often quite intuitive and easy to understand. For this reason, many examples of copulas encountered in this book are exchangeable. The respective definition is the following.

Definition 1.9 (Exchangeability)

(1) A random vector $(X_1, \ldots, X_d)$ on a probability space $(\Omega, \mathcal{F}, \mathbb{P})$ is called exchangeable *if for all permutations $\pi : \{1, \ldots, d\} \to \{1, \ldots, d\}$ one has that*

$$(X_1, \ldots, X_d) \stackrel{d}{=} (X_{\pi(1)}, \ldots, X_{\pi(d)}).$$

(2) An infinite sequence $\{X_k\}_{k \in \mathbb{N}}$ of random variables on a probability space $(\Omega, \mathcal{F}, \mathbb{P})$ is called exchangeable *if for each $d \in \mathbb{N}$ and $1 \leq i_1 < i_2 < \ldots < i_d$ the random vector $(X_{i_1}, \ldots, X_{i_d})$ is exchangeable.*

(3) A distribution function F is called exchangeable, *if it is invariant with respect to permutations of its arguments, i.e. if for all permutations $\pi : \{1, \ldots, d\} \to \{1, \ldots, d\}$ one has that*

$$F(x_1, \ldots, x_d) = F(x_{\pi(1)}, \ldots, x_{\pi(d)}), \quad x_1, \ldots, x_d \in \mathbb{R}.$$

Using the notation of Sklar's theorem (see Theorem 1.2), this means that the margins $F_1, \ldots, F_d$ are identical and the copula C is permutation invariant.

A bivariate copula C is exchangeable if and only if $C(u_1, u_2) = C(u_2, u_1)$ for all $u_1, u_2 \in [0, 1]$. Moreover, exchangeability of a bivariate copula means that a scatterplot exhibits a great level of symmetry with respect to the diagonal $\{u_1 = u_2\}$. The notion of an infinite exchangeable sequence is included, since it plays an important role in the subsequent paragraph. It is immediately clear that a random vector $(X_1, \ldots, X_d)$ on a probability space $(\Omega, \mathcal{F}, \mathbb{P})$ is exchangeable if and only if its distribution function F (or equivalently its survival function $\bar{F}$) is invariant with respect to permutations of its arguments. This also explains part (3) in Definition 1.9. Intuitively, $(X_1, \ldots, X_d)$ is exchangeable if and only if all margins are identical, all two-margins are identical, all three-margins are identical, and so on. Here is an example.

Example 1.15 (Exchangeable Normal Distribution)
Let $\boldsymbol{X} := (X_1, \ldots, X_d) \sim \mathcal{N}_d(\boldsymbol{\mu}, \Sigma)$ be multivariate normally distributed with mean vector $\boldsymbol{\mu} := (\mu_1, \ldots, \mu_d)$ and positive definite covariance matrix Σ. Denote as usual the diagonal elements of Σ by $\sigma_1^2 := \Sigma_{11}, \ldots, \sigma_d^2 := \Sigma_{dd}$ and the off-diagonal elements by $\rho_{ij}\, \sigma_i\, \sigma_j := \Sigma_{ij}$, $i \neq j$, where $\rho_{ij} :=$ Corr(X_i, X_j). Since the multivariate normal distribution is characterized by $\boldsymbol{\mu}$ and Σ, a check for exchangeability requires one to only consider one- and two-dimensional margins. First, $X_1 \stackrel{d}{=} \ldots \stackrel{d}{=} X_d$ is equivalent to $\mu_1 = \ldots = \mu_d$, $\sigma_1^2 = \ldots = \sigma_d^2$. Provided we have this, $(X_i, X_j) \stackrel{d}{=} (X_1, X_2)$ is equivalent to $\rho_{ij} = \rho_{12}$ for all $i \neq j$. This finally implies that

$$\boldsymbol{X} \text{ is exchangeable} \Leftrightarrow \mu_i =: \mu \in \mathbb{R},\ \sigma_i^2 =: \sigma^2 > 0,\ \rho_{ij} =: \rho.$$

If this is the case, one can additionally show that $\rho > -1/(d-1)$ in order to guarantee that Σ is positive definite (see Lemma 1.8). To summarize, the d-dimensional exchangeable multivariate normal distribution is parameterized by three parameters $(\mu, \sigma^2, \rho) \in \mathbb{R} \times (0, \infty) \times (-1/(d-1), 1]$.

The dependence structure of an exchangeable random vector is limited to some extent. For instance, for exchangeable random vectors $(X_1, \ldots, X_d)$ with an existing covariance matrix, one can show that their pairwise correlation $\rho := \text{Corr}(X_i, X_j)$ (which is independent of i, j) is bounded below by $-1/(d-1)$. In particular, it follows that an infinite exchangeable sequence can never exhibit negative correlation between pairs.

Lemma 1.8 (Positive Definiteness of Exchangeable Covariance Matrix)
Let $\boldsymbol{X}$ be an exchangeable random vector with an existing covariance matrix Σ. Denote the pairwise correlation coefficient by $\rho \in [-1,1]$. Then $\rho \geq -1/(d-1)$. If Σ is positive definite, we obtain $\rho > -1/(d-1)$.

Proof. Denote the covariance matrix by Σ and recall that all off-diagonal entries are identical by exchangeability. Σ is positive semi-definite, hence it follows for every $\boldsymbol{x} = (x_1, \ldots, x_d) \in \mathbb{R}^d \setminus \{\boldsymbol{0}\}$ that

$$\boldsymbol{x}\,\Sigma\,\boldsymbol{x}' \geq 0 \Leftrightarrow \rho \geq -\frac{\sum_{i=1}^{d} x_i^2}{\sum_{i=1}^{d} \left(x_i \sum_{j\neq i} x_j\right)} =: -f(x_1, \ldots, x_d).$$

Let $x_1 = \ldots = x_d =: x \in \mathbb{R} \setminus \{0\}$ to obtain

$$\rho \geq -f(x, \ldots, x) = -\frac{d\,x^2}{d\,(d-1)\,x^2} = -\frac{1}{d-1}.$$

If Σ is positive definite, all "$\geq$" signs are replaced by "$>$" signs. □

1.2.3 *Homogeneous Mixture Models*

If all components of a random vector $\boldsymbol{X} := (X_1, \ldots, X_d)$ are independent and identically distributed with common (univariate) distribution function F, then simulation is trivial. We simply have to call d times the univariate sampling algorithm which returns a sample of F. One popular and convenient approach to extend this procedure to a case with dependent components is to randomize the function F. For example, the function F may be chosen randomly from a pre-specified class of distribution functions. As an example, take any parametric family of distribution functions and randomly draw the parameter(s) in a first step. Such a random choice of F makes the components of $\boldsymbol{X}$ dependent. Here is an example.

Example 1.16 (First Example of a Homogeneous Mixture)
On a probability space $(\Omega, \mathcal{F}, \mathbb{P})$ let $Y \sim Bin(1, 1/2)$ denote the outcome of a coin toss. If $Y = 1$, then we draw $(X_1, \ldots, X_d)$ i.i.d. from the $U[-1,0]$-distribution, and if $Y = 0$ we draw them i.i.d. from a $U[0,1]$-distribution. It is then clear that

$$\mathbb{P}(X_1 < 0, X_2 > 0) = \mathbb{P}(Y = 1, Y = 0) = 0,$$
$$\mathbb{P}(X_1 < 0)\,\mathbb{P}(X_2 > 0) = \mathbb{P}(Y = 1)\,\mathbb{P}(Y = 0) = \frac{1}{2} \cdot \frac{1}{2} = \frac{1}{4},$$

implying that X_1 and X_2 are dependent. In mathematical terms, this example may be constructed on a probability space $(\Omega, \mathcal{F}, \mathbb{P})$ supporting $d+1$ independent random variables $Y \sim Bin(1, 1/2)$, $U_1, \ldots, U_d \sim U[0,1]$ as follows:

(1) Define the (random) distribution function $F = \{F_x\}_{x \in \mathbb{R}}$ by

$$F_x := \begin{cases} (1+x)\,\mathbb{1}_{\{-1 \le x \le 0\}} + \mathbb{1}_{\{x>0\}} & , Y = 1 \\ x\,\mathbb{1}_{\{0 \le x \le 1\}} + \mathbb{1}_{\{x>1\}} & , Y = 0 \end{cases}, \quad x \in \mathbb{R}.$$

(2) Sample $X_1, \ldots, X_d$ i.i.d. from F. This might be accomplished by setting

$$X_k := \begin{cases} -U_k & , Y = 1 \\ U_k & , Y = 0 \end{cases}, \quad k = 1, \ldots, d.$$

In general, a generic simulation algorithm for such random vectors is given in Algorithm 1.3.

Algorithm 1.3 (Generic Sampling of Homogeneous Mixtures)
The input for the algorithm is a stochastic model for $F = \{F_x\}_{x \in \mathbb{R}}$.

(1) Sample a realization of the distribution function $F = \{F_x\}_{x \in \mathbb{R}}$.
(2) Draw i.i.d. samples $X_1, \ldots, X_d$ from F.

We say that a random vector $(X_1, \ldots, X_d)$, which is constructed as in Algorithm 1.3, is obtained as a *homogeneous mixture*. This nomenclature indicates that all possible paths $\{F(\omega)\}_{\omega \in \Omega}$ of F are "mixed" with a certain probability measure. This mixture is "homogeneous", since all components of $(X_1, \ldots, X_d)$ are affected by F in the very same manner. On the one hand, such dependence structures are very convenient for sampling applications, since a simulation algorithm can be performed in two consecutive steps. On the other hand, the class of multivariate distributions obtained by this approach is of a very special and limited form (see Theorem 1.4). Nevertheless, many important examples of copulas are motivated by such an approach (see, e.g., Sections 2.2 and 3.3).

In the sequel, the possible kinds of dependence structures that can be simulated via Algorithm 1.3 are determined. In this regard it is useful to note that there is an intimate connection between exchangeability and the aforementioned mixture approach. It is obvious that the random vector $(X_1, \ldots, X_d)$ constructed in Algorithm 1.3 is exchangeable. However, it even satisfies a stronger condition. To this end, the notion of extendibility is important.

Definition 1.10 (Extendibility)

(1) *An exchangeable random vector* $(X_1, \ldots, X_d)$ *on a probability space* $(\Omega, \mathcal{F}, \mathbb{P})$ *is called* extendible, *if there exists a probability space* $(\tilde{\Omega}, \tilde{\mathcal{F}}, \tilde{\mathbb{P}})$ *supporting an infinite exchangeable sequence* $\{\tilde{X}_k\}_{k \in \mathbb{N}}$ *such that* $(X_1, \ldots, X_d) \stackrel{d}{=} (\tilde{X}_1, \ldots, \tilde{X}_d)$.

(2) *An exchangeable distribution function* F *(resp. a copula* C*) is called* extendible, *if on some probability space* $(\Omega, \mathcal{F}, \mathbb{P})$ *a random vector* $(X_1, \ldots, X_d) \sim F$ *(resp.* $(U_1, \ldots, U_d) \sim C$*) is extendible.*

Unlike in the case of exchangeability, there is no immediate analytic way to check or define the extendibility of a distribution function F (or a copula C). Whereas it is easy to check whether a given distribution function is exchangeable, it is typically very difficult to check whether it is extendible or not. For examples see Chapters 2 and 3.

Remark 1.4 (Exchangeability vs. Extendibility)
If a random vector is extendible, this immediately implies that it is also exchangeable by definition. Conversely, not every exchangeable random vector is extendible. In fact, we will encounter several such examples in later chapters. To provide a very simple example, consider a probability space $(\Omega, \mathcal{F}, \mathbb{P})$ *supporting a standard normally distributed random variable* $X \sim \mathcal{N}(0,1)$. *Then the bivariate random vector* $(X, -X)$ *is clearly exchangeable, since* $X \stackrel{d}{=} -X$. *However, Lemma 1.9 shows that* $(X, -X)$ *cannot be extendible, since* $Corr(X, -X) = -1 < 0$.

It is immediately clear that a random vector $\boldsymbol{X} = (X_1, \ldots, X_d)$ which is constructed as in Algorithm 1.3 is not only exchangeable, but also extendible: instead of drawing d samples $X_1, \ldots, X_d$ in step (2), we could as well draw these samples in perturbed order (exchangeability), and we could as well draw infinitely many samples $X_1, X_2, X_3, \ldots$ (extendibility). Conversely, it can also be shown that every exchangeable and extendible random vector can be simulated[9] by Algorithm 1.3. The proof of this fact relies on a seminal structural result by Bruno de Finetti and can be stated as follows in Theorem 1.4.

[9] However, it is not always easy to find the random distribution function $\{F_x\}_{x \in \mathbb{R}}$.

Theorem 1.4 (De Finetti's Theorem)
An infinite sequence of random variables $\{\tilde{X}_k\}_{k\in\mathbb{N}}$ *on a probability space* $(\tilde{\Omega}, \tilde{\mathcal{F}}, \tilde{\mathbb{P}})$ *is exchangeable if and only if it is conditionally i.i.d., i.e. there exists a sub-*σ*-algebra* $\tilde{\mathcal{G}} \subset \tilde{\mathcal{F}}$ *such that*

$$\tilde{\mathbb{P}}(\tilde{X}_1 \leq x_1, \ldots, \tilde{X}_d \leq x_d \,|\, \tilde{\mathcal{G}}) = \prod_{k=1}^{d} \tilde{\mathbb{P}}(\tilde{X}_1 \leq x_k \,|\, \tilde{\mathcal{G}})$$

holds for all $x_1, \ldots, x_d \in \mathbb{R}$.

Proof. Originally due to de Finetti (1937). An elegant proof can be given as an application of the so-called reversed martingale convergence theorem (see Aldous (1985, Theorem 3.1)). □

Assume $(X_1, \ldots, X_d)$, defined on $(\Omega, \mathcal{F}, \mathbb{P})$, is exchangeable and extendible. Let $\{\tilde{X}_k\}_{k\in\mathbb{N}}$ be an extension of $(X_1, \ldots, X_d)$ on $(\tilde{\Omega}, \tilde{\mathcal{F}}, \tilde{\mathbb{P}})$ and let $\tilde{\mathcal{G}}$ be as in Theorem 1.4. Since $(X_1, \ldots, X_d) \stackrel{d}{=} (\tilde{X}_1, \ldots, \tilde{X}_d)$, sampling $(\tilde{X}_1, \ldots, \tilde{X}_d)$ is equivalent to sampling $(X_1, \ldots, X_d)$. A sample of $(\tilde{X}_1, \ldots, \tilde{X}_d)$ can be obtained using Algorithm 1.3 with input $F = \{F_x\}_{x\in\mathbb{R}}$, where

$$F_x := \tilde{\mathbb{P}}(\tilde{X}_1 \leq x \,|\, \tilde{\mathcal{G}}), \quad x \in \mathbb{R}.$$

Note that F_x is a $\tilde{\mathcal{G}}$-measurable random variable, and so is particularly random! There are in fact some examples where the latter conditional probability can be computed explicitly and sampling from the respective $\{F_x\}_{x\in\mathbb{R}}$ is possible. Here is an example.

Example 1.17 (Extendible Multivariate Normal Distribution)
Let $\boldsymbol{X} = (X_1, \ldots, X_d)$ *be exchangeable and multivariate normally distributed, with a positive definite covariance matrix. Example 1.15 shows that the distribution of* $\boldsymbol{X}$ *is parameterized by* $(\mu, \sigma^2, \rho) \in \mathbb{R} \times (0, \infty) \times (-1/(d-1), 1]$. *It is shown in the sequel that* $\boldsymbol{X}$ *is extendible if and only if* $\rho \geq 0$. *The necessity is clear, since* $-1/(d-1) \to 0$ *as* $d \to \infty$. *To see the sufficiency, consider a probability space* $(\tilde{\Omega}, \tilde{\mathcal{F}}, \tilde{\mathbb{P}})$ *supporting i.i.d. standard normally distributed random variables* $\epsilon_0, \epsilon_1, \epsilon_2, \ldots$. *Define the random variables*

$$\tilde{X}_k := \mu + \sigma\,(\sqrt{\rho}\,\epsilon_0 + \sqrt{1-\rho}\,\epsilon_k), \quad k \in \mathbb{N}.$$

Then $\{\tilde{X}_k\}_{k\in\mathbb{N}}$ *is an infinite exchangeable sequence, which satisfies*

$$\begin{pmatrix} \tilde{X}_1 \\ \tilde{X}_2 \\ \vdots \\ \tilde{X}_d \end{pmatrix} = \begin{pmatrix} \mu \\ \mu \\ \vdots \\ \mu \end{pmatrix} + \begin{pmatrix} \sigma\sqrt{\rho} & \sigma\sqrt{1-\rho} & 0 & \ldots & 0 \\ \sigma\sqrt{\rho} & 0 & \sigma\sqrt{1-\rho} & \ldots & 0 \\ & \vdots & & \ddots & \vdots \\ \sigma\sqrt{\rho} & 0 & 0 & \ldots & \sigma\sqrt{1-\rho} \end{pmatrix} \begin{pmatrix} \epsilon_0 \\ \epsilon_1 \\ \vdots \\ \epsilon_d \end{pmatrix},$$

implying that $(\tilde{X}_1, \ldots, \tilde{X}_d) \stackrel{d}{=} (X_1, \ldots, X_d)$ *(see Lemma 4.3(2)). Hence,* $\{\tilde{X}_k\}_{k \in \mathbb{N}}$ *is an extension of* $\boldsymbol{X}$*. Note in particular that* $\rho \geq 0$ *is important for this extension, since otherwise* $\sqrt{\rho}$ *is not well defined. Using the language of de Finetti's theorem,* $\tilde{X}_1, \tilde{X}_2, \ldots$ *are i.i.d. conditioned on the sigma algebra generated by* ϵ_0*, i.e.* $\tilde{\mathcal{G}} := \sigma(\epsilon_0)$*, with univariate distribution function (*Φ *is the distribution function of* $\mathcal{N}(0,1)$*)*

$$F_x = \tilde{\mathbb{P}}(\tilde{X}_1 \leq x | \tilde{\mathcal{G}}) = \Phi\Big(\frac{x - \mu - \sigma\sqrt{\rho}\,\epsilon_0}{\sigma\sqrt{1-\rho}}\Big), \quad x \in \mathbb{R}.$$

Note that F_x *is random, since it depends on* ϵ_0*.*

This book is especially concerned with the simulation of copulas. In order to guarantee that the multivariate distribution function obtained from a homogeneous mixture as in Algorithm 1.3 is a copula, one has to guarantee that the marginal distributions of the components are uniformly distributed on the unit interval. This means that one has to ensure that $F = \{F_x\}_{x \in \mathbb{R}}$ satisfies $F_0 = 0$, $F_1 = 1$, and $\mathbb{P}(X_i \leq x) = \mathbb{E}[F_x] = x$ for $x \in [0,1]$. In this case, we denote as usual $(U_1, \ldots, U_d) := (X_1, \ldots, X_d)$ and $F = \{F_u\}_{u \in [0,1]}$. Some specific examples of such algorithms can be found in Chapters 2 and 3. Here is a simple example.

Example 1.18 (Maresias Copula)
Let $G, H : [0,1] \to [0,1]$ *be two distribution functions on* $[0,1]$ *satisfying* $G(u) = 2u - H(u)$*,* $u \in [0,1]$*. Valid choices for* G *are, for instance,* $G_1(u) = \mathbb{1}_{\{u > 1/2\}}(2u - 1)$*,* $G_2(u) = u^\alpha$ *for* $\alpha \in [1,2]$*, or*

$$G_3(u) = \begin{cases} u/\alpha & , u < 1/2 \\ (2 - 1/\alpha)\,u - (1 - 1/\alpha) & , u \geq 1/2 \end{cases}, \quad u \in [0,1],$$

for a parameter $\alpha \geq 1$*. Consider a probability space* $(\Omega, \mathcal{F}, \mathbb{P})$ *supporting a Bernoulli-distributed random variable* $Y \sim Bin(1, 1/2)$*. Define the random distribution function*

$$F_u := \begin{cases} G(u) & , Y = 1 \\ H(u) & , Y = 0 \end{cases}, \quad u \in [0,1].$$

It is then easy to observe that $\mathbb{E}[F_u] = 1/2\,\big(G(u) + H(u)\big) = u$*,* $u \in [0,1]$*. Hence, if we first draw* Y *in order to determine* F*, and subsequently draw random variables* $U_1, U_2, \ldots, U_d$ *independently from* F*, they are all uniformly distributed on* $[0,1]$*. Hence, their joint distribution function is a copula, given by*

$$C_G(u_1, \ldots, u_d) = \frac{1}{2}\Big(\prod_{k=1}^{d} G(u_k) + \prod_{k=1}^{d} H(u_k)\Big), \quad u_1, \ldots, u_d \in [0,1].$$

Copulas obtained from this construction are called Maresias copulas, *because this example was constructed there. Figure 1.5 illustrates the functions G and H, as well as a scatterplot of the resulting bivariate copula C_G, for the case $G = G_1$.*

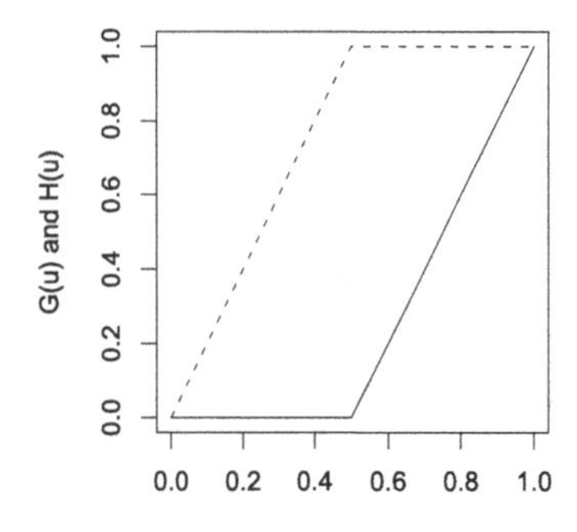

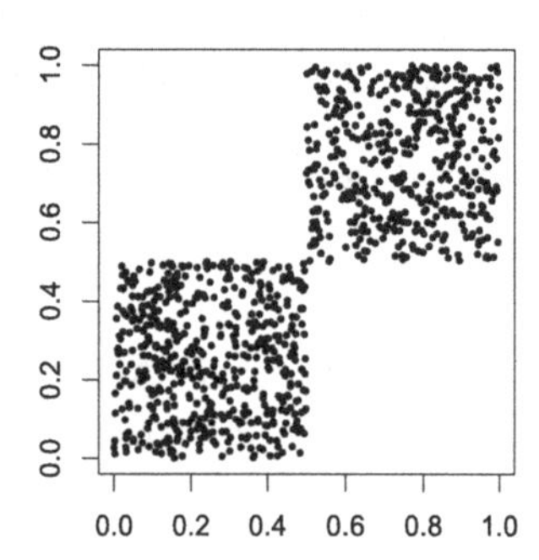

Fig. 1.5 Function plot of G_1 and the corresponding $H_1(u) := 2\,u - G_1(u)$ (left). Scatterplot of 1 000 samples from the Maresias copula corresponding to G_1 (right).

Example 1.19 (Extendibility of Bivariate Cuadras–Augé Copula)
Recall the copula C_α from Example 1.4. Since $C_\alpha(u_1, u_2) = C_\alpha(u_2, u_1)$ for all $u_1, u_2 \in [0, 1]$, a random vector $(U_1, U_2) \sim C_\alpha$ on $(\Omega, \mathcal{F}, \mathbb{P})$ is exchangeable. It can be shown furthermore that it is also extendible (see Section 3.3 for more details). This chapter also shows that an extension does not need to be unique.

Random vectors that are constructed by homogeneous mixtures exhibit a somewhat limited dependence structure. For instance, the distribution (function) of an arbitrary k-margin $(X_{i_1}, \ldots, X_{i_k})$ of $(X_1, \ldots, X_d)$ does only depend on k, and not on the specific subset $\{i_1, \ldots, i_k\} \subset \{1, \ldots, d\}$. This simply means that extendibility implies exchangeability. A further restriction is the fact that such random vectors in some sense cannot exhibit negative association (see, e.g., the following lemma).

Lemma 1.9 (Pairwise Correlation in Homogeneous Mixture Models)
Let $(X_1, \ldots, X_d)$ be an (exchangeable and) extendible random vector on a probability space $(\Omega, \mathcal{F}, \mathbb{P})$. Furthermore, assume that all (co-) variances of $(X_1, \ldots, X_d)$ exist. Then for each $1 \leq i < j \leq d$ it holds that $Corr(X_i, X_j) \geq 0$.

Proof. This follows directly from Lemma 1.8 by considering $d \to \infty$. However, with de Finetti's theorem at hand, we can provide an alternative proof. Without loss of generality, let $\mathbb{E}[X_i] = 0$ for all $i = 1, \ldots, d$. Otherwise consider $Y_i := X_i - \mathbb{E}[X_i]$, $i = 1, \ldots, d$. Denote by $\{\tilde{X}_k\}_{k \in \mathbb{N}}$ an extension of $(X_1, \ldots, X_d)$ on $(\tilde{\Omega}, \tilde{\mathcal{F}}, \tilde{\mathbb{P}})$. Moreover, denote by $\tilde{\mathcal{G}}$ the σ-algebra of Theorem 1.4, and define $F_x := \tilde{\mathbb{P}}(\tilde{X}_1 \leq x \,|\, \tilde{\mathcal{G}})$, $x \in \mathbb{R}$. We then have independently of the index i that $\tilde{\mathbb{E}}[\tilde{X}_i \,|\, \tilde{\mathcal{G}}] = \int_{\mathbb{R}} x \, dF_x$. Using the conditional i.i.d. structure one therefore obtains

$$\mathbb{E}[X_i X_j] = \tilde{\mathbb{E}}[\tilde{X}_i \tilde{X}_j] = \tilde{\mathbb{E}}\big[\tilde{\mathbb{E}}[\tilde{X}_i \,|\, \tilde{\mathcal{G}}] \, \tilde{\mathbb{E}}[\tilde{X}_j \,|\, \tilde{\mathcal{G}}]\big] = \tilde{\mathbb{E}}\Big[\Big(\int_{\mathbb{R}} x \, dF_x\Big)^2\Big] \geq 0.$$

This implies the claim. □

Non-negative pairwise correlations are the most common (but only one particular) way of measuring positive association between the components of a random vector. In a non-normal setup, correlations might not be the best measures of association, or might not even exist. One alternative notion of positive association (among numerous) is the following: a random vector $\boldsymbol{X} := (X_1, \ldots, X_d)$ on a probability space $(\Omega, \mathcal{F}, \mathbb{P})$ exhibits *positive orthant dependency (POD)* if for all $x_1, \ldots, x_d \in \mathbb{R}$ it satisfies[10] positive upper and lower orthant dependency, i.e.

$$\text{(PUOD)} \qquad \mathbb{P}(X_1 > x_1, \ldots, X_d > x_d) \geq \prod_{k=1}^{d} \mathbb{P}(X_k > x_k), \quad \text{and}$$

$$\text{(PLOD)} \qquad \mathbb{P}(X_1 \leq x_1, \ldots, X_d \leq x_d) \geq \prod_{k=1}^{d} \mathbb{P}(X_k \leq x_k). \tag{1.13}$$

In terms of a survival copula $\hat{C}$ and a copula C of $\boldsymbol{X}$, (1.13) is equivalent to the fact that $\hat{C} \geq \Pi$ and $C \geq \Pi$ pointwise. Clearly, if the components of $\boldsymbol{X}$ are independent, then (1.13) is satisfied with equality. We have seen in Theorem 1.4 that extendible random vectors have components that are conditionally i.i.d. given some information. Thus, one might wonder whether extendible random vectors always exhibit POD. However, in general this conjecture is wrong, as the following toy example shows.

[10]For a detailed study of different measures of association see Müller and Stoyan (2002).

Example 1.20 (Homogeneous Mixtures Need Not Exhibit POD) *Consider a probability space $(\Omega, \mathcal{F}, \mathbb{P})$ supporting a random variable $U \sim U[0,1]$. Define the (random) survival function*

$$\bar{F}_x := \begin{cases} 1 & , x \in (-\infty, 1-U) \\ \frac{1}{2} & , x \in [1-U, 1+U) \\ 0 & , x \in [1+U, \infty) \end{cases}, \quad x \in \mathbb{R}.$$

Now let the random vector (X_1, X_2) be defined as a homogeneous mixture from $\{F_x\}_{x \in \mathbb{R}}$, where $F_x := 1 - \bar{F}_x$, $x \in \mathbb{R}$. Then

$$\mathbb{P}\Big(X_1 > \frac{1}{2}, X_2 > \frac{3}{2}\Big) = \mathbb{E}\big[\bar{F}_{\frac{1}{2}}\, \bar{F}_{\frac{3}{2}}\big] = 0 + \frac{1}{2} \cdot \frac{1}{2}\, \mathbb{P}(U > 1/2) = \frac{1}{8}.$$

However, we also find

$$\mathbb{P}\Big(X_1 > \frac{1}{2}\Big)\, \mathbb{P}\Big(X_1 > \frac{3}{2}\Big) = \mathbb{E}\big[\bar{F}_{\frac{1}{2}}\big]\, \mathbb{E}\big[\bar{F}_{\frac{3}{2}}\big] = \frac{3}{4} \cdot \frac{1}{4} = \frac{3}{16} > \frac{1}{8}.$$

Therefore, (X_1, X_2) does not exhibit POD, not even PUOD, though it is a conditionally i.i.d. model.

1.2.4 *Heterogeneous Mixture Models/Hierarchical Models*

In many practical applications one faces situations where exchangeable dependence structures are not sufficiently flexible. For instance, the assumption of exchangeability implies that all bivariate subvectors of the random vector in question are equal in distribution. In reality, this is clearly not always satisfied. As Theorem 1.4 shows, exchangeability is often the consequence of a stochastic model in which a latent source of dependence (namely $F = \{F_x\}_{x \in \mathbb{R}}$ in the case of a homogeneous mixture) affects all components in the very same manner. However, it is often more realistic to assume that there are multiple latent factors. Considering, e.g., a portfolio with loans or stocks issued by d companies, it is natural to subdivide the portfolio into groups according to, e.g., industrial branches, geographic regions, or some other economic criterion, and to associate a latent factor with each subgroup. Given such a partition it might be intuitive to assume that, e.g., the pharmaceutical branch is affected by different market factors than the financial branch. Nevertheless, there might be a global factor which affects all industrial branches, so the branches cannot be modeled independently of each other. The result of such a construction is a dependence structure (induced by various latent factors) where each subgroup (being affected in the same way by the latent factors) is extendible. The subgroups, however, are dependent. To model such structures without losing the advantages of the mixture approach with regard to efficient simulation, one often extends the approach to a *heterogeneous mixture*. A generic probabilistic construction of such a stochastic model is outlined in Algorithm 1.4.

Algorithm 1.4 (Generic Sampling of Heterogeneous Mixture Models)

This algorithm provides a generic sampling scheme for heterogeneous mixture models. As input one has the dimension $d \geq 2$ and a partition of the dimension $d_1 + d_2 + \ldots + d_J = d$ into J groups, where $d_1, \ldots, d_J \in \mathbb{N}$. Moreover, a stochastic model for J possibly dependent distribution functions $F^{(j)} = \{F_x^{(j)}\}_{x \in \mathbb{R}}$ is required, $j = 1, \ldots, J$. Sampling is then accomplished as follows:

(1) Sample the paths of the groups' distribution functions $F^{(1)}, \ldots, F^{(J)}$.
(2) For each $j = 1, \ldots, J$ draw d_j i.i.d. samples $X_{j,1}, \ldots, X_{j,d_j}$ from $F^{(j)}$ and return the random vector:

$$\big(\underbrace{X_{1,1}, \ldots, X_{1,d_1}}_{group\ 1}, \underbrace{X_{2,1}, \ldots, X_{2,d_2}}_{group\ 2}, \ldots, \underbrace{X_{J,1}, \ldots, X_{J,d_J}}_{group\ J}\big).$$

In case the copula behind the model is required, the unconditional marginal distribution functions must be derived and applied to the components $X_{j,l}$, $l = 1, \ldots, d_j$, $j = 1, \ldots, J$.

Clearly, if $F^{(1)}, \ldots, F^{(J)}$ are independent, the simulation boils down to J independent homogeneous mixtures. So the crucial point for such heterogeneous models is the incorporation of a dependence structure between the random distribution functions $F^{(1)}, \ldots, F^{(J)}$. Often, it is reasonable to assume that all components are affected by some global factor and subgroups are additionally affected by independent and group-specific factors. The result is a *hierarchical* dependence structure. A general construction principle based on independent stochastic objects is outlined here. The resulting copulas of this construction principle are termed *h-extendible* in Mai and Scherer (2011b), reflecting the interpretation of a hierarchical and extendible structure. This construction is introduced here and is illustrated with several specific examples in later chapters.

Definition 1.11 (H-Extendible Copulas: Recursive Definition)

Consider the probability space $(\Omega, \mathcal{F}, \mathbb{P})$. A random vector $(X_1, \ldots, X_d)$ is called h-extendible with $1 \leq n \leq d$ levels of hierarchy, if the following holds: for $n = 1$, h-extendibility corresponds to the previously defined notion of *extendibility. For $n \geq 2$, there exists some σ-algebra $\mathcal{G} \subset \mathcal{F}$ and a partition $d_1 + \ldots + d_J = d$, $d_j \in \mathbb{N}$, such that conditioned on $\mathcal{G}$:*

(1) $(X_1, \ldots, X_d)$ splits into J independent subvectors,
(2) each subvector is h-extendible with at most $n-1$ levels, and
(3) at least one subvector has $n-1$ levels.

We call a copula h-extendible with n levels of hierarchy if it is the distribution function of an h-extendible random vector $(U_1, \ldots, U_d)$ whose univariate marginal laws are uniform on $[0,1]$. If, additionally, the dependence structure within all subgroups is of the same parametric family as the original vector, the copula is called h-extendible with respect to the family in question.

This recursive definition is simple and at the same time flexible (from a theoretical point of view). An iterative reformulation, however, might be better suited for later stating a general construction principle.

Remark 1.5 (H-Extendible Copulas: Iterative Definition)
Again, consider the probability space $(\Omega, \mathcal{F}, \mathbb{P})$. The vector $(U_1, \ldots, U_d)$ has an h-extendible copula with n levels of hierarchy as its distribution function if the univariate marginal laws are uniform, i.e. $U_k \sim U[0,1]$, $k = 1, \ldots, d$, and there exists an increasing sequence of σ-algebras $\mathcal{G}_1 \subsetneq \ldots \subsetneq \mathcal{G}_n \subset \mathcal{F}$ such that:

(1) Conditioned on $\mathcal{G}_1$, $(U_1, \ldots, U_d)$ splits into J independent groups of sizes $d_1 + \ldots + d_J = d$. Assuming that the subvector $(U_{d_1+\ldots+d_{j-1}+1}, \ldots, U_{d_1+\ldots+d_j})$ contains the elements of group $j \in \{1, \ldots, J\}$, $d_0 := 0$, one has

$$\mathbb{P}(U_1 \leq u_1, \ldots, U_d \leq u_d \,|\, \mathcal{G}_1) = \prod_{j=1}^{J} \mathbb{P}\big(U_i \leq u_i, \forall\, U_i \in \textit{group } j \,|\mathcal{G}_1\big).$$

(2) Conditioned on $\mathcal{G}_2$, the groups of level (1) further split into independent subgroups. Each d_j is further partitioned as $d_j = d_{j,1} + \ldots + d_{j,J_j}$ and

$$\mathbb{P}(U_1 \leq u_1, \ldots, U_d \leq u_d \,|\, \mathcal{G}_2) =$$
$$\prod_{j=1}^{J} \prod_{l=1}^{d_{j,J_j}} \mathbb{P}\big(U_i \leq u_i,\, \forall\, U_i \in \textit{subgroup } l \textit{ of group } j \,|\, \mathcal{G}_2\big).$$

(k) For $2 < k < n$ we further iterate the definition. Given $\mathcal{G}_k$, level $(k-1)$ subgroups split into independent subgroups.

(n) *Conditioned on $\mathcal{G}_n$, we have: (a) all components are independent (but not necessarily identically distributed) and (b) inside each final level (n) subgroup components are independent and identically distributed, i.e.*

$$\mathbb{P}(U_1 \leq u_1, \ldots, U_d \leq u_d \,|\, \mathcal{G}_n) \overset{(a)}{=} \prod_{i=1}^{d} \mathbb{P}(U_i \leq u_i \,|\, \mathcal{G}_n),$$

$$\mathbb{P}(U_i \leq u_i, \forall\, U_i \in \textit{final subgroup } l \,|\, \mathcal{G}_n) \overset{(b)}{=} \prod_{i\,:\,U_i \in \textit{final subgroup } l} \mathbb{P}(U_k \leq u_i \,|\, \mathcal{G}_n),$$

for U_k arbitrarily chosen from the final subgroup l.

We now provide a general construction principle based on Remark 1.5. For $n = 1$ we assume that the dependence structure of an extendible copula model $(U_1, \ldots, U_d) \sim C_\theta$ is induced by a global stochastic object $M^{(0)}$, affecting i.i.d. components alike. The stochastic object $M^{(0)}$ can be a random variable, a random vector, a stochastic process, or any other stochastic object defined on $(\Omega, \mathcal{F}, \mathbb{P})$. The components $(U_1, \ldots, U_d)$ are defined as $U_k := f(\epsilon_k, M^{(0)})$, $k = 1, \ldots, d$, where $\epsilon_1, \ldots, \epsilon_d$ is an i.i.d. sequence of random variables independent of the random object $M^{(0)}$ and f is a real-valued measurable functional. Going one step further with a factor model in mind, we divide the random vector into J groups of sizes $d_1, \ldots, d_J$, respectively, where $d_1 + \ldots + d_J = d$. For notational convenience, assume that $U_1, \ldots, U_{d_1}$ belong to the first group, $U_{d_1+1}, \ldots, U_{d_1+d_2}$ to the second group, and so on. We now generalize the extendible model by introducing additional independent group-specific stochastic objects $M^{(1)}, \ldots, M^{(J)}$ such that

$$\begin{aligned} U_k &:= f_1\big(\epsilon_k, M^{(0)}, M^{(1)}\big), \quad k = 1, \ldots, d_1, \\ &\vdots \\ U_k &:= f_J\big(\epsilon_k, M^{(0)}, M^{(J)}\big), \quad k = d_1 + \ldots + d_{J-1} + 1, \ldots, d. \end{aligned}$$

This construction requires real-valued measurable functionals $f_1, \ldots, f_J$ and independent stochastic objects $M^{(1)}, \ldots, M^{(J)}$. The result is a random vector $(U_1, \ldots, U_d)$ satisfying the hierarchical axioms of Definition 1.5. With

$$\mathcal{G}_1 := \sigma\big(M^{(0)}\big) \subsetneq \mathcal{G}_2 := \sigma\big(M^{(0)}, M^{(1)}, \ldots, M^{(J)}\big)$$

one has (abbreviating $d_0 := 0$)

$$\mathbb{P}(U_1 \leq u_1, \ldots, U_d \leq u_d \,|\, \mathcal{G}_1) =$$

$$\prod_{j=1}^{J} \mathbb{P}\big(U_{d_1+\ldots+d_{j-1}+1} \leq u_{d_1+\ldots+d_{j-1}+1}, \ldots, U_{d_1+\ldots+d_j} \leq u_{d_1+\ldots+d_j} \,\big|\, \mathcal{G}_1\big)$$

and

$$\mathbb{P}(U_{d_1+\ldots+d_{j-1}+1} \leq u_{d_1+\ldots+d_{j-1}+1}, \ldots, U_{d_1+\ldots+d_j} \leq u_{d_1+\ldots+d_j} \,|\, \mathcal{G}_2) =$$

$$\prod_{l=d_1+\ldots+d_{j-1}+1}^{d_1+\ldots+d_j} \mathbb{P}\big(U_k \leq u_l \,\big|\, \mathcal{G}_2\big), \quad j = 1, \ldots, J,$$

where k is an arbitrary index in $\{d_1 + \ldots + d_{j-1} + 1, \ldots, d_1 + \ldots + d_j\}$. So far, it is not guaranteed that the components $U_1, \ldots, U_d$ have a uniform distribution, so we have an h-extendible random vector but not necessarily an h-extendible copula. Ensuring $U_1, \ldots, U_d \sim U[0,1]$, and therefore obtaining an h-extendible copula, the functionals $f_1, \ldots, f_J$ and the stochastic objects $M^{(1)}, \ldots, M^{(J)}$ must be chosen in a well-considered way, which might be quite tricky. It is even more complicated to design the model in a way that ensures a particular parametric class of extendible copulas $\{C_\theta\}$ in each step.

1.2.5 *Extreme-Value Copulas*

Some copulas studied later on have a very nice analytic property, which plays a central role in extreme-value theory.

Definition 1.12 (Extreme-Value Copula)
A copula $C : [0,1]^d \to [0,1]$ *is called an* extreme-value copula *if it satisfies the extreme-value property* $C(u_1^t, \ldots, u_d^t) = C(u_1, \ldots, u_d)^t$ *for all* $t > 0$, $u_1, \ldots, u_d \in [0,1]$.

Specific examples of extreme-value copulas include the independence copula Π, the upper Fréchet–Hoeffding bound M, as well as Marshall–Olkin copulas as discussed in Chapter 3. Such copulas occur in multivariate extreme-value theory as possible limit copulas. More precisely, let F be a multivariate distribution function with continuous margins $F_1, \ldots, F_d$. Consider a probability space $(\Omega, \mathcal{F}, \mathbb{P})$ supporting i.i.d. random vectors $\{(X_1^{(n)}, \ldots, X_d^{(n)})\}_{n \in \mathbb{N}}$ with distribution function F. For each $n \in \mathbb{N}$ one considers the vector of componentwise maxima $(M_{1:n}, \ldots, M_{d:n})$, where $M_{k:n} := \max\{X_k^{(1)}, \ldots, X_k^{(n)}\}$, $k = 1, \ldots, d$. If there exist sequences

$\{a_{1:n}\}_{n\in\mathbb{N}},\ldots,\{a_{d:n}\}_{n\in\mathbb{N}}$, $\{b_{1:n}\}_{n\in\mathbb{N}},\ldots,\{b_{d:n}\}_{n\in\mathbb{N}}$ and a random vector $(X_1,\ldots,X_d)$ such that

$$\Big(\frac{M_{1:n}-a_{1:n}}{b_{1:n}},\ldots,\frac{M_{d:n}-a_{d:n}}{b_{d:n}}\Big)\xrightarrow{d}(X_1,\ldots,X_d),\quad \text{as } n\to\infty,\quad (1.14)$$

then the copula of the limit random vector $(X_1,\ldots,X_d)$ must be of the extreme-value kind. A proof of this fact can be found, e.g., in Joe (1997, p. 172ff). Conversely, given an extreme-value copula C, it may occur as the copula of a limit random vector as in (1.14): consider i.i.d. random vectors $\{(U_1^{(n)},\ldots,U_d^{(n)})\}_{n\in\mathbb{N}}$ with joint distribution function C on a probability space $(\Omega,\mathcal{F},\mathbb{P})$. For each $n\in\mathbb{N}$ denote by $M_{k:n}:=\max\{U_k^{(1)},\ldots,U_k^{(n)}\}$ the componentwise maxima, for $k=1,\ldots,d$. Choosing $a_{k:n}\equiv 1$ and $b_{k:n}=1/n$ for all $k\in\{1,\ldots,d\}$ and $n\in\mathbb{N}$, it is observed for $t_1,\ldots,t_d\geq 0$ that

$$\begin{aligned}
&\mathbb{P}\Big(\frac{M_{1:n}-1}{1/n}\leq -t_1,\ldots,\frac{M_{d:n}-1}{1/n}\leq -t_d\Big)\\
&=\mathbb{P}\Big(U_1^{(1)}\leq -\frac{t_1}{n}+1,\ldots,U_d^{(1)}\leq -\frac{t_d}{n}+1\Big)^n\\
&=C\Big(-\frac{t_1}{n}+1,\ldots,-\frac{t_d}{n}+1\Big)^n\\
&=C\Big(\Big(-\frac{t_1}{n}+1\Big)^n,\ldots,\Big(-\frac{t_d}{n}+1\Big)^n\Big)\quad\xrightarrow{n\to\infty}\quad C\big(e^{-t_1},\ldots,e^{-t_d}\big).
\end{aligned}$$

The third equality in the computation above requires the extreme-value property of C. Notice that the entries $\exp(-t_k)$, $k=1,\ldots,d$, of C in the limit constitute univariate extreme-value distribution functions of the Weibull kind as t_k ranges in $[0,\infty)$ (see, e.g., Joe (1997, p. 170)). Hence, C occurs as a possible limit copula in (1.14).

From an analytical perspective, each extreme-value copula admits a so-called *Pickands representation*. This characterizes an extreme-value copula by a measure on the d-dimensional unit simplex subject to certain boundary conditions.

Theorem 1.5 (De Haan and Resnick (1977), Pickands (1981)) *A d-dimensional copula C is an extreme-value copula if and only if there exists a (positive) finite measure δ on the d-dimensional unit simplex*

$$A_d:=\big\{(u_1,\ldots,u_d)\in[0,1]^d\,\big|\,u_1+\ldots+u_d=1\big\},$$

subject to the conditions $\int_{A_d}u_j\,\delta(du_1,\ldots,du_d)=1$, $j=1,\ldots,d$, *such that*

$$C(u_1,\ldots,u_d)=\Big(\prod_{i=1}^d u_i\Big)^{P\left(\frac{\log u_1}{\sum_{k=1}^d\log u_k},\ldots,\frac{\log u_d}{\sum_{k=1}^d\log u_k}\right)},$$

where P, called the Pickands dependence function, *is defined on A_d and given by*

$$P(w_1, \ldots, w_d) = \int_{A_d} \max\{u_1\, w_1, \ldots, u_d\, w_d\}\, \delta(du_1, \ldots, du_d).$$

Proof. This theorem is named after Pickands (1981), even though it is in turn based on de Haan and Resnick (1977). A full proof (using the language of extreme-value distributions rather than extreme-value copulas) can be retrieved from Resnick (1987, Proposition 5.11, p. 268ff). □

Theorem 1.5 can be used to derive expressions for tail dependence parameters of general extreme-value copulas based on the measure δ (see Li (2009)). Moreover, Theorem 1.5 is useful for constructing parametric families of extreme-value copulas.

Example 1.21 (Pickands Dependence Function of Cuadras–Augé Copulas)
Consider the bivariate copula C_α from Example 1.4. The extreme-value property is easily verified to hold for C_α, so in regard of Theorem 1.5 it is natural to ask what the measure δ and the corresponding dependence function P look like. It is derived in Falk et al. (2004, Example 4.3.2, p. 124) that δ is a discrete measure, whose mass is concentrated on three points. More precisely, it is determined by

$$\delta(\{(1,0)\}) = \delta(\{(0,1)\}) = 1 - \alpha, \quad \delta(\{(1/2,1/2)\}) = 2\,\alpha.$$

One verifies that the boundary conditions are valid, i.e.

$$\int_{A_2} u_1\, \delta(du_1, du_2) = \int_{A_2} u_2\, \delta(du_1, du_2) = (1-\alpha)\cdot 1 + 2\,\alpha \cdot \frac{1}{2} = 1.$$

Moreover, the Pickands dependence function P is computed to have the form

$$\begin{aligned} P(w_1, w_2) &= \int_{A_2} \max\{u_1\, w_1, u_2\, w_2\}\, \delta(du_1, du_2) \\ &= (1-\alpha)\, w_1 + (1-\alpha)\, w_2 + 2\,\alpha\, \frac{1}{2} \max\{w_1, w_2\} \\ &= \max\{w_1, w_2\} + (1-\alpha) \min\{w_1, w_2\}. \end{aligned}$$

Indeed, one may check that for $u_1, u_2 \in (0,1)$

$$\begin{aligned} P\Big(\frac{\log u_1}{\log(u_1\, u_2)}, \frac{\log u_2}{\log(u_1\, u_2)}\Big) = &\frac{\log\big(\min\{u_1, u_2\}\big)}{\log(u_1\, u_2)} \\ &+ (1-\alpha)\,\frac{\log\big(\max\{u_1, u_2\}\big)}{\log(u_1\, u_2)}. \end{aligned}$$

Hence, it is verified that

$$(u_1\,u_2)^{P\left(\frac{\log u_1}{\log(u_1\,u_2)},\frac{\log u_2}{\log(u_1\,u_2)}\right)} = e^{\log\left(\min\{u_1,u_2\}\right)+\log\left(\max\{u_1,u_2\}^{1-\alpha}\right)}$$
$$= C_\alpha(u_1,u_2).$$

Since the Pickands dependence function $P(w_1,w_2)$ is defined only for non-negative w_1,w_2 with $w_1+w_2=1$, one may alternatively parameterize it by $w\in[0,1]$ setting $\tilde{P}(w):=P(w,1-w)$. The function $\tilde{P}$ can then easily be visualized for different choices of $\alpha\in[0,1]$ (see Figure 1.6). The fact that $\tilde{P}$ is not differentiable in $w=1/2$ for $\alpha>0$ indicates that C_α has a singular component (see Joe (1997, Theorem 6.5, p. 176)).

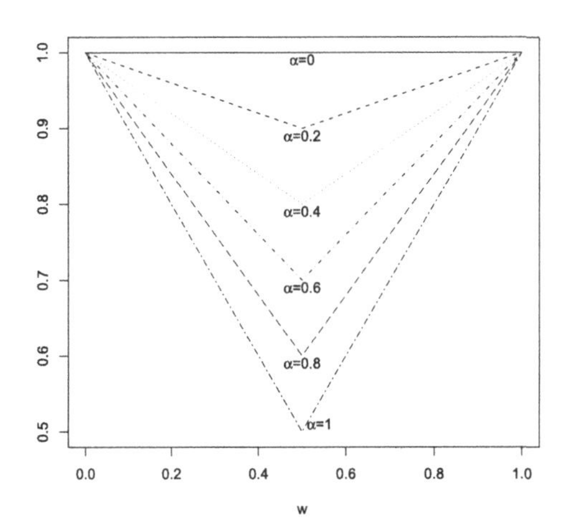

Fig. 1.6 The Pickands dependence function $\tilde{P}(w)=\max\{w,1-w\}+(1-\alpha)\min\{w,1-w\}$ of a bivariate Cuadras–Augé copula is illustrated for $\alpha=0,0.2,0.4,0.6,0.8,1$.

Example 1.22 (Copulas of Class BC_2)

In Example 1.9 we introduced the family of copulas

$$C_{a,b}(u_1,u_2)=\min\{u_1^a,u_2^b\}\,\min\{u_1^{1-a},u_2^{1-b}\},\quad u_1,u_2\in[0,1],$$

where $a,b\in(0,1)$. It is left as an exercise for the reader to show that BC_2 copulas (1) are extreme-value copulas, and (2) have a discrete Pickands measure δ with at most two atoms. Scatterplots of this family are given in Figure 1.7 below.

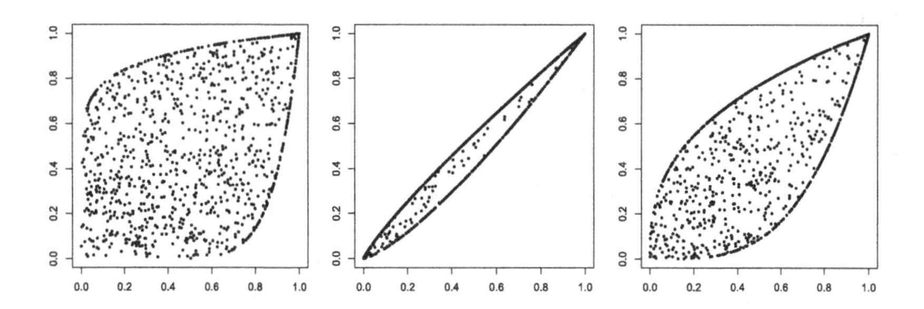

Fig. 1.7 Scatterplots of 1 000 samples from the copula BC_2 with parameters $(a, b) = (0.9, 0.1)$ (left), $(0.4, 0.3)$ (middle), and $(0.3, 0.5)$ (right).

Besides being the limit copulas of componentwise maxima, extreme-value copulas have a second natural interpretation. They are precisely the copulas of so-called *min-stable multivariate exponential distributions* (see Joe (1997, p. 174–175)). This means that combined with exponential margins they form a class of multivariate distribution functions F such that $(X_1, \ldots, X_d) \sim F$, $1 \leq i_1 < \ldots < i_k \leq d \Rightarrow \min\{X_{i_1}, \ldots, X_{i_k}\} \sim Exp(\lambda)$ for some $\lambda > 0$.

Chapter 2

Archimedean Copulas

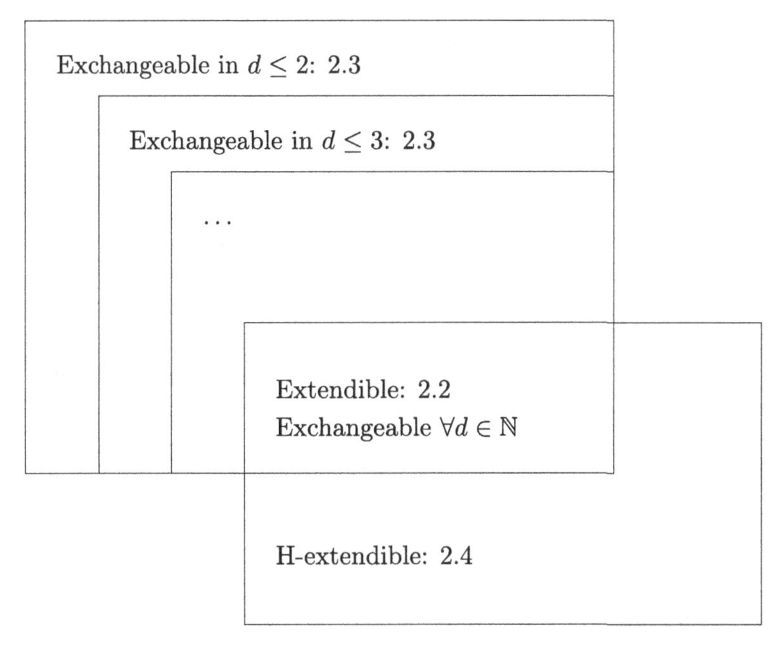

Fig. 2.1 Classification of families of Archimedean copulas, including the sections where these are discussed in this chapter.

This chapter on Archimedean copulas is organized as follows (see Figure 2.1). After a brief motivation in Section 2.1, Section 2.2 studies extendible Archimedean copulas. Those allow for a stochastic representation with components that are conditionally i.i.d. in the sense of de Finetti's theorem (see Theorem 1.4). Subsequently, Section 2.3 studies the more general family of exchangeable Archimedean copulas. This family additionally allows for negative association between components. Intuitively, this family "shrinks" in the dimension d, with the extendible class as limit for $d \to \infty$. Finally, Section 2.4 shows how to construct and simulate h-extendible (hierarchical) Archimedean copulas. This is important in order to take into account non-exchangeable dependence patterns without losing some advantages of the (extendible) Archimedean class.

2.1 Motivation

Let $(\Omega, \mathcal{F}, \mathbb{P})$ be a probability space supporting a list of i.i.d. exponential random variables $E_1, \ldots, E_d$ with mean $\mathbb{E}[E_1] = 1$. Independent of this list, let M be a positive random variable. Define the vector of random variables $(X_1, \ldots, X_d)$ by

$$(X_1, \ldots, X_d) := \Big(\frac{E_1}{M}, \ldots, \frac{E_d}{M}\Big). \tag{2.1}$$

What can we conclude about this vector?

(1) **Mixture model interpretation:** Dividing an exponential random variable by a positive number corresponds to dividing its mean by exactly this value, or, similarly, corresponds to multiplying its intensity parameter by this number. In the above construction we start with E_k's having an intensity of 1, which are then multiplied by the realization of $1/M$. In this sense, the above construction can equivalently be interpreted as a two-step experiment. In a first step, the intensity M is drawn. In a second step, a d-dimensional vector of i.i.d. $Exp(M)$ variables is drawn. Using the notation of Section 1.2.3, we have a conditionally i.i.d. model with random distribution function $\{F_x\}_{x\in\mathbb{R}}$, $F_x := \big(1 - \exp(-M\,x)\big)\mathbb{1}_{\{x>0\}}$.

(2) **Margins:** For each margin $k \in \{1, \ldots, d\}$, one has

$$\begin{aligned}\bar{F}_k(x) &= \mathbb{P}(X_k > x) = \mathbb{P}(E_k > x\,M)\\ &= \mathbb{E}\big[\mathbb{E}[\mathbb{1}_{\{E_k > x\,M\}}|M]\big] = \mathbb{E}\big[e^{-x\,M}\big] =: \varphi(x), \quad x \geq 0,\end{aligned}$$

where φ is the Laplace transform of M. To conclude, the marginal laws are related to the mixing variable M via its Laplace transform.

(3) **Copula:** As long as M is not deterministic, the X_k's are dependent, since they are all affected by the random variable M. Intuitively speaking, whenever the realization of M is large (small), all realizations of X_k are more likely to be small (large). From this

regard, it is natural that the resulting copula is parameterized by the Laplace transform of M. This indeed holds, as we observe that the survival copula of the random vector $(X_1, \ldots, X_d)$ is given by

$$\varphi(\varphi^{-1}(u_1) + \ldots + \varphi^{-1}(u_d)) := C_\varphi(u_1, \ldots, u_d), \qquad (2.2)$$

where $u_1, \ldots, u_d \in [0,1]$ and φ is the Laplace transform of M. Copulas of this structure are called *Archimedean copulas.* Written differently, with C_φ given by Equation (2.2), it holds that

$$\mathbb{P}\big(X_1 > x_1, \ldots, X_d > x_d\big) = C_\varphi\big(\varphi(x_1), \ldots, \varphi(x_d)\big),$$

where $x_1, \ldots, x_d \geq 0$. The copula C_φ in Equation (2.2) is obviously a symmetric function in its arguments, showing that the vector of X_k's has an exchangeable distribution (which also follows from the construction as a homogeneous mixture model).

Proof. The joint survival function of $(X_1, \ldots, X_d)$ is given by

$$\begin{aligned}
\bar{F}(x_1, \ldots, x_d) &= \mathbb{P}(X_1 > x_1, \ldots, X_d > x_d) \\
&= \mathbb{E}\big[\,\mathbb{E}\big[\mathbb{1}_{\{E_1 > x_1 M\}} \cdots \mathbb{1}_{\{E_d > x_d M\}} \big| M\big]\,\big] \\
&= \mathbb{E}\big[e^{-M \sum_{k=1}^{d} x_k}\big] = \varphi\Big(\sum_{k=1}^{d} x_k\Big),
\end{aligned}$$

where $(x_1, \ldots, x_d) \in [0, \infty)^d$. The proof is established by transforming the marginals to $U[0,1]$ and by using the survival version of Sklar's theorem. □

Instead of the constructive way, most textbooks introduce Archimedean copulas analytically. A d-dimensional copula is called *Archimedean,*[1] if it admits the functional form[2] $C_\varphi(u_1, \ldots, u_d) := \varphi\big(\varphi^{-1}(u_1) + \ldots + \varphi^{-1}(u_d)\big)$, where $u_1, \ldots, u_d \in [0,1]$. In this case, the function $\varphi : [0, \infty) \to [0,1]$ is called the *generator* of C_φ and φ^{-1} denotes its inverse. We shall see that this parametric family allows for a large spectrum of dependence structures, and, at the same time, remains analytically tractable due to the convenient form (2.2). This renders the class of Archimedean copulas a very popular class for applications. However, both the constructive and the analytical approach evoke some questions:

[1]The wording *Archimedean copula* stems from a purely analytical property, related to the classical Archimedean property, and was first used in Ling (1965).

[2]Let us briefly remark that some authors change the order of φ and φ^{-1}. Analytically, this is of course identical, since $(\varphi^{-1})^{-1} = \varphi$. Based on the probabilistic construction, however, we interpret φ as a Laplace transform and therefore prefer the initially used definition.

(1) Construction (2.1) and the earlier considerations show that each positive random variable M induces a copula of form (2.2). The question at hand is whether all copulas of form (2.2) can be constructed in this way. Formulated analytically, what kind of functions φ define[3] an Archimedean copula? Sampling Archimedean copulas requires a probabilistic model. At this point it is open if model (2.1) contains all Archimedean copulas (which is not the case) or if there exists a more general probabilistic model (which is the case, see Section 2.3).

(2) Given that Archimedean copulas can be parameterized by positive random variables M (via φ), it is natural to ask what dependence properties of C_φ can be related to stochastic properties of M and to analytical properties of φ.

(3) An immediate observation is that C_φ (seen as a function) is symmetric, meaning that C_φ is invariant under permutations of its arguments. Given that φ is chosen such that C_φ is a copula, the associated random vector $(U_1, \ldots, U_d) \sim C_\varphi$ is exchangeable. Since this might not be justified in all applications, a natural question is to ask for non-exchangeable generalizations. Such extensions, their probabilistic models, and sampling schemes are provided in Section 2.4.

(4) Different Archimedean generators φ can imply the same copula C_φ. More precisely, let φ define the copula C_φ and let $c > 0$ be some arbitrary positive constant. Then, the function $\varphi_c(x) := \varphi(c\,x)$ has inverse $\varphi_c^{-1}(x) = \varphi^{-1}(x)/c$ and implies the very same Archimedean copula when used in Equation (2.2), i.e. $C_\varphi = C_{\varphi_c}$. Given that $\varphi(x) = \mathbb{E}[\exp(-x\,M)]$ is the Laplace transform of some positive random variable M, $\varphi_c(x) = \varphi(c\,x) = \mathbb{E}[\exp(-x\,c\,M)]$ is the Laplace transform of $c\,M$. We conclude that such a deterministic shift in construction (2.1) only affects the univariate marginal laws but not the resulting dependence structure. The same holds when $E_1, \ldots, E_d$ are distributed as $Exp(\lambda)$ for some $\lambda > 0$, so taking unit exponentials in the first place is just a convenient standardization.

[3] For instance, one easily verifies that $\varphi(x) := \exp(-x)$ corresponds to the independence copula Π, whereas the choice $\varphi(x) = x$ does not imply a copula.

2.2 Extendible Archimedean Copulas

The basis for this section is construction (2.1). Notice that the crucial step in the derivation of the survival copula of $(X_1, \ldots, X_d) = (E_1/M, \ldots, E_d/M)$ is that, conditioned on M, the random variables X_k are i.i.d. exponential with intensity M. The exchangeable d-dimensional random vector $(X_1, \ldots, X_d)$ can be extended to arbitrary dimensions by simply adding further components $X_k := E_k/M$, $k > d$, where the additional E_k's are $Exp(1)$-distributed and independent of all previously defined objects. Hence, this construction naturally provides an extendible sequence where the conditional independence construction related to de Finetti's theorem (see Theorem 1.4), namely construction (2.1), is explicitly known. Formulating this construction as an algorithm, which was first done by Marshall and Olkin (1988), yields a very convenient sampling strategy for extendible Archimedean copulas.

Algorithm 2.1 (Sampling Extendible Archimedean Copulas)

(1) Sample i.i.d. $E_1, \ldots, E_d$, *where* $E_k \sim Exp(1)$, $k \in \{1, \ldots, d\}$.
(2) Sample a positive random variable M *with Laplace transform* φ.
(3) Return $(U_1, \ldots, U_d)$, *where* $U_k := \varphi\big(E_k/M\big)$, $k \in \{1, \ldots, d\}$.

Proof. Recall that the survival function of $(E_1/M, \ldots, E_d/M)$ is given by $\varphi(x_1 + \ldots + x_d) = C_\varphi\big(\varphi(x_1), \ldots, \varphi(x_d)\big)$. Moreover, the marginal survival functions are φ, since for $x \geq 0$ one has $\mathbb{P}(E_k/M > x) = \mathbb{E}[\exp(-x\,M)] = \varphi(x)$, $k = 1, \ldots, d$. Therefore, $\big(\varphi(E_1/M), \ldots, \varphi(E_d/M)\big) \sim C_\varphi$, since φ is continuous and strictly decreasing, which is precisely the vector that is returned in the last step of Algorithm 2.1. □

We observe that two things are required for using Algorithm 2.1: (1) the functional form of the Laplace transform φ, and (2) a sampling scheme for the positive random variable M. This means that starting with a random variable M on the one hand, it is required to compute its Laplace transform and to have a sampling scheme available. On the other hand, starting from a Laplace transform φ, we have to identify the underlying random variable M which then also has to be sampled.[4] Obviously, the best case is to have

[4]Starting with φ it might be difficult to identify the random variable whose Laplace transform is φ. Helpful in this regard are tables of Laplace transforms (see Oberhettinger and Badii (1973)). Alternatively, one might address this issue numerically, which is done, e.g., in Hofert (2011). Sampling bivariate Archimedean copulas using the analytical form of the generator is done in Section 2.5.1.

a repertoire of parametric families of random variables, along with their Laplace transforms and some efficient sampling schemes, at hand. This is provided in Section 2.2.4.

An important advantage of Algorithm 2.1 is the low effort to sample a high-dimensional Archimedean copula. More precisely, the additional effort of one more dimension is of order $\mathcal{O}(1)$. Only one additional exponential random variable is to be drawn and the transformation of the last step in Algorithm 2.1 has to be applied.[5] Overall, only $d+1$ random variables have to be drawn for the simulation of a d-dimensional random vector.

2.2.1 *Kimberling's Result and Bernstein's Theorem*

Let us switch to the analytical perspective on Archimedean copulas and introduce the notion of an (Archimedean) generator.

Definition 2.1 (Archimedean Generator)
An (Archimedean) generator *is a function* $\varphi : [0,\infty) \to [0,1]$ *with the following properties: (1)* $\varphi(0) = 1$ *and* $\lim_{x\nearrow\infty}\varphi(x) = 0$*, (2)* φ *is continuous, (3)* φ *is decreasing on* $[0,\infty)$ *and strictly decreasing on* $\big[0, \inf\{x > 0 : \varphi(x) = 0\}\big)$*, where* $\inf\emptyset := \infty$.

This definition represents the necessary requirements on φ such that, if applied in Equation (2.2), the resulting function C_φ fulfills the properties of groundedness and of uniform margins: φ^{-1} is well defined on $(0,1]$ because of (2) and (3). Required is the convention $\varphi^{-1}(0) := \inf\{x > 0 : \varphi(x) = 0\}$ for generators that reach 0 at some point. Such generators are called *non-strict*, in contrast to *strict* generators that remain strictly positive on $[0,\infty)$. We check uniform margins and groundedness via

$$\begin{aligned} C_\varphi(1,\ldots,1,u_j,1,\ldots,1) &= \varphi\big(\varphi^{-1}(1)+\ldots+\varphi^{-1}(u_j)+\ldots+\varphi^{-1}(1)\big) \\ &\overset{(1)}{=} \varphi\big(0+\ldots+0+\varphi^{-1}(u_j)+0+\ldots+0\big) = u_j \end{aligned}$$

and

$$\begin{aligned} &C_\varphi(u_1,\ldots,u_{j-1},0,u_{j+1},\ldots,u_d) = \\ &\qquad \varphi\big(\underbrace{\varphi^{-1}(u_1)+\ldots+\varphi^{-1}(0)+\ldots+\varphi^{-1}(u_d)}_{\geq\varphi^{-1}(0)=\inf\{x>0:\varphi(x)=0\}}\big) = 0. \end{aligned}$$

[5] This is typical for extendible models when the conditional independence construction is known and used for sampling.

Still, in order to define a copula via Equation (2.2), it remains to ensure that each d-volume is positive.[6] This problem was solved by Schweizer and Sklar (1963) for the bivariate case; see also Moynihan (1978), Schweizer and Sklar (1983, Section 6.3), or Nelsen (2006, p. 111). One can show that C_φ is a copula in dimension $d = 2$ if and only if φ is a convex generator. However, there exist convex Archimedean generators that do not imply Archimedean copulas in dimension $d \geq 3$. A simple example is the function $\varphi : [0, \infty) \to [0, 1]$, with $x \mapsto (1 - x)\mathbb{1}_{\{x \in [0,1]\}}$, which is a convex generator. Considering the resulting function C_φ, we observe that this is precisely the lower Fréchet–Hoeffding bound of Lemma 1.3,

$$C_\varphi(u_1, \ldots, u_d) = \max\{u_1 + \ldots + u_d - d + 1, 0\}, \quad (u_1, \ldots, u_d) \in [0, 1]^d.$$

This function, however, is only a copula for dimension $d = 2$. In higher dimensions it assigns negative mass to certain rectangles.[7]

The next step is to investigate the problems raised in the introduction to this chapter from a probabilistic view. We know from the probabilistic model (2.1) that each positive random variable M defines an Archimedean copula: the corresponding Laplace transform φ defines an Archimedean copula C_φ in all dimensions $d \geq 2$. Hence, Laplace transforms of positive random variables must be suitable generators. Before going further, let us recall the required notion of complete monotonicity.

Definition 2.2 (Complete Monotonicity)
An (Archimedean) generator φ is completely monotone *(c.m.) if it has derivatives of all orders on $(0, \infty)$ and*

$$(-1)^j \varphi^{(j)}(x) \geq 0, \quad \forall x > 0,\ j \in \mathbb{N}_0,$$

where $\varphi^{(j)}$ denotes the jth order derivative of φ.

Note that this condition involves φ being (1) continuous on $(0, \infty)$ (since it is differentiable), (2) decreasing (since the first derivative is non-positive), and (3) convex (since the second derivative is non-negative). Examples of completely monotone functions are $\varphi(x) = \exp(-\vartheta\, x)$ and $\varphi(x) = (1+x)^{-\vartheta}$ for $\vartheta > 0$. We can now formulate Bernstein's theorem, which links completely monotone generators to Laplace transforms, and therefore establishes a probabilistic link to such functions.

[6] An example for a generator that does not imply an Archimedean copula is provided by McNeil and Nešlehová (2009). The function $\varphi : [0, \infty) \to [0, 1]$, with $x \mapsto (1 - x)\mathbb{1}_{\{x \in [0,1/2]\}} + (3/2 - 2x)\mathbb{1}_{\{x \in (1/2,3/4]\}}$ is a generator in the sense of Definition 2.1, but the resulting function C_φ (in dimension $d = 2$) assigns negative mass to the rectangle $[11/16, 13/16] \times [11/16, 13/16]$.

[7] See, e.g., Embrechts et al. (2003) or McNeil et al. (2005, p. 200): the C_φ volume of $[1/2, 1]^d$ is $1 - d/2$.

Theorem 2.1 (Bernstein's Theorem)
A generator $\varphi : [0, \infty) \to [0, 1]$ is completely monotone if and only if φ is the Laplace transform of a positive random variable M, i.e. $\varphi(x) = \mathbb{E}[\exp(-x\,M)]$ and $\mathbb{P}(M > 0) = 1$.

Proof. Originally in Bernstein (1929). See also Widder (1946, Theorem 12a, p. 160) or Feller (1966, Theorem 1, p. 439). □

First of all, we recall that distinct distributions have distinct Laplace transforms (see Feller (1966, Theorem 1, p. 430)). Combining the fact that each positive random variable M implies a copula (for all $d \geq 2$) with Bernstein's theorem, which provides a relation between Laplace transforms of positive random variables and completely monotone generators, we have solved the classification of extendible Archimedean copulas halfway. We now know that a completely monotone generator φ implies a copula C_φ in all dimensions. The reverse direction is established by Kimberling's theorem (see Kimberling (1974)). This establishes necessary and sufficient conditions on a generator φ to define copulas C_φ in all dimensions $d \geq 2$.

Theorem 2.2 (Kimberling's Theorem: Extendible C_φ)
Equation (2.2) defines a copula in all dimensions $d \geq 2$ if and only if the generator φ is completely monotone.

Proof. The seminal reference is Kimberling (1974); see also Alsina et al. (2006, Theorem 4.4.6). The result follows from the more general result of Theorem 2.3. □

Let us briefly summarize the key facts of this section: (1) There is an intimate correspondence between the class of extendible Archimedean copulas (defining a copula in all dimensions $d \geq 2$) and the class of positive random variables. (2) They are related via the Laplace transform φ and construction (2.1). Technically, this involves the notion of complete monotonicity, Bernstein's theorem, and Kimberling's theorem. Hence, extendible Archimedean copulas are precisely the ones with completely monotone generators. (3) All extendible Archimedean copulas can therefore be conveniently sampled using Algorithm 2.1. Required, however, is a sampling scheme for M and an analytical expression of its Laplace transform φ.

2.2.2 *Properties of Extendible Archimedean Copulas*

This section lists several properties of extendible Archimedean copulas. In most references, these properties are inferred from analytical properties of φ. We aim at relating these properties (whenever possible) to the mixture model that generates this class of copulas to provide alternative proofs and to gain a deeper understanding of the induced dependence structure.

2.2.2.1 *Tail Dependence*

Both upper- and lower-tail dependence coefficients of Archimedean copulas can directly be inferred from their definition and the simple functional form of Archimedean copulas. For the second to last step in (2.3) and (2.4), the substitution $\varphi^{-1}(u) := x$ is used. The last step is an application of de l'Hospital's rule.

$$
\begin{aligned}
UTD_{C_\varphi} &= 2 - \lim_{u \nearrow 1} \frac{1 - \varphi\big(2\,\varphi^{-1}(u)\big)}{1 - u} \\
&= 2 - \lim_{x \searrow 0} \frac{1 - \varphi(2\,x)}{1 - \varphi(x)} = 2 - 2 \lim_{x \searrow 0} \frac{\varphi'(2\,x)}{\varphi'(x)}, \qquad (2.3)
\end{aligned}
$$

$$
LTD_{C_\varphi} = \lim_{u \searrow 0} \frac{\varphi\big(2\,\varphi^{-1}(u)\big)}{u} = \lim_{x \nearrow \infty} \frac{\varphi(2\,x)}{\varphi(x)} = 2 \lim_{x \nearrow \infty} \frac{\varphi'(2\,x)}{\varphi'(x)}. \qquad (2.4)
$$

The choice of mixing variable M in construction (2.1) obviously affects the tail dependence coefficients of the resulting Archimedean copula. Some general statements in this regard are listed in the following lemma; a deeper analysis is provided in Li (2009) and Charpentier and Segers (2009). Examples where the lower-tail dependence coefficient does not exist are presented in Larsson and Nešlehová (2011).

Lemma 2.1 (Tail Dependence of Archimedean Copulas)
Consider an extendible Archimedean copula C_φ as in (2.1). Then:

(1) If the mixing variable M has existing expectation, i.e. $\mathbb{E}[M] < \infty$, the upper-tail dependence coefficient of the resulting copula is 0, i.e. $UTD_{C_\varphi} = 0$.
(2) If the mixing variable M is bounded away from 0, i.e. $\mathbb{P}\big(M \in [0, \epsilon)\big) = 0$ for some $\epsilon > 0$, then $LTD_{C_\varphi} = 0$.

Proof. Without loss of generality let M be non-deterministic. The assumption $\mathbb{E}[M] < \infty$ implies that $\mathbb{E}[M] = -\lim_{x \searrow 0} \varphi'(x)$ exists in $(0, \infty)$. Part (1) then follows from (2.3). The assumption $\mathbb{P}\big(M \in [0, \epsilon)\big) = 0$ for some $\epsilon > 0$ implies that $\varphi(2\,x) = \mathbb{E}[\exp(-2\,x\,M)] \leq \mathbb{E}[\exp(-x\,M - x\,\epsilon)] = \exp(-x\,\epsilon)\,\varphi(x)$. This establishes (2), since

$$0 \leq LTD_{C_\varphi} = \lim_{x \nearrow \infty} \frac{\varphi(2\,x)}{\varphi(x)} \leq \lim_{x \nearrow \infty} \frac{e^{-x\,\epsilon}\varphi(x)}{\varphi(x)} = 0. \qquad \square$$

Archimedean copulas are flexible in the sense that one can construct families without tail dependence, with (only) upper-tail dependence, with (only) lower-tail dependence, and with both upper- and lower-tail dependence. To discuss this further, Lemma 2.1 is useful in two regards. First, for several parametric families we can immediately conclude that the upper- and/or lower-tail dependence coefficient is 0.[8] Second, it suggests that if we want to construct an Archimedean copula with lower-tail dependence, the mixing variable M must not be bounded away from 0. Similarly, a heavy-tailed mixing variable is required in order to have upper-tail dependence. The converse statement, however, is not true, as the following example shows.

Example 2.1 (Infinite Expectation of $M \not\Rightarrow UTD_{C_\varphi} > 0$)
Let $M \sim$ Pareto$(1, 1)$, i.e. let M have density $f(x) = x^{-2}\mathbb{1}_{\{x \geq 1\}}$. The Laplace transform of M is

$$\varphi(x) = \mathbb{E}\big[e^{-x\,M}\big] = \int_1^\infty e^{-x\,m} m^{-2} dm = x \int_x^\infty e^{-m} m^{-2} dm, \quad x > 0,$$

using a change of variables to verify the last equality. Consequently,

$$\varphi'(x) = \int_x^\infty e^{-m} m^{-2} dm - e^{-x} x^{-1}.$$

Using de l'Hospital's rule, this yields

$$\lim_{x \searrow 0} \frac{\varphi'(2\,x)}{\varphi'(x)} = \lim_{x \searrow 0} \frac{\int_{2\,x}^\infty e^{-m} m^{-2} dm - e^{-2\,x}/(2\,x)}{\int_x^\infty e^{-m} m^{-2} dm - e^{-x}/x} = \ldots = \lim_{x \searrow 0} \frac{e^{-2\,x}}{e^{-x}} = 1.$$

Applied in Equation (2.3), this shows that $UTD_{C_\varphi} = 0$.

2.2.2.2 *Kendall's Tau*

Kendall's tau of an Archimedean copula can be derived as a functional of the generator φ (see Nelsen (2006, p. 163) and Joe (1997, p. 91)). We find

$$\tau_{C_\varphi} = 1 + 4 \int_0^1 \frac{\varphi^{-1}(x)}{(\varphi^{-1})'(x)} dx = 1 - 4 \int_0^\infty u\big(\varphi'(u)\big)^2 du. \qquad (2.5)$$

[8]When we later consider parametric families of Archimedean copulas, we shall see that most of them do not possess upper-tail dependence, which results exactly from the fact that they are defined via a light-tailed mixing variable M. Moreover, all parametric families with M taking values in $\mathbb{N}$ do not have lower-tail dependence.

In terms of applications, Kendall's tau is often used to estimate one-parametric bivariate Archimedean copulas: one sets the parameter such that the empirical version of Kendall's tau[9] agrees with the theoretical value.

Example 2.2 (Clayton Copula)
The Clayton family, introduced in Equation (2.11), has the generator $\varphi(x) = (1+x)^{-1/\vartheta}$ *for* $\vartheta \in (0,\infty)$. *Hence,* $\varphi^{-1}(x) = x^{-\vartheta} - 1$ *and* $(\varphi^{-1})'(x) = -\vartheta\, x^{-\vartheta-1}$. *Consequently*

$$\begin{aligned}\tau_{C_\varphi} &= 1 + 4\int_0^1 \frac{x^{-\vartheta}-1}{-\vartheta\, x^{-\vartheta-1}}dx \\ &= 1 + 4\int_0^1 \frac{x^{\vartheta+1}-x}{\vartheta}dx = \ldots = \frac{\vartheta}{\vartheta+2} \in (0,1).\end{aligned}$$

2.2.2.3 *Density of Archimedean Copulas*

The density of an Archimedean copula with a completely monotone generator can be computed by taking the d-partial derivatives with respect to $u_1, \ldots, u_d$ (see McNeil et al. (2005, p. 197)). One obtains

$$c_\varphi(u_1,\ldots,u_d) = \varphi^{(d)}\big(\varphi^{-1}(u_1) + \ldots + \varphi^{-1}(u_d)\big)(\varphi^{-1})'(u_1)\cdots(\varphi^{-1})'(u_d).$$

The above formula looks quite innocuous but becomes numerically challenging when the dimension d is large. This issue is addressed in Hofert et al. (2011) and explicit formulas for $\varphi^{(d)}$ are given for some Archimedean families. Note that interpreting φ as the Laplace transform of M might be beneficial in this situation, since then $\varphi^{(d)}(x) = (-1)^d\,\mathbb{E}[M^d \exp(-x\,M)]$, which can be solved explicitly or by numerical integration; depending on the respective family φ.

2.2.2.4 *Positive Lower Orthant Dependence (PLOD)*

Archimedean copulas with completely monotone generators φ are positive lower orthant dependent (PLOD), i.e. $\Pi(u_1,\ldots,u_d) \le C_\varphi(u_1,\ldots,u_d)$ for all $(u_1,\ldots,u_d) \in [0,1]^d$ (see Joe (1997, p. 20)), where Π is the independence copula. This property is quite natural given the construction (2.1), where independent components are affected in the very same way (and are thus made "positively dependent") by the random variable M. Note that

[9]Given n samples from a bivariate distribution, the empirical Kendall's tau is defined as the number of concordant pairs minus the number of discordant pairs divided by the total number of pairs (see McNeil et al. (2005, p. 229)).

this does not necessarily hold for Archimedean copulas without completely monotone generators, where, however, the stochastic model is also quite different (see construction (2.14)).

Proof. We have to show that for all $(u_1, \ldots, u_d) \in [0,1]^d$ one has

$$\Pi(u_1, \ldots, u_d) \leq \varphi\big(\varphi^{-1}(u_1) + \ldots + \varphi^{-1}(u_d)\big). \tag{2.6}$$

Using $\varphi_\Pi(x) := \exp(-x)$ and $\varphi_\Pi^{-1}(x) := -\log(x)$, Π can be written as an Archimedean copula. We set $\varphi^{-1}(u_k) =: x_k$ for $k = 1, \ldots, d$ and let $f(x) := -\log\big(\varphi(x)\big)$. Note that complete monotonicity of φ implies that $\varphi > 0$, making this transformation admissible. Inequality (2.6) can then be rearranged to

$$f(x_1) + \ldots + f(x_d) \geq f(x_1 + \ldots + x_d), \quad \forall\, (x_1, \ldots, x_d) \in [0, \infty)^d. \tag{2.7}$$

By induction, it is enough to show that $f(x_1) + f(x_2) \geq f(x_1 + x_2)$, for $(x_1, x_2) \in [0, \infty)^2$. We observe that $f(0) = -\log\big(\varphi(0)\big) = 0$. Moreover, using the complete monotonicity of φ, the second derivative $f^{(2)}(x) = \big(\varphi'(x)^2 - \varphi^{(2)}(x)\varphi(x)\big)/\varphi^2(x)$ is less than or equal to 0. This follows from a connection of absolutely monotone functions with Hankel determinants, establishing

$$\det\begin{pmatrix} \varphi(x) & -\varphi'(x) \\ -\varphi'(x) & (-1)^2\varphi^{(2)}(x) \end{pmatrix} = \varphi(x)\varphi^{(2)}(x) - \varphi'(x)\varphi'(x) \geq 0, \quad \forall\, x \geq 0$$

(see Widder (1946, Corollary 16, p. 167)). Hence, f is concave. Without loss of generality, let $x_1 + x_2 > 0$ and write

$$x_1 = \frac{x_1}{x_1 + x_2} \cdot (x_1 + x_2) + \frac{x_2}{x_1 + x_2} \cdot 0,$$
$$x_2 = \frac{x_1}{x_1 + x_2} \cdot 0 + \frac{x_2}{x_1 + x_2} \cdot (x_1 + x_2).$$

Using that f is concave and $f(0) = 0$, we obtain

$$f(x_1) \geq \frac{x_1}{x_1 + x_2} \cdot f(x_1 + x_2), \qquad f(x_2) \geq \frac{x_2}{x_1 + x_2} \cdot f(x_1 + x_2).$$

Combining both inequalities implies the required subadditivity of f. □

Let us remark that a closely related result on the ordering of $C_{\varphi_1} \leq C_{\varphi_2}$ is provided in Genest and MacKay (1986) (see also Nelsen (2006, p. 136)). There, it is shown that $C_{\varphi_1}(u_1, \ldots, u_d) \leq C_{\varphi_2}(u_1, \ldots, u_d)$ for all $(u_1, \ldots, u_d) \in [0,1]^d$ if and only if $f(x) := \varphi_1^{-1}\big(\varphi_2(x)\big)$ is subadditive.

2.2.3 *Constructing Multi-Parametric Families*

Outer/exterior power family: It is known (see Nelsen (2006, Theorem 4.5.1, p. 141)) that for a completely monotone generator φ and for $\alpha \in (0,1]$, a new completely monotone generator $\varphi_\alpha(x) := \varphi(x^\alpha)$ can be defined.[10] This statement can be shown either analytically (which is done in the above reference) or via the following probabilistic interpretation. Consider an α-stable Lévy subordinator[11] $\Lambda = \{\Lambda_t\}_{t\geq 0}$ and an independent positive random variable M with Laplace transform φ. Stopping the process Λ at time M yields another positive random variable, denoted $\hat{M} := \Lambda_M$. Its Laplace transform is given by $\mathbb{E}[\exp(-x\,\Lambda_M)] = \mathbb{E}[\,\mathbb{E}[\exp(-x\,\Lambda_M)|M]\,] = \mathbb{E}[\exp(-M\,x^\alpha)] = \varphi(x^\alpha)$, where we used the fact that the Laplace exponent of an α-stable Lévy subordinator is $\Psi(x) = -\log(\mathbb{E}[\exp(-x\,\Lambda_t)])/t = x^\alpha$. Generators defined in this way have one additional parameter and are therefore more flexible. Note that the simulation of such structures is easy if M and Λ can be simulated. Moreover, constructions of this kind turn out to be useful for defining h-extendible (hierarchical) Archimedean copulas via Lévy subordinators (see Section 2.4).

Inner/interior power family: It is also known (see Nelsen (2006, Theorem 4.5.1, p. 141)) that for a completely monotone generator φ and for $\beta \geq 1$, a new completely monotone generator $\varphi_\beta(x) := \varphi(x)^\beta$ can be defined. Here, we only have a probabilistic argument for $\beta \in \mathbb{N}$. In this case, consider β i.i.d. copies $M_1, \ldots, M_\beta$ of the positive random variable M whose Laplace transform is φ. Then, the sum $M_1 + \ldots + M_\beta$ is again a positive random variable and its Laplace transform is the product of the individual Laplace transforms, hence φ^β. Again, this is straightforward to simulate if a simulation scheme for M is available.

More on outer/inner families can be found in Nelsen (1997).

2.2.4 *Parametric Families*

This section introduces some of the Archimedean families that have been studied in the literature. We purposely selected families that have specific properties and describe those. The included scatterplots indicate the large flexibility of Archimedean copulas and might be used to visually understand the various dependence structures. Recall that each positive random vari-

[10]The reader might be surprised by the wording "outer/exterior". The explanation for this is that the original source (see Oakes (1994)) parameterizes Archimedean copulas by means of φ^{-1} instead of φ, where the power is then taken "outside".

[11]An introduction to Lévy processes is given in the Appendix.

able defines an Archimedean copula via the mixture model (2.1). Therefore, our list of random variables cannot be complete. A battery of additional generator functions including their properties can be found in Nelsen (2006, p. 116).

2.2.4.1 *Ali–Mikhail–Haq Family*

The generator of the *Ali–Mikhail–Haq (AMH) family* is given by

$$\varphi(x) = \frac{1-\vartheta}{e^x - \vartheta}, \quad \varphi^{-1}(x) = \log\Big(\frac{1-\vartheta}{x} + \vartheta\Big), \quad \vartheta \in [0,1). \tag{2.8}$$

This corresponds to a discrete mixing variable M with Geo$(1-\vartheta)$-distribution,[12] i.e.

$$\mathbb{P}(M = m) = (1-\vartheta)\,\vartheta^{m-1}, \quad m \in \mathbb{N}.$$

A simple representation[13] of such a distribution is $M \stackrel{d}{=} \lceil \log(U)/\log(\vartheta) \rceil$, where $U \sim U[0,1]$ and $\lceil x \rceil$ is the ceiling function $\lceil x \rceil := \min\{m \in \mathbb{Z} : m \geq x\}$ (see Devroye (1986, p. 499)). In the bivariate case, the range of ϑ can be extended to $[-1,1)$, but the interpretation of φ as the Laplace transform of a Geometric distribution is only available on $[0,1)$. When ϑ decreases to 0, the limit is the independence copula (see Figure 2.2). This is quite natural from the perspective of the mixing model, since for $\vartheta = 0$ we have $\mathbb{P}(M = 1) = 1$. The Ali–Mikhail–Haq family does not have tail dependence. Kendall's tau for this family is $\tau_{C_\varphi} = 1 - 2\big(\vartheta + (1-\vartheta)^2 \log(1-\vartheta)\big)/(3\,\vartheta^2)$. This illustrates that the Ali–Mikhail–Haq family is not able to imply strong dependence, since Kendall's tau is increasing in ϑ and converges to $1/3$ for $\vartheta \nearrow 1$. A further characterization of this family, based on quotients of polynomials, is provided in Nelsen (2006, p. 148). The name of this family stems from Ali et al. (1978).

2.2.4.2 *Frank Family*

The generator of the *Frank family* is given by

$$\varphi(x) = -\frac{1}{\vartheta} \log\big(e^{-x}(e^{-\vartheta} - 1) + 1\big), \quad \varphi^{-1}(x) = -\log\Big(\frac{e^{-\vartheta\,x} - 1}{e^{-\vartheta} - 1}\Big), \tag{2.9}$$

[12]For $x \geq 0$: $\mathbb{E}\big[e^{-x\,M}\big] = \sum_{m=1}^{\infty} e^{-x\,m}(1-\vartheta)\,\vartheta^{m-1} = e^{-x}(1-\vartheta)\sum_{m=0}^{\infty}(e^{-x}\,\vartheta)^m = (1-\vartheta)/(e^x - \vartheta)$.

[13]Note: $\mathbb{P}(\lceil \frac{\log(U)}{\log(\vartheta)} \rceil = m) = \mathbb{P}(\vartheta^{m-1} > U \geq \vartheta^m) = \vartheta^{m-1} - \vartheta^m = (1-\vartheta)\,\vartheta^{m-1}$.

where $\vartheta \in (0, \infty)$. This corresponds to a discrete mixing variable M with distribution[14]

$$\mathbb{P}(M = m) = \frac{(1 - e^{-\vartheta})^m}{m\,\vartheta}, \quad m \in \mathbb{N}.$$

Note that this distribution might be seen as a *logarithmic distribution*[15] with parameter $p = \big(1 - \exp(-\vartheta)\big)$. The Frank family does not have tail dependence. Kendall's tau for this family is $\tau_{C_\varphi} = 1 + 4\big(\int_0^\vartheta t(\vartheta \exp(t) - \vartheta)^{-1} dt - 1\big)/\vartheta$. Its limiting cases are the independence copula for $\vartheta \searrow 0$ and the comonotonicity copula for $\vartheta \nearrow \infty$. The distinct property of the Frank family is that, in the bivariate case,[16] it is the only Archimedean family which is radially symmetric, i.e. its copula and its survival copula are the same. This statement was shown in Frank (1979) and is visualized in Figure 2.3.

2.2.4.3 *Joe Family*

The generator of the Joe family is given by

$$\varphi(x) = 1 - (1 - e^{-x})^{1/\vartheta}, \quad \varphi^{-1}(x) = -\log\big(1 - (1 - x)^{\vartheta}\big), \tag{2.10}$$

where $\vartheta \in [1, \infty)$. This corresponds to a discrete mixing variable M with a Sibuya distribution[17]

$$\mathbb{P}(M = m) = (-1)^{m+1} \binom{1/\vartheta}{m}, \quad m \in \mathbb{N}.$$

The sampling of general discrete random variables with a given probability distribution function is treated in Chapter 6. A more specific algorithm to sample from M in the present situation is presented in Hofert (2011). This algorithm is repeated below for the readers' convenience.

Algorithm 2.2 (Sampling M for Joe's Family: Hofert (2011))
Define $\alpha := 1/\vartheta$ and the functions

$$F(n) := 1 - \frac{1}{n\,B(n, 1 - \alpha)}, \quad n \in \mathbb{N},$$

$$G^{-1}(y) := \big((1 - y)\,\Gamma(1 - \alpha)\big)^{-\vartheta}, \quad y \in [0, 1],$$

---

[14]For $x \geq 0$: $\mathbb{E}\big[e^{-x\,M}\big] = \sum_{m=1}^{\infty} e^{-x\,m} \frac{(1-e^{-\vartheta})^m}{m\,\vartheta} = \frac{1}{\vartheta} \sum_{m=1}^{\infty} \frac{(e^{-x}(1-e^{-\vartheta}))^m}{m} = -\frac{1}{\vartheta} \log(e^{-x}(e^{-\vartheta} - 1) + 1)$.

[15]$X \sim \mathrm{Log}(p)$ if $\mathbb{P}(X = m) = p^m/\big(-m\log(1-p)\big)$ for $m \in \mathbb{N}$ and $p \in (0, 1)$. Sampling such distributions is considered in Kemp (1981).

[16]In the trivariate case, one can quickly find numerically a point $(u_1, u_2, u_3) \in [0, 1]^3$ satisfying $C(u_1, u_2, u_3) \neq \hat{C}(u_1, u_2, u_3)$.

[17]For $x \geq 0$: $\mathbb{E}\big[e^{-x\,M}\big] = \sum_{m=1}^{\infty} e^{-x\,m} (-1)^{m+1} \binom{1/\vartheta}{m} = 1 - \sum_{m=0}^{\infty} \binom{1/\vartheta}{m} (-e^{-x})^m = 1 - (1 - e^{-x})^{1/\vartheta}$.

where Γ *and* B *abbreviate the Gamma and Beta functions,*[18] *respectively. Then,* M *can be sampled via the following algorithm:*

(1) Sample $U \sim U[0,1]$.
(2) If $U \leq \alpha$ *then return* 1.
(3) If $F(\lfloor G^{-1}(U) \rfloor) < U$ *then return* $\lceil G^{-1}(U) \rceil$. *Otherwise* $\lfloor G^{-1}(U) \rfloor$,

where $\lfloor x \rfloor := \max\{m \in \mathbb{Z} : m \leq x\}$ *and* $\lceil x \rceil := \min\{m \in \mathbb{Z} : m \geq x\}$.

The Joe family does not have lower-tail dependence but has upper-tail dependence equal to $2 - 2^{\frac{1}{\vartheta}}$. Kendall's tau for this family is $\tau_{C_\varphi} = 1 - 4 \sum_{k=1}^{\infty} \big(k\,(\vartheta\, k + 2)\,(\vartheta\,(k-1) + 2)\big)^{-1}$. The limiting cases of this family are the independence copula for $\vartheta = 1$ (where $\mathbb{P}(M = 1) = 1$) and the comonotonicity copula for $\vartheta \nearrow \infty$ (see Figure 2.4). This family appears, e.g., in Joe (1993).

2.2.4.4 *Clayton Family*

The wording[19] *Clayton family* stems from the reference Clayton (1978). The generator of the Clayton family is given by

$$\varphi(x) = (1+x)^{-1/\vartheta}, \quad \varphi^{-1}(x) = x^{-\vartheta} - 1, \quad \vartheta \in (0,\infty). \tag{2.11}$$

This corresponds to a Gamma-distributed[20] mixing variable $M \sim \Gamma(1/\vartheta, 1)$. In the bivariate case, the range of ϑ can even be extended to $[-1,0) \cup (0,\infty)$. Then, however, the interpretation of the copula as the result of a mixture model involving a Gamma-distributed random variable is lost and the generator function is not completely monotone anymore. The Clayton family has lower-tail dependence $2^{-\frac{1}{\vartheta}}$ but does not have upper-tail dependence. Kendall's tau for this family is $\tau_{C_\varphi} = \vartheta/(\vartheta+2)$ (see Example 2.2). The

[18]The Beta function is defined as $B(x,y) := \Gamma(x)\,\Gamma(y)/\Gamma(x+y)$.

[19]H. Joe names this copula the MTCJ copula and provides as additional references Mardia (1962), Takahasi (1965), Kimeldorf and Sampson (1975), and Cook and Johnson (1981), where this copula appears explicitly or implicitly.

[20]Using the substitution $\tilde{m} := m\,(1+x)$ and the definition of the Gamma function, we find

$$\begin{aligned}
\mathbb{E}\big[e^{-x\,M}\big] &= \int_0^\infty e^{-xm} \frac{m^{1/\vartheta - 1} e^{-m}}{\Gamma(1/\vartheta)}\, dm = \frac{1}{\Gamma(1/\vartheta)} \int_0^\infty e^{-m(x+1)} m^{1/\vartheta - 1}\, dm \\
&= \frac{1}{\Gamma(1/\vartheta)} \int_0^\infty e^{-\tilde{m}} \Big(\frac{\tilde{m}}{1+x}\Big)^{1/\vartheta - 1} \frac{1}{1+x}\, d\tilde{m} \\
&= \frac{1}{\Gamma(1/\vartheta)(1+x)^{1/\vartheta}} \int_0^\infty e^{-\tilde{m}} \tilde{m}^{1/\vartheta - 1}\, d\tilde{m} = (1+x)^{-1/\vartheta}, \quad x \geq 0.
\end{aligned}$$

limiting cases are the independence copula when $\vartheta \searrow 0$ and the comonotonicity copula when $\vartheta \nearrow \infty$ (see Figure 2.5).

The Clayton copula is hidden in the following application (see Schmitz (2004)). Consider a list of i.i.d. random variables $X_1, \ldots, X_n$ with continuous distribution function F. Let $X_{(1)} \leq \ldots \leq X_{(n)}$ denote the order statistic of this sample. We can compute the copula of $(-X_{(1)}, X_{(n)})$ via $\mathbb{P}(-X_{(1)} \leq x, X_{(n)} \leq y) = \mathbb{P}\big(\bigcap_{i=1}^n \{X_i \in [-x, y]\}\big) = \big(F(y) - F(-x)\big)^n$ for $-x \leq y$ and 0 otherwise. Considering the marginals of $-X_{(1)}$ and $X_{(n)}$, respectively, note that $F_{X_{(n)}}(y) = F(y)^n$ and $F_{-X_{(1)}}(x) = \big(1 - F(-x)\big)^n$. Sklar's theorem yields the copula $C_{-X_{(1)},X_{(n)}}(u, v) = \max\{0, (u^{1/n} + v^{1/n} - 1)^n\}$, which is the bivariate Clayton copula with parameter $\vartheta = 1/n$. The copula of $X_{(1)}$ and $X_{(n)}$ can then be found by using the fact that $-X_{(1)}$ is a strictly decreasing transformation of $X_{(1)}$. Therefore (see Nelsen (2006, Theorem 2.4.4, p. 26) or Embrechts et al. (2003, Example 2.3)), the copula of the smallest and the largest sample is $C_{X_{(1)},X_{(n)}}(u, v) = v - C_{-X_{(1)},X_{(n)}}(1 - u, v)$. This shows that $X_{(1)}$ and $X_{(n)}$ are asymptotically independent.

A second example where the Clayton family naturally occurs is presented in McNeil et al. (2005, Example 5.12). Here, it is shown that the (bivariate) Clayton copula is the survival copula of the (bivariate) Pareto distribution (see also Mardia (1962)).

2.2.4.5 *Gumbel Family*

This family was introduced in Gumbel (1960a). The generator of the *Gumbel family* is given by

$$\varphi(x) = e^{-x^{1/\vartheta}}, \quad \varphi^{-1}(x) = \big(-\log(x)\big)^\vartheta, \quad \vartheta \in [1, \infty). \tag{2.12}$$

This corresponds to a $1/\vartheta$-stable-distributed mixing variable $M \sim S(1/\vartheta)$. The Gumbel family has no lower-tail dependence but has upper-tail dependence $2 - 2^{\frac{1}{\vartheta}}$. Kendall's tau for this family is $\tau_{C_\varphi} = (\vartheta - 1)/\vartheta$. The limiting cases for $\vartheta \searrow 1$ and $\vartheta \nearrow \infty$ are the independence and the comonotonicity copula, respectively (see Figure 2.6). The Gumbel family plays a distinct role as the only Archimedean family that is at the same time an extreme-value copula (see Definition 1.12). Writing out the functional form of C_φ one easily verifies $\varphi\big(\varphi^{-1}(u_1^t) + \ldots + \varphi^{-1}(u_d^t)\big) = \varphi\big(\varphi^{-1}(u_1) + \ldots + \varphi^{-1}(u_d)\big)^t$ for all $t \geq 0$. The proof that there are no other Archimedean families with this property is due to Genest and Rivest (1989) and can also be found in Nelsen (2006, p. 143).

2.2.4.6 *Inverse Gaussian Family*

The generator of the inverse Gaussian (IG) family is given by

$$\varphi(x) = e^{(1-\sqrt{1+2\vartheta^2 x})/\vartheta}, \quad \varphi^{-1}(x) = \frac{\big(1-\vartheta\log(x)\big)^2-1}{2\,\vartheta^2}, \tag{2.13}$$

where $\vartheta \in (0,\infty)$. This corresponds to a continuous mixing variable $M \sim IG(1, 1/\vartheta)$. The IG family has no tail dependence. We do not have a closed-form expression for Kendall's tau, but it is not difficult to numerically compute it from Equation (2.5). Scatterplots based on different parameter values are given in Figure 2.7.

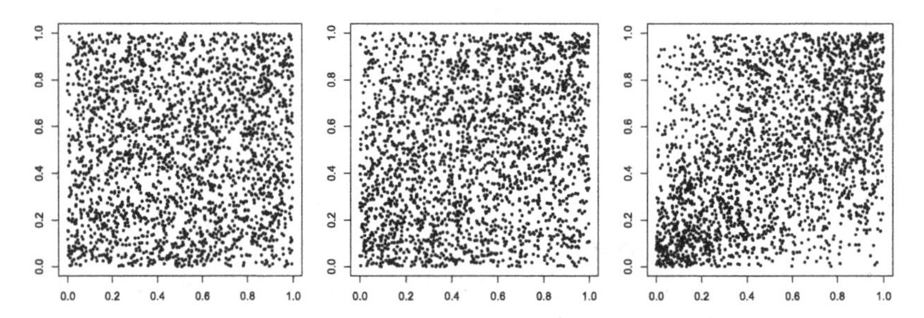

Fig. 2.2 Scatterplot of 2 500 samples from the AMH copula with $\vartheta = 0.1$ (left), 0.5 (middle), and 0.9 (right). Note that the dependence is increasing with ϑ, and the limiting case $\vartheta = 0$ corresponds to independence.

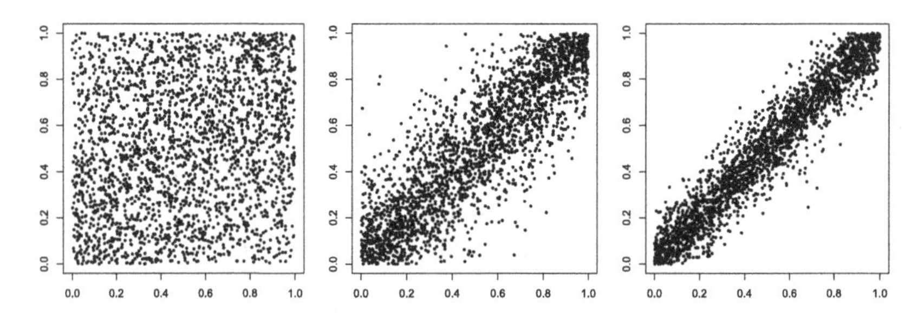

Fig. 2.3 Scatterplot of 2 500 samples from the Frank copula with $\vartheta = 1$ (left), 10 (middle), and 20 (right). Note that the dependence is increasing with ϑ, the limiting case $\vartheta \searrow 0$ corresponds to independence, and the case $\vartheta \nearrow \infty$ to comonotonicity. Also note that the scatterplot has two axes of symmetry: (1) symmetry around the first diagonal (corresponding to exchangeability) and (2) symmetry around the second diagonal (corresponding to the fact that the bivariate Frank copula is radially symmetric).

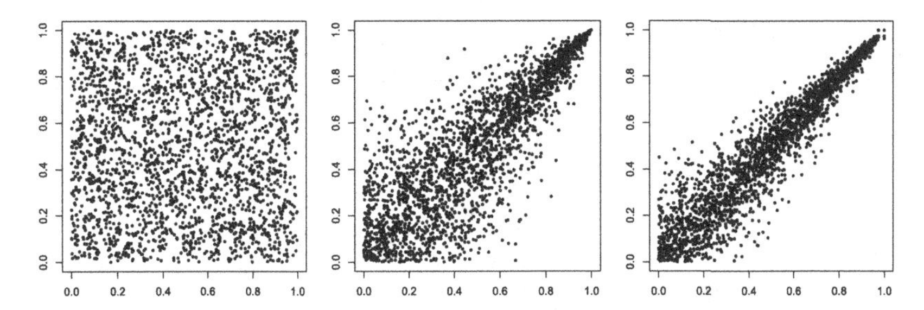

Fig. 2.4 Scatterplot of 2 500 samples from the Joe copula with $\vartheta = 1.1$ (left), 5 (middle), and 10 (right). Note that the dependence is increasing with ϑ, the limiting case $\vartheta = 1$ corresponds to independence, and $\vartheta \nearrow \infty$ corresponds to the comonotonicity copula.

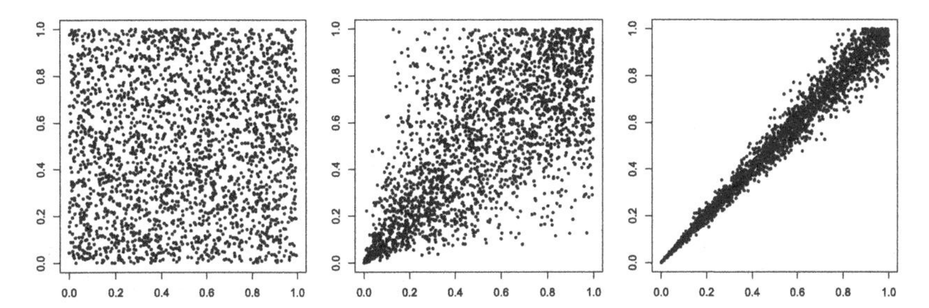

Fig. 2.5 Scatterplot of 2 500 samples from the Clayton copula with $\vartheta = 0.1$ (left), 2 (middle), and 20 (right). Note that the dependence is increasing with ϑ, the limiting case $\vartheta \searrow 0$ corresponds to independence, and the case $\vartheta \nearrow \infty$ to comonotonicity.

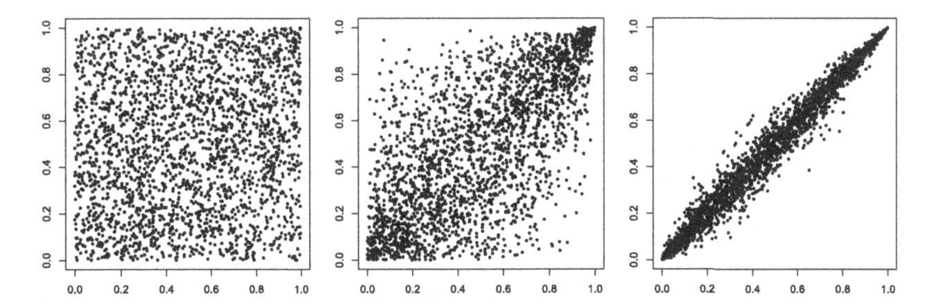

Fig. 2.6 Scatterplot of 2 500 samples from the Gumbel copula with $\vartheta = 1.1$ (left), 2 (middle), and 10 (right). Note that the dependence is increasing with ϑ, the limiting case $\vartheta \searrow 1$ corresponds to independence, and the case $\vartheta \nearrow \infty$ to comonotonicity.

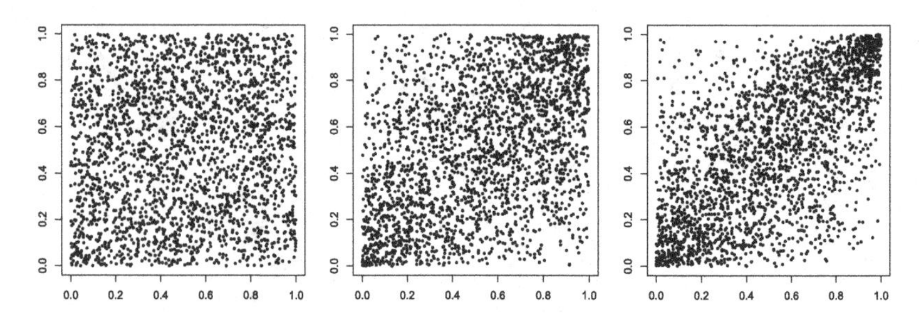

Fig. 2.7 Scatterplot of 2 500 samples from the IG copula with $\vartheta = 0.1$ (left), 2 (middle), and 20 (right). Note that the dependence is increasing with ϑ.

2.3 Exchangeable Archimedean Copulas

Recalling Kimberling's theorem from Section 2.2, on the one hand it is established that φ defines a copula C_φ in all dimensions $d \geq 2$ if and only if φ is a completely monotone generator. On the other hand, the function $x \mapsto (1-x)\mathbb{1}_{\{x\in[0,1]\}}$ provides an example of a generator that defines a copula C_φ in dimension $d = 2$ but not for $d \geq 3$. This leaves us with the following questions: (1) How do we characterize generators that define copulas C_φ for some fixed $d > 2$ (recall that the bivariate case was solved by Schweizer and Sklar (1963))? (2) Given such a characterization, is there some probabilistic model that might be used to sample from C_φ?

Sketching a roadmap for this subsection, it shall turn out that (somewhat similar to Bernstein's theorem) the notion of d-monotonicity can be related to the so-called Williamson d-transform of a positive random variable R. Moreover, C_φ is a proper copula in dimension d if the generator φ is d-monotone (note the similarity to Kimberling's theorem). Finally, a stochastic model for general exchangeable Archimedean copulas in dimension d is constructed, involving the random variable R that is related to φ via the Williamson d-transform.

To address the issues (1) and (2) raised above, let us start with some very general thoughts. If C_φ is an Archimedean copula for some fixed dimension d, the symmetry of C_φ manifests that the underlying distribution is exchangeable. Moreover, we have already classified all extendible Archimedean copulas, so we are looking for an exchangeable Archimedean copula that is not extendible. What is special about extendibility (and can thus be ruled out)? We know that the components of extendible random vectors have non-negative pairwise correlation (in the Archimedean case, they are PLOD, see Section 2.2.2.4). This puts us on the right track, as the example for a generator that defines an exchangeable Archimedean copula for $d = 2$ but not for $d \geq 3$ was the generator that corresponds to the countermonotonicity copula in dimension $d = 2$ (see Example 1.3). Hence, it might be worthwhile to investigate exchangeable models that are flexible enough to allow for negative dependence. One such model is the following. On a probability space $(\Omega, \mathcal{F}, \mathbb{P})$ consider a list $E_1, \ldots, E_d$ of i.i.d. exponentially distributed random variables with mean 1 and define

$$\mathbb{S}_d := \Big(\frac{E_1}{E_1 + \ldots + E_d}, \ldots, \frac{E_d}{E_1 + \ldots + E_d}\Big).$$

Intuitively, it is clear that the components defined in the above way are negatively dependent.[21] More precisely, one can show the following lemma. The density of $\mathbb{S}_d$ is provided in Fang et al. (1990, Theorem 5.2(2), p. 115).

Lemma 2.2 (Properties of $\mathbb{S}_d$)

(1) The sum $E_1 + \ldots + E_d$ follows an Erlang distribution, i.e. a Gamma distribution with scale parameter d and rate parameter 1, *i.e. with density*

$$f(x) = \mathbb{1}_{\{x>0\}} \frac{x^{d-1} e^{-x}}{(d-1)!}, \quad x \in \mathbb{R}.$$

(2) $\mathbb{S}_d$ is uniformly distributed on the d-dimensional unit simplex, i.e. on the set

$$A_d := \{(u_1, \ldots, u_d) \in [0,1]^d : u_1 + \ldots + u_d = 1\}.$$

(3) $\mathbb{S}_d$ and $E_1 + \ldots + E_d$ are independent random variables.

(4) The survival copula $\hat{C}$ of $\mathbb{S}_d$ is of the Archimedean kind, given by

$$\begin{aligned}\hat{C}(u_1, \ldots, u_d) &= \big(u_1^{\frac{1}{d-1}} + \ldots + u_d^{\frac{1}{d-1}} - (d-1)\big)_+^{d-1} \\ &= C_{\varphi_{d-1}}(u_1, \ldots, u_d),\end{aligned}$$

with $\varphi_{d-1}(x) := (1-x)_+^{d-1} = \max(1-x, 0)^{d-1}$.

(5) If R is independent of $\mathbb{S}_d$ and Erlang distributed with parameter d, then

$$R\,\mathbb{S}_d \stackrel{d}{=} (E_1, \ldots, E_d).$$

[21] Consider, e.g., the case $d = 2$ where this is especially apparent, since $E_1/(E_1 + E_2) = 1 - E_2/(E_1 + E_2)$.

Proof.

(1) The statement is obtained by means of a standard computation and it also follows as a byproduct from the proof of (2).
(2) Note that $(E_1, \ldots, E_d)$ has joint probability density function

$$(x_1, \ldots, x_d) \mapsto \mathbb{1}_{\{x_1>0,\ldots,x_d>0\}} e^{-\sum_{i=1}^d x_i}.$$

Define the transformation $h : (0,\infty)^d \to (0,1)^{d-1} \times (0,\infty)$ by

$$(x_1, \ldots, x_d) \mapsto \Big(\frac{x_1}{\sum_{i=1}^d x_i}, \ldots, \frac{x_{d-1}}{\sum_{i=1}^d x_i}, \sum_{i=1}^d x_i\Big).$$

Then, $h^{-1} : (0,1)^{d-1} \times (0,\infty) \to (0,\infty)^d$ is given by

$$(y_1, \ldots, y_{d-1}, y) \mapsto \Big(y y_1, y y_2, \ldots, y y_{d-1}, y\big(1 - \sum_{i=1}^{d-1} y_i\big)\Big).$$

It follows that

$$(h^{-1})' = \begin{bmatrix} \frac{\partial}{\partial y_1} h_1^{-1} & \ldots & \frac{\partial}{\partial y_1} h_{d-1}^{-1} & \frac{\partial}{\partial y_1} h_d^{-1} \\ \vdots & \ddots & \vdots & \vdots \\ \frac{\partial}{\partial y_{d-1}} h_1^{-1} & \ldots & \frac{\partial}{\partial y_{d-1}} h_{d-1}^{-1} & \frac{\partial}{\partial y_{d-1}} h_d^{-1} \\ \frac{\partial}{\partial y} h_1^{-1} & \ldots & \frac{\partial}{\partial y} h_{d-1}^{-1} & \frac{\partial}{\partial y} h_d^{-1} \end{bmatrix}$$
$$= \begin{bmatrix} y & 0 & \ldots & 0 & -y \\ 0 & y & \ldots & 0 & -y \\ \vdots & \vdots & \ddots & \vdots & \vdots \\ 0 & 0 & \ldots & y & -y \\ y_1 & y_2 & \ldots & y_{d-1} & \big(1 - \sum_{i=1}^{d-1} y_i\big) \end{bmatrix},$$

where h_i abbreviates $h_i(x_1, \ldots, x_d) := (h(x_1, \ldots, x_d))_i$, the ith component of h. By either using Laplace expansion along the last row, or by alternatively using the fact that the determinant is a multi-linear functional that is invariant when one row or column is added to another one, we obtain the determinant $\det\big((h^{-1})'\big) = y^{d-1}$. Density transformation (see, e.g., Czado and Schmidt (2011, Satz 1.3, p. 6)) provides that the joint probability density function of $h(E_1, \ldots, E_d)$ is given by

$$\begin{aligned}
&f_{h(E_1,\ldots,E_d)}\big(y_1, \ldots, y_{d-1}, y\big) \\
&= f_{(E_1,\ldots,E_d)}\big(h^{-1}(y_1, \ldots, y_{d-1}, y)\big) \cdot \big|\det\big((h^{-1})'\big)\big| \\
&= \mathbb{1}_{\{yy_1>0,\ldots,yy_{d-1}>0,y\left(1-\sum\limits_{i=1}^{d-1} y_i\right)>0\}} e^{-y}y^{d-1} \\
&\overset{(*)}{=} \mathbb{1}_{\{y_1>0,\ldots,y_{d-1}>0,y>0,\sum\limits_{i=1}^{d-1} y_i<1\}} e^{-y}y^{d-1} \\
&= \underbrace{\left(\mathbb{1}_{\{y>0\}}e^{-y}y^{d-1}\frac{1}{(d-1)!}\right)}_{(**)} \cdot \underbrace{\left((d-1)!\mathbb{1}_{\{y_1>0,\ldots,y_{d-1}>0,\sum\limits_{i=1}^{d-1} y_i<1\}}\right)}_{(***)},
\end{aligned}$$

where $(**)$ equals the density of $h_d(E_1, \ldots, E_d) = E_1 + \ldots + E_d$, i.e. an Erlang distribution with scale parameter d, and $(***)$ is the density of $(h_1, \ldots, h_{d-1})(E_1, \ldots, E_d) = \big(E_1/\sum_{i=1}^d E_i, \ldots, E_{d-1}/\sum_{i=1}^d E_i\big)$, which is uniformly distributed on the set in the indicator. To justify $(*)$, first consider $yy_1 > 0, \ldots, yy_{d-1} > 0, y\big(1 - \sum_{i=1}^{d-1} y_i\big) > 0$ and assume $y < 0$. It follows that $y_1 < 0, \ldots, y_{d-1} < 0$ as $yy_1, \ldots, yy_{d-1}$ should be positive. Hence, $y\big(1 - \sum_{i=1}^{d-1} y_i\big) < 0$. This contradiction shows that the assumption is wrong and $y > 0$. Thus, it follows from $yy_1 > 0, \ldots, yy_{d-1} > 0, y\big(1 - \sum_{i=1}^{d-1} y_i\big) > 0$ that $y_1 > 0, \ldots, y_{d-1} > 0, y > 0, \sum_{i=1}^{d-1} y_i < 1$. The other direction is trivial.

It follows that $\big(E_1/\sum_{i=1}^d E_i, \ldots, E_{d-1}/\sum_{i=1}^d E_i\big)$ is uniformly distributed on the set $\{y_1 > 0, \ldots, y_{d-1} > 0, \sum_{i=1}^{d-1} y_i < 1\}$ and consequently

$$\left(\frac{E_1}{\sum_{i=1}^d E_i}, \ldots, \frac{E_{d-1}}{\sum_{i=1}^d E_i}, \frac{E_d}{\sum_{i=1}^d E_i}\right)$$

follows the uniform distribution on A_d, as

$$\frac{E_d}{\sum_{i=1}^d E_i} = 1 - \frac{\sum_{i=1}^{d-1} E_i}{\sum_{i=1}^d E_i}.$$

(3) The claim follows directly from the proof of (2), since $E_1 + \ldots + E_d$ and

$$\left(\frac{E_1}{\sum_{i=1}^d E_i}, \ldots, \frac{E_{d-1}}{\sum_{i=1}^d E_i}\right)$$

are independent and $\mathbb{S}_d$ is a function of the latter.

(4) Consider $u \in (0,1)$ and $k \in \{1,\dots,d\}$ and compute

$$\mathbb{P}\Big(\frac{E_k}{E_1+\ldots+E_d} > u\Big) = \mathbb{P}\Big(E_k > \frac{u}{1-u}\sum_{i\neq k} E_i\Big) \overset{(*)}{=} \mathbb{E}\Big[e^{-\frac{u}{1-u}\sum_{i\neq k} E_i}\Big]$$
$$\overset{(**)}{=} \Big(1+\frac{\frac{u}{1-u}}{1}\Big)^{-(d-1)} = \big(\frac{1}{1-u}\big)^{-(d-1)}$$
$$= (1-u)^{d-1} =: \bar{F}_k(u).$$

Thus, $\bar{F}_k^{-1}(v) = 1 - v^{\frac{1}{d-1}}$ for $v \in (0,1)$. Observe that to justify $(*)$ we condition on $\sum_{i\neq k} E_i$ and in $(**)$ we exploit the fact that $\sum_{i\neq k} E_i \sim$ Erlang$(d-1) = \Gamma(1, d-1)$, which is why the Laplace transform is known (see Section A.2.2).

Moreover,[22] by substituting the density of $(h_1,\dots,h_{d-1})(E_1,\dots,E_d)$ obtained in the proof of (2), we deduce using $u_i \in (0,1)$ that

$$\begin{aligned}
&\mathbb{P}\big(\mathbb{S}_d > (u_1,\dots,u_d)\big)\\
&= \mathbb{P}\Big(h_1 > u_1,\dots,h_{d-1} > u_{d-1}, 1 - \sum_{i=1}^{d-1} h_i > u_d\Big)\\
&= \int_{u_1}^1 \int_{u_2}^1 \cdots \int_{u_{d-1}}^1 \mathbb{1}_{\{\sum_{i=1}^{d-1} x_i < 1-u_d\}} (d-1)!\, dx_{d-1}\ldots dx_2 dx_1\\
&= \iint_{[0,1]^{d-1}} \mathbb{1}_{\{x_1>u_1\}}\cdots \mathbb{1}_{\{x_{d-2}>u_{d-2}\}} \mathbb{1}_{\{\sum_{i=1}^{d-1} x_i - (x_1+\ldots+x_{d-2}) > u_{d-1}\}} \times\\
&\qquad \times \mathbb{1}_{\{\sum_{i=1}^{d-1} x_i < 1-u_d\}} (d-1)!\, d(x_1,\dots,x_{d-1}).
\end{aligned}$$

We substitute $(x_1,\dots,x_{d-1}) \overset{f}{\mapsto} (x_1,\dots,x_{d-2},\sum_{i=1}^{d-1} x_i)$ with $\det(f') = 1$ and transform further

$$\begin{aligned}
&\mathbb{P}(\mathbb{S}_d > (u_1,\dots,u_d))\\
&= \iint_{[0,1]^{d-2}\times[0,d-1]} \mathbb{1}_{\{x_1>u_1\}}\cdots \mathbb{1}_{\{x_{d-2}>u_{d-2}\}} \mathbb{1}_{\{x-(x_1+\ldots+x_{d-2})>u_{d-1}\}} \times\\
&\qquad \times \mathbb{1}_{\{x<1-u_d\}} (d-1)!\, d(x_1,\dots,x_{d-2},x)\\
&= (d-1)! \int_{u_1}^1 \cdots \int_{u_{d-2}}^1 \int_0^{d-1} \mathbb{1}_{\{1-u_d > x > u_{d-1}+(x_1+\ldots+x_{d-2})\}}\, dx\, d(x_1,\dots,x_{d-2})\\
&= (d-1)! \int_{u_1}^1 \cdots \underbrace{\int_{u_{d-2}}^1 \Big(1 - u_d - \big(u_{d-1} + (x_1+\ldots+x_{d-2})\big)\Big)_+ dx_{d-2}}_{\substack{= -\frac{1}{2}\left(1-(u_d+u_{d-1}+x_1+\ldots+x_{d-2})\right)_+^2 \Big|_{x_{d-2}=u_{d-2}}^{x_{d-2}=1} \\ = \frac{1}{2}\left(1-(u_d+u_{d-1}+u_{d-2}+x_1+\ldots+x_{d-3})\right)_+^2}} \ldots dx_1.
\end{aligned}$$

[22]Operators such as ">" are understood componentwise and we abbreviate $h_i := h_i(E_1,\dots,E_d)$.

By applying iteratively the same computations, we obtain

$$\begin{aligned}\mathbb{P}\big(\mathbb{S}_d > (u_1,\ldots,u_d)\big) &= (d-1)!\frac{1}{(d-1)!}\Big[1-(u_d+\ldots+u_1)\Big]_+^{d-1}\\ &= \Big[1-(u_d+\ldots+u_1)\Big]_+^{d-1}\\ &= \max\big(1-(u_d+\ldots+u_1),0\big)^{d-1}.\end{aligned}$$

Now Sklar's theorem provides that

$$\begin{aligned}\hat{C}_{\mathbb{S}_d}(u_1,\ldots,u_d) &= \Big[1-\Big(\big(1-u_d^{\frac{1}{d-1}}\big)+\ldots+\big(1-u_1^{\frac{1}{d-1}}\big)\Big)\Big]_+^{d-1}\\ &= \big[u_d^{\frac{1}{d-1}}+\ldots+u_1^{\frac{1}{d-1}}-(d-1)\big]_+^{d-1}\\ &= C_{\varphi_{d-1}}(u_1,\ldots,u_d),\end{aligned}$$

with $\varphi_{d-1}(x) := (1-x)_+^{d-1} = \max(1-x,0)^{d-1}$.

(5) Consider a random variable $R \sim \mathit{Erlang}(d)$ (note $\mathbb{P}(R>0)=1$), independent of $\mathbb{S}_d$, and let $x_1>0,\ldots,x_d>0$. Then, because of (4), one has

$$\begin{aligned}\mathbb{P}\big(R\mathbb{S}_d > (x_1,\ldots,x_d)\big) &= \mathbb{E}\Big[\mathbb{P}\big(\mathbb{S}_d > \frac{1}{R}(x_1,\ldots,x_d)\mid R\big)\Big]\\ &= \mathbb{E}\Big[\big(1-\frac{1}{R}(x_1+\ldots+x_d)\big)_+^{d-1}\Big]\\ &= \int_0^\infty \frac{1}{(d-1)!}e^{-x}x^{d-1}\Big(1-\sum_{i=1}^d x_i/x\Big)_+^{d-1}dx\\ &= \int_{\sum_{i=1}^d x_i}^\infty \frac{1}{(d-1)!}e^{-x}\Big(x-\sum_{i=1}^d x_i\Big)^{d-1}dx\\ &= \int_0^\infty \frac{1}{(d-1)!}e^{-y-\sum_{i=1}^d x_i}y^{d-1}dy\\ &= e^{-\sum_{i=1}^d x_i}\underbrace{\int_0^\infty \frac{1}{(d-1)!}e^{-y}y^{d-1}dy}_{\substack{=1 \text{ as integral over density}\\ \text{of Erlang distribution}}} = \underbrace{e^{-\sum_{i=1}^d x_i}}_{\text{surv. func. of }(E_1,\ldots,E_d)},\end{aligned}$$

where we have substituted $y := x-\sum_{i=1}^d x_i$. The claim follows. □

2.3.1 Constructing Exchangeable Archimedean Copulas

Now consider a positive random variable R, independent of $\mathbb{S}_d$. Define the random vector

$$(X_1, \ldots, X_d) := R\,\mathbb{S}_d = \Big(\frac{R E_1}{\sum_{k=1}^d E_k}, \ldots, \frac{R E_d}{\sum_{k=1}^d E_k} \Big). \tag{2.14}$$

Let us start with two important cases:

(1) The simplest case in (2.14) is to take $R = 1$. Then, the random vector boils down to a uniform distribution on the d-dimensional simplex, as seen in Lemma 2.2(2).
(2) It is also possible to relate this model to the initial mixture model of extendible Archimedean copulas. It holds that $\mathbb{S}_d$ and $\sum_{k=1}^d E_k$ are independent (see Lemma 2.2(3)). This allows us to define $R := (\sum_{k=1}^d E_k)/M$ and the model agrees with the model for extendible Archimedean copulas, given in Equation (2.1). Therefore, (2.14) is a proper generalization of (2.1).

In general, we can interpret the model (2.14) as mixing a uniform distribution on the unit d-dimensional simplex using a random radius R. Conditioned on the factor R, the induced survival copula equals the Archimedean copula with generator $\varphi(x) = \max(1-x, 0)^{d-1} = (1-x)_+^{d-1}$, being a pointwise lower bound for all Archimedean copulas (see McNeil and Nešlehová (2009, Proposition 4.6)). In this regard, instead of conditional independence (as in the case of model (2.1)) one could heuristically speak of "*conditional countermonotonicity*". As opposed to the constructions in the completely monotone case (where an infinite sequence $\{X_k\}_{k\in\mathbb{N}}$ with $X_k := E_k/M$ is constructed), the present construction might not be extended to dimension $d+1$. The model (2.14) is again related to Archimedean copulas via its survival copula. It holds that $(X_1, \ldots, X_d)$ has an Archimedean survival copula, where the generator is the Williamson d-transform of the random variable R, denoted $\varphi_{d,R}$. Since this is a little-known transform, let us state the definition.

Definition 2.3 (Williamson d-Transform)
The Williamson d-transform (see Williamson (1956)) of a positive random variable R is defined as

$$\varphi_{d,R}(x) := \mathbb{E}\Big[\max\big(1 - \frac{x}{R}, 0\big)^{d-1} \Big], \qquad x \geq 0, \tag{2.15}$$

where $d \in \mathbb{N}$ with $d \geq 2$.

It is worth noting that the distribution of a positive random variable is characterized uniquely by its Williamson d-transform. Moreover, there exists an inversion formula which is surprisingly simple.[23] More precisely, the distribution function F_R of a positive random variable R with Williamson d-transform $\varphi_{d,R}$ is given by

$$F_R(x) = 1 - \sum_{k=0}^{d-2} \frac{(-1)^k x^k \varphi_{d,R}^{(k)}(x)}{k!} - \frac{(-1)^{d-1} x^{d-1} \varphi_{d,R+}^{(d-1)}(x)}{(d-1)!}, \qquad (2.16)$$

where $x \geq 0$ and $\varphi_{d,R+}^{(d-1)}$ denotes the $(d-1)$-fold derivative of $\varphi_{d,R}$, whenever non-negative.

Example 2.3 (Independence Copula in Dimension $d \geq 2$)
The d-dimensional independence copula $\Pi(u_1, \ldots, u_d) = u_1 \cdots u_d$ is of the Archimedean kind with generator $\varphi_{d,R}(x) = \exp(-x)$. Unlike in the extendible model (2.1), where it is easily seen that $M \equiv$ const implies the independence copula, in construction (2.14) it is not immediate how R must be chosen in order to obtain independent components. To determine the distribution of R, we use the inversion formula (2.16) and find with $\exp(-x)^{(k)} = (-1)^k \exp(-x)$ that

$$F_R(x) = 1 - \sum_{k=0}^{d-2} \frac{x^k\, e^{-x}}{k!} - \frac{x^{d-1}\, e^{-x}}{(d-1)!} = 1 - e^{-x} \sum_{k=0}^{d-1} \frac{x^k}{k!}, \qquad x \geq 0.$$

This is precisely the distribution function of an Erlang distribution with d degrees of freedom. This example also illustrates that the choice of R depends on the dimension and the result is clearly consistent with Lemma 2.2(5).

The main theorem of this section is now stated. It characterizes d-dimensional Archimedean copulas via d-monotone generator functions. These, in turn, are related to positive random variables via Williamson's d-transform. This relation then manifests a probabilistic model for each such copula.

Theorem 2.3 (Exchangeable Archimedean Copulas)
Consider the model (2.14) for a positive random variable R.

(1) The survival copula of (2.14) is an Archimedean copula whose generator is given by the Williamson d-transform of R.

[23] Having in mind the non-trivial inversion formula(s) of the Laplace transform (see, e.g., Widder (1946, Chapter II, Section 7)).

(2) *The Williamson d-transform of R is d-*monotone *on $[0,\infty)$, i.e. (a) it is differentiable on $(0,\infty)$ up to the order $d-2$ and the derivatives satisfy*

$$(-1)^j\,\varphi_{d,R}^{(j)}(x) \geq 0, \quad \forall\, x>0,\ j=0,1,\ldots,d-2,$$

(b) $(-1)^{d-2}\,\varphi_{d,R}^{(d-2)}$ is non-increasing and convex on $(0,\infty)$, and (c) $\varphi_{d,R}$ is continuous at 0.

(3) *The set of d-monotone functions on $[0,\infty)$ starting from 1 agrees with the set of Williamson d-transforms of positive random variables.*

(4) *All (exchangeable) Archimedean copulas can be obtained by this approach.*

(5) *The extendible subset is obtained by choosing $R := T/M$, where $T \sim Erlang(d)$ is independent of the positive random variable M.*

Proof.

(1) The proof involves two steps: (a) From Lemma 2.2, it follows that

$$\begin{aligned}\mathbb{P}\big(R\,\mathbb{S}_d > (x_1,\ldots,x_d)\big) &= \mathbb{P}\big(\mathbb{S}_d > \frac{1}{R}(x_1,\ldots,x_d)\big)\\ &= \mathbb{E}\Big[\mathbb{P}\big(\mathbb{S}_d > \frac{1}{R}(x_1,\ldots,x_d)\mid R\big)\Big]\\ &= \mathbb{E}\Big[\big(1-\frac{1}{R}(x_1+\ldots+x_d)\big)_+^{d-1}\Big]\\ &= \varphi_{d,R}(x_1+\ldots+x_d).\end{aligned}$$

(b) Moreover,

$$\begin{aligned}\mathbb{P}\Big(R\frac{E_k}{\sum_{i=1}^d E_i} > x\Big) &= \mathbb{P}\Big(E_k > \frac{x/R}{1-x/R}\sum_{i\neq k} E_i, \frac{x}{R}<1\Big)\\ &\overset{(*)}{=} \mathbb{E}\Big[\mathbb{1}_{\{\frac{x}{R}<1\}} e^{-\frac{x/R}{1-x/R}\sum_{i\neq k} E_i}\Big]\\ &\overset{(**)}{=} \mathbb{E}\Big[\mathbb{1}_{\{\frac{x}{R}<1\}}\big(1+\frac{x/R}{1-x/R}\big)^{-(d-1)}\Big]\\ &= \mathbb{E}\Big[\Big(1-\frac{x}{R}\Big)_+^{d-1}\Big] = \varphi_{d,R}(x).\end{aligned}$$

Note that in $(*)$ we compute the probability conditioned on $\sum_{i\neq k} E_i$ and R and to justify $(**)$ we use the fact that $\sum_{i\neq k} E_i \sim \text{Erlang}(d-1)$ and condition on R. The statement follows by applying Sklar's theorem.

(2) See McNeil and Nešlehová (2009, Proposition 3.1(i)).

(3) To prove (3), for $\varphi_{d,R}^{(d-2)}$ a suitable random variable is constructed in McNeil and Nešlehová (2009, Appendix B).

(4) The claim follows from McNeil and Nešlehová (2009, Theorem 2.2) and (1), (2), and (3).
(5) Choose $R := T/M$ with independent random variables $T \sim$ Erlang(d) and M, the latter with Laplace transform φ. Then

$$\begin{aligned}
&\varphi_{d,R}(x) \\
&= \mathbb{E}\Big[\mathbb{E}\Big[\Big(1-\frac{x\,M}{T}\Big)_+^{d-1}\,\Big|\,M\Big]\Big] = \mathbb{E}\Big[\int_0^\infty \Big(1-\frac{x\,M}{u}\Big)_+^{d-1}\frac{u^{d-1}e^{-u}}{(d-1)!}du\Big] \\
&= \mathbb{E}\Big[\int_0^\infty (u-x\,M)_+^{d-1}\frac{e^{-u}}{(d-1)!}du\Big] = \mathbb{E}\Big[\int_{x\,M}^\infty \frac{(u-x\,M)^{d-1}\,e^{-u}}{(d-1)!}du\Big] \\
&\overset{(*)}{=} \mathbb{E}\Big[\int_0^\infty \frac{y^{d-1}e^{-(y+x\,M)}}{(d-1)!}dy\Big] \overset{(**)}{=} \mathbb{E}\Big[e^{-x\,M}\Big]\int_0^\infty \frac{e^{-y}y^{d-1}}{(d-1)!}dy = \varphi(x),
\end{aligned}$$

where $(*)$ requires the substitution $y = u - x\,M$ and for $(**)$ we observe that we integrate out the density of an Erlang distribution. This implies the claim.

□

To conclude, for a fixed dimension $d \geq 2$ a generator implies an Archimedean copula if and only if the generator is d-monotone.[24] Each such generator corresponds uniquely to a positive random variable R with the Williamson d-transform equal to the generator. Finally, the respective Archimedean copula appears as the survival copula of construction (2.14), which might be used to sample the Archimedean copula.

2.3.2 *Sampling Exchangeable Archimedean Copulas*

The following algorithm samples an Archimedean copula via construction (2.14).

Algorithm 2.3 (Sampling Exchangeable Archimedean Copulas)
Fix $d \geq 2$ and let R be a positive random variable with Williamson d-transform $\varphi_{d,R}$. Sampling a d-dimensional Archimedean copula with generator $\varphi_{d,R}$ is possible via the following steps.

(1) Sample i.i.d. $E_1, \ldots, E_d$, where $E_1 \sim Exp(1)$.
(2) Sample $R > 0$, whose Williamson d-transform is $\varphi_{d,R}$.
(3) Compute $X_k := R\,\frac{E_k}{E_1+\ldots+E_d}$ for $k = 1, \ldots, d$.
(4) Return $(U_1, \ldots, U_d)$, where $U_k := \varphi_{d,R}(X_k)$.

[24] From this, Kimberling's theorem (see Theorem 2.2) follows as an immediate corollary.

We observe that Algorithm 2.3 is similarly convenient to implement as Algorithm 2.1. Moreover, this algorithm is equally fast: only $d+1$ random variables have to be drawn and one transform must be applied to each coordinate. However, there are some subtle differences. While Algorithm 2.1 is independent of the dimension (only the Laplace transform of the mixing variable is required), the Williamson d-transform must be available in the respective dimension. In particular, $\varphi_{d,R} \neq \varphi_{d+1,R}$ in general. The consequences are (1) we cannot simply add further dimensions, and (2) we must compute the Williamson d-transform for each dimension in which we want to sample[25] the Archimedean copula.

Example 2.4 (Mixing with a Die)
Let R be the outcome of rolling a standard die, i.e. $\mathbb{P}(R = i) = 1/6$ for $i = 1, \ldots, 6$. The Williamson d-transform of R is easily found to be

$$\varphi_{d,R}(x) = \frac{1}{6} \sum_{i=1}^{6} \max\left(1 - \frac{x}{i}, 0\right)^{d-1}, \qquad x \geq 0.$$

Using this to simulate a bivariate and trivariate Archimedean copula with Algorithm 2.3 yields Figure 2.8. Note the singular component of the resulting copula. Its specific form results from the fact that we mix the countermonotonicity copula with the six possible outcomes of the die experiment (each having the same probability).

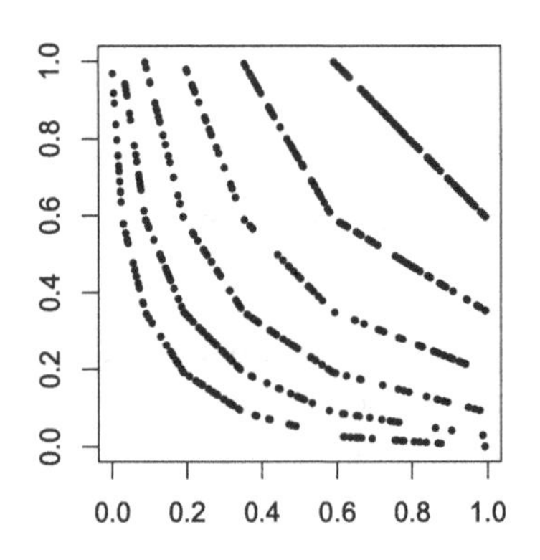

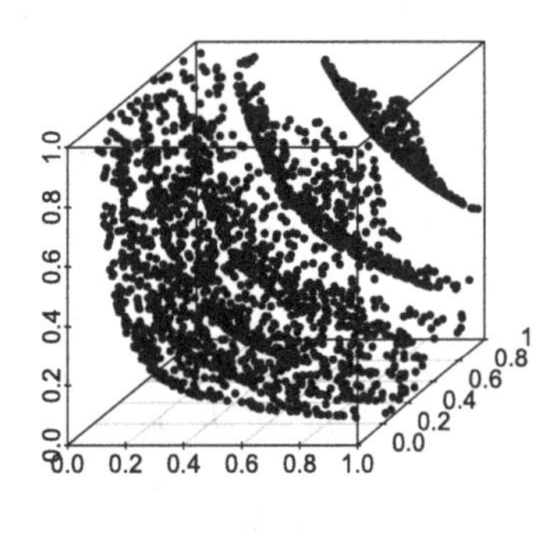

Fig. 2.8 Scatterplot of 500 samples from the bivariate Archimedean copula of Example 2.4 (left) and 2 500 samples from the trivariate copula (right).

[25]It is often much easier to find this transform for $d = 2$ (compared to $d > 2$), since for $d = 2$ the power in the definition of the Williamson d-transform vanishes.

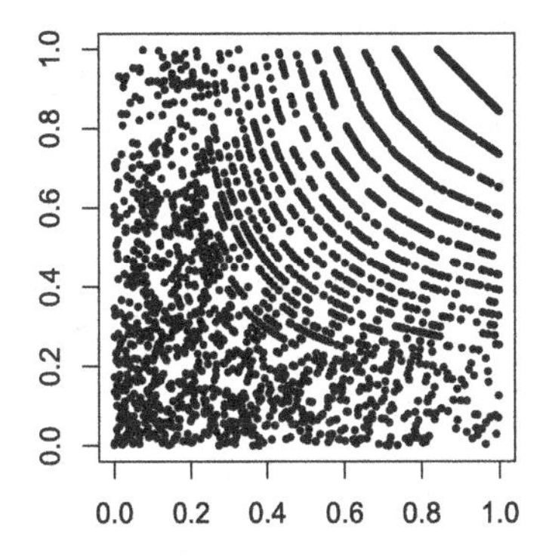

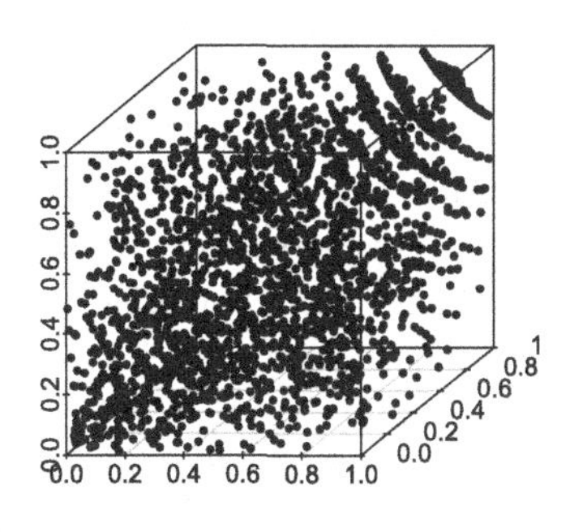

Fig. 2.9 Scatterplot of 2 500 samples from the Archimedean copula of Example 2.5 with $\vartheta = 0.05$ for the bivariate (left) and trivariate case (right).

Example 2.5 (Mixing with a ***Geo*****(ϑ)-Distribution)**
In this example, taken from McNeil and Nešlehová (2009), let R follow a Geo(ϑ)-distribution, $\vartheta \in (0,1)$. The Williamson d-transform of R is

$$\varphi_{d,R}(x) = \sum_{i=1}^{\infty} \max\big(1 - \frac{x}{i}, 0\big)^{d-1} \vartheta(1-\vartheta)^{i-1}, \qquad x \geq 0,$$

and a scatterplot is provided in Figure 2.9.

2.3.3 *Properties of Exchangeable Archimedean Copulas*

In this section, we briefly review some properties of non-extendible Archimedean copulas. The results (2.3) and (2.4) on tail dependence and Kendall's tau (2.5) are identical to the case of extendible Archimedean copulas.

2.3.3.1 *Exchangeable C_φ are not necessarily PLOD*

In contrast to extendible Archimedean copulas, the class of exchangeable Archimedean copulas is no longer PLOD. To verify this, simply consider the bivariate countermonotonicity copula $W(u_1, u_2) = (u_1 + u_2 - 1)\mathbb{1}_{\{u_1+u_2 \geq 1\}}$, corresponding to the two-monotone generator $\varphi_{2,R\equiv 1}(x) = (1-x)\mathbb{1}_{\{x \in [0,1]\}}$, as an example. More generally, observe that a generator defined via the Williamson d-transform (2.15) might be non-strict, i.e. $\varphi_{d,R}(x) = 0$ for suitable x, d, and R (see, e.g., Example 2.4). Archimedean copulas defined via non-strict generators φ can never be PLOD. To justify this statement, observe that the support of C_φ is the set $\{(u_1,\ldots,u_d) \in [0,1]^d : \varphi^{-1}(u_1) + \ldots + \varphi^{-1}(u_d) \leq \varphi^{-1}(0)\}$. For non-strict φ we can select $(\hat{u}_1,\ldots,\hat{u}_d) \in (0,1]^d$ such that $C_\varphi(\hat{u}_1,\ldots,\hat{u}_d) = 0 < \Pi(\hat{u}_1,\ldots,\hat{u}_d)$. Hence, C_φ is not PLOD. However, based on construction (2.14) it is possible to guess an alternative lower bound, which is proven in McNeil and Nešlehová (2009, Proposition 13). Construction (2.14) starts with negatively dependent components $\mathbb{S}_d$ which are multiplied by the random variable R. Since this latter multiplication (with random R) increases the dependence among the components, a lower bound is obtained for $R \equiv 1$ (or any other positive constant R). More precisely, for all d-monotone functions $\varphi_{d,R}$ one has for all $(u_1,\ldots,u_d) \in [0,1]^d$, $d \geq 2$,

$$C_{\varphi_{d,R\equiv 1}}(u_1,\ldots,u_d) \leq C_{\varphi_{d,R}}(u_1,\ldots,u_d), \tag{2.17}$$

where $\varphi_{d,R\equiv 1}(x) = (1-x)_+^{d-1}$.

2.3.3.2 *Density*

Since a d-monotone generator is not necessarily d times differentiable, the existence and form of a copula density is a difficult issue in the present situation. A solution (including statements on a singular component) is provided in McNeil and Nešlehová (2009, Propositions 8 and 9), which we repeat below.

Theorem 2.4 (Density of $C_{\varphi_{d,R}}$)
Consider a d-dimensional Archimedean copula with d-monotone generator $\varphi_{d,R}$. Further, denote by H the distribution function of the random vector $R\,\mathbb{S}_d$. Then:

(1) $C_{\varphi_{d,R}}$ is absolutely continuous if and only if H is.
(2) Sufficient for $C_{\varphi_{d,R}}$ being absolutely continuous is $\varphi_{d,R}$ being $(d+1)$-monotone.
(3) $C_{\varphi_{d,R}}$ is absolutely continuous if and only if $\varphi_{d,R}^{(d-1)}$ exists and is absolutely continuous on $(0,\infty)$. In this case, the density of $C_{\varphi_{d,R}}$ is given for almost every $(u_1,\ldots,u_d) \in (0,1)^d$ by

$$c(u_1,\ldots,u_d) = \frac{\varphi_{d,R}^{(d)}\big(\varphi_{d,R}^{-1}(u_1) + \ldots + \varphi_{d,R}^{-1}(u_d)\big)}{\varphi_{d,R}'(\varphi_{d,R}^{-1}(u_1)) \cdots \varphi_{d,R}'(\varphi_{d,R}^{-1}(u_d))}.$$

2.4 Hierarchical (H-Extendible) Archimedean Copulas

The simple construction principles (2.1) and (2.14) and the convenient parametric form (2.2) renders the family of Archimedean copulas a very tractable class of copulas. Being parameterized by real-valued functions, i.e. by infinite-dimensional objects, Archimedean copulas allow us to model a vast spectrum of dependence structures. This is illustrated by the examples in the previous sections. However, classical Archimedean copulas are symmetric functions, and the related random vectors are exchangeable. This property is often not justified in applications. To overcome this shortcoming without giving up the convenient structure of Archimedean copulas, the class of hierarchical (or nested) Archimedean copulas was introduced. Loosely speaking, one might think of this class as Archimedean copulas connecting certain groups. These groups are then connected by some other copula, again of the Archimedean family. The functional form of such structures is given by

$$C_{\varphi_0}\big(C_{\varphi_1}(u_{1,1},\ldots,u_{1,d_1}),\ldots,C_{\varphi_J}(u_{J,1},\ldots,u_{J,d_J})\big). \tag{2.18}$$

Structures with deeper levels of hierarchy are defined recursively, and the respective inner copula is then further nested. Definition (2.18) is based on $J+1$ generator functions: φ_0 for the outer Archimedean family, φ_j for group $j=1,\ldots,J$. However, it is difficult to decide if for a certain choice of generators the resulting function (2.18) is a proper copula. It is typically not possible to arbitrarily combine $J+1$ generators in the above way. However, a quite convenient sufficient condition is known. Denoting by Φ_∞ the set of all completely monotone Archimedean generators, it is possible to construct h-extendible copulas with respect to the family $\{C_\varphi\}_{\varphi\in\Phi_\infty}$ of extendible Archimedean copulas, as outlined in this section. The following sufficient condition on the involved generators to define a copula is given in Joe (1997, p. 88) for three- and four-dimensional copulas and may be found in McNeil (2008) for more general structures: if for each $j=0,\ldots,J$ the function φ_j is a c.m. generator, and if additionally

$$(\varphi_0^{-1}\circ\varphi_j)' \text{ is c.m.}, \; j\in\{1,\ldots,J\}, \tag{2.19}$$

then (2.18) defines a proper copula, which is h-extendible with respect to the family $\{C_\varphi\}_{\varphi\in\Phi_\infty}$. The function φ_0 is referred to as the *outer generator* and the functions φ_j, for $j=1,\ldots,J$, as the *inner generators*. The involved generators are called *compatible* if condition (2.19) holds. A sampling algorithm for such structures was first stated by McNeil (2008). An implementation of several families of classical and hierarchical Archimedean copulas can be found in the R package nacopula (see Hofert and Mächler (2011)).

This section focuses on a construction principle for hierarchical Archimedean copulas, based on the reference Hering et al. (2010). More precisely, we show how a random vector with a hierarchical Archimedean survival copula (2.18) is constructed. Such an approach is useful, since it provides a sampling strategy for hierarchical Archimedean copulas. The construction principle turns out to be flexible enough to comprise all copulas of form (2.18) whose generators satisfy (2.19).

2.4.1 *Compatibility of Generators*

The (difficult-to-verify) nesting condition (2.19) is now investigated further. Specific examples for generators satisfying (2.19) are given in Joe (1997), McNeil (2008), Hofert (2008), and Hofert (2011). Here, we rather fix a c.m. outer generator φ_0 and identify the set

$$M_{\varphi_0} := \{\varphi \text{ c.m. generator} \,|\, (\varphi_0^{-1} \circ \varphi)' \text{ c.m.}\}$$

of inner generators which are compatible with φ_0 (see Theorem 2.5 below). In order to establish this result, the notion of Lévy subordinators, i.e. non-decreasing Lévy processes, is required. Such processes are introduced in Section A.2 of the Appendix. Bridging the gap to copulas, an application of the Lévy–Khinchin theorem (see Theorem A.4) allows us to determine the set M_{φ_0}. Theorem 2.5 shows that for a given (outer) generator φ_0, all compatible (inner) generators can be parameterized by φ_0, a drift constant $\mu \geq 0$, and a Lévy measure ν on $(0, \infty)$.

Theorem 2.5 (Compatible Generators)
Let φ_0 be a c.m. generator. Then

$$M_{\varphi_0} = \Big\{\varphi \,\Big|\, \varphi(x) = \varphi_0\Big(\mu x + \int_{(0,\infty)} (1 - e^{-xt})\,\nu(dt)\Big),$$

$$\textit{where } \mu \geq 0 \textit{ and } \nu \textit{ is a measure on } (0,\infty) \textit{ satisfying (A.4)},$$

$$\textit{and either } \mu > 0, \textit{ or } \nu\big((0,1)\big) = \infty, \textit{ or both}\Big\}.$$

Proof. It is shown in Feller (1966, p. 450) that $\Psi : [0, \infty) \to [0, \infty)$ is the Laplace exponent of a classical Lévy subordinator if and only if $\Psi(0) = 0$, $\lim_{x \downarrow 0} \Psi(x) = 0$, and Ψ has a c.m. derivative on $(0, \infty)$. It follows that for two c.m. generators φ_0 and φ, $(\varphi_0^{-1} \circ \varphi)'$ is c.m. if and only if $\varphi_0^{-1} \circ \varphi =: \Psi$

is the Laplace exponent of a classical Lévy subordinator. Fixing φ_0, this implies that all possible inner generators φ are given by $\varphi = \varphi_0 \circ \Psi$ for the Laplace exponent Ψ of a classical Lévy subordinator Λ. More clearly, φ is the Laplace transform of Λ_M, where Λ is a classical Lévy subordinator and M is an independent positive random variable with Laplace transform φ_0. Note that such a random variable M exists due to Bernstein's theorem (see Theorem 2.1). Since $\lim_{x\to\infty} \varphi(x) = 0$ is necessary to obtain a generator, one must choose Laplace exponents Ψ satisfying $\lim_{x\to\infty} \Psi(x) = \infty$. This excludes Lévy subordinators Λ with the property that $\mathbb{P}(\Lambda_t = 0) > 0$ for some $t > 0$, which are compound Poisson subordinators with drift $\mu = 0$ and Lévy measure ν satisfying $\nu\big((0,1)\big) < \infty$. The claim is finally established by the Lévy–Khinchin representation. □

2.4.2 *Probabilistic Construction and Sampling*

It is shown in the previous section that if the nesting condition (2.19) holds, the functions $\varphi_0^{-1} \circ \varphi_j$, for $j = 1, \ldots, J$, are Laplace exponents of classical Lévy subordinators. On the other side, one can start with a suitable combination of Lévy subordinators to arrive at a probabilistic model for hierarchical Archimedean copulas. In this sense, the sampling algorithm of McNeil (2008) for hierarchical Archimedean copulas of form (2.18) can be rewritten using classical Lévy subordinators. To do so, consider a probability space $(\Omega, \mathcal{F}, \mathbb{P})$ supporting i.i.d. exponential random variables with mean 1, denoted $\{E_{j,i}\}_{j=1,\ldots,J,\, i=1,\ldots,d_j}$. Moreover, independent of these random variables let $M > 0$ be a positive random variable (which might be interpreted as a random time) with Laplace transform $\varphi_0(x) = \mathbb{E}[\exp(-x\,M)]$. Independent of all $E_{j,i}$ and M, let $\Lambda^{(1)}, \ldots, \Lambda^{(J)}$ be J independent, classical Lévy subordinators with Laplace exponents $\Psi_1, \ldots, \Psi_J$, respectively. These Laplace exponents are assumed to satisfy $\lim_{x\to\infty} \Psi_j(x) = \infty$, $j = 1, \ldots, J$, which corresponds to $\Lambda_t^{(j)} > 0$ a.s. for all $t > 0$ and $j = 1, \ldots, J$. We now construct the random vector

$$\Bigg(\underbrace{\frac{E_{1,1}}{\Lambda_M^{(1)}}, \ldots, \frac{E_{1,d_1}}{\Lambda_M^{(1)}}}_{\text{group } 1}, \underbrace{\frac{E_{2,1}}{\Lambda_M^{(2)}}, \ldots, \frac{E_{2,d_2}}{\Lambda_M^{(2)}}}_{\text{group } 2}, \ldots\ldots, \underbrace{\frac{E_{J,1}}{\Lambda_M^{(J)}}, \ldots, \frac{E_{J,d_J}}{\Lambda_M^{(J)}}}_{\text{group } J}\Bigg). \tag{2.20}$$

An interpretation of this construction is the following. Originally i.i.d. exponential random variables are partitioned into J groups. Our first aim is for each group to have an Archimedean dependence structure. We know from construction (2.1) that this can be achieved by dividing all members of the group by the same positive random variable. However, we additionally aim at an Archimedean dependence structure between the groups. Both aims are achieved by taking as the positive random variable affecting group j the Lévy subordinator $\Lambda^{(j)}$ stopped at the random time M. The dependence structure within this group is of the Archimedean kind with the generator given by the Laplace transform of $\Lambda_M^{(j)}$. On the other side, we use the fact that all group-specific Lévy subordinators are stopped at the same random time M to introduce dependence between the groups. We shall later see that the induced dependence between members of a group is at least as large as the dependence between members of different groups. The precise form of the hierarchical Archimedean copula induced by construction (2.20) is computed in Theorem 2.6.

Theorem 2.6 (Probabilistic Model with Lévy Subordinators) *The survival copula of the random vector constructed in (2.20) admits the form (2.18). The outer generator φ_0 is the Laplace transform of the positive random variable M. The group-specific generators are given by $\varphi_j := \varphi_0 \circ \Psi_j$, for $j = 1, \ldots, J$. They are the Laplace transforms of $\Lambda_M^{(j)}$. Moreover, the univariate survival functions of the components are*

$$\mathbb{P}\big(E_{j,i}/\Lambda_M^{(j)} > x\big) = (\varphi_0 \circ \Psi_j)(x), \; x > 0, \; j = 1, \ldots, J, \; i = 1, \ldots, d_j.$$

Proof. The joint survival function of the random vector constructed in (2.20) is

$$\begin{aligned}
\mathbb{P}\left(\frac{E_{j,i}}{\Lambda_M^{(j)}} > x_{j,i}, \forall j, i\right) &= \mathbb{E}\Big[e^{-\sum_{j=1}^{J} \Lambda_M^{(j)} \sum_{i=1}^{d_j} x_{j,i}}\Big] \\
&= \mathbb{E}\Big[\prod_{j=1}^{J} e^{-M\,\Psi_j\left(\sum_{i=1}^{d_j} x_{j,i}\right)}\Big] \\
&= \mathbb{E}\Big[e^{-M \sum_{j=1}^{J} \Psi_j\left(\sum_{i=1}^{d_j} x_{j,i}\right)}\Big] \\
&= \varphi_0\left(\sum_{j=1}^{J} \varphi_0^{-1} \circ (\varphi_0 \circ \Psi_j)\left(\sum_{i=1}^{d_j} x_{j,i}\right)\right).
\end{aligned}$$

The component $E_{j,i}/\Lambda_M^{(j)}$ has survival function

$$\mathbb{P}\big(E_{j,i}/\Lambda_M^{(j)} > x\big) = \mathbb{E}\Big[e^{-x\Lambda_M^{(j)}}\Big] = \mathbb{E}\Big[e^{-M\,\Psi_j(x)}\Big] = (\varphi_0 \circ \Psi_j)(x).$$

Hence, the survival copula has the claimed form. □

Composing hierarchical Archimedean copulas using Lévy subordinators implies that the copula is specified by an arbitrary positive random variable M

and J quite arbitrary classical Lévy subordinators. The nesting condition (2.19) holds by construction, i.e. Theorem 2.6 ensures that the resulting generators are compatible. Moreover, Equation (2.20) suggests a convenient sampling strategy which is formulated as Algorithm 2.4.

Algorithm 2.4 (Sampling Hierarchical Archimedean Copulas)

(1) Sample i.i.d. $E_{j,i} \sim Exp(1)$, $j = 1, \ldots, J$, $i = 1, \ldots, d_j$.
(2) Sample[26] $M > 0$.
(3) For each group $j = 1, \ldots, J$, *sample the subordinator* $\Lambda^{(j)}$ *at time* M, *i.e. sample*[27] *the random variable* $\Lambda_M^{(j)}$.
(4) Return $(U_{1,1}, \ldots, U_{J,d_J})$, *where*

$$U_{j,i} := (\varphi_0 \circ \Psi_j)\big(E_{j,i}/\Lambda_M^{(j)}\big), \quad j = 1, \ldots, J, \, i = 1, \ldots, d_j.$$

Note that for sampling a hierarchical Archimedean copula in dimension d with J groups, only $d + J + 1$ random variables have to be simulated, which is extremely fast. Efficient sampling strategies for various subordinators are known (see Cont and Tankov (2004, p. 171ff) and the references therein). There exist approximate sampling strategies for general Lévy subordinators (see, e.g., Bondesson (1982) and Damien et al. (1995)). Some examples for subordinators are found in the Appendix.

2.4.3 *Properties*

The hierarchical Archimedean copulas constructed earlier have the following h-extendible structure. Let $(U_{1,1}, \ldots, U_{1,d_1}, \ldots, U_{J,1}, \ldots, U_{J,d_J})$ have distribution function (2.18). If $1 \leq j_1 < \ldots < j_i \leq J$ are indices of i distinct groups, then $(U_{j_1,1}, \ldots, U_{j_i,1}) \sim C_{\varphi_0}$. Moreover, within the jth group, it holds that $(U_{j,1}, \ldots, U_{j,d_j}) \sim C_{\varphi_j}$. This shows that the construction is truly h-extendible with respect to the family $\{C_\varphi\}_{\varphi \in \Phi_\infty}$. Considering bivariate marginals of (2.18), it is easily observed that any two components from different groups are coupled via the Archimedean copula C_{φ_0}. On the other hand, any two components from some group j are coupled via the Archimedean copula C_{φ_j}. Therefore, the dependence properties of bivariate marginals can be traced back from known results on bivariate Archimedean copulas. However, appealing to construction (2.20), the inner generators φ_j have the specific form $\varphi_j = \varphi_0 \circ \Psi_j$, where Ψ_j is the Laplace exponent of the

[26] M has Laplace transform φ_0 (the outer generator).
[27] $\Lambda_M^{(j)}$ has Laplace transform $\varphi_0 \circ \Psi_j$ (the inner generator of group j).

Lévy subordinator of group j and φ_0 is the Laplace transform of M. The construction suggests that components from the same group should be at least as dependent as components from different groups. This presumption holds and is made precise in Lemma 2.3.

Lemma 2.3 (Concordance Ordering)
Let φ_0 be a completely monotone Archimedean generator and let Ψ be the Laplace exponent of a classical Lévy subordinator with $\lim_{x\to\infty}\Psi(x)=\infty$. Then, it holds that $C_{\varphi_0\circ\Psi}(u,v)\geq C_{\varphi_0}(u,v)$ for all $u,v\in[0,1]$.

Proof. See Joe (1997, Corollary 4.2, p. 90). □

Note that Lemma 2.3 may be applied to show that various measures of association (e.g. Kendall's tau and the upper- and lower-tail dependence coefficient) are ordered in the same way for pairs within a group compared to pairs from different groups. The probabilistic model based on Lévy subordinators allows one to obtain a deeper understanding of copula (2.18). Based upon the choice of random variable M and Lévy subordinators, it is possible to draw conclusions about implied dependence measures. As an example, pairwise upper-tail dependence coefficients are treated. Recall from Equation (2.3) that (given existence) $UTD_{C_\varphi}=2-2\lim_{x\downarrow 0}\varphi'(2\,x)/\varphi'(x)$. Applied to the present situation (a pair of random variables from different groups having the Archimedean survival copula C_{φ_0}, a pair from the same group j having survival copula $C_{\varphi_0\circ\Psi_j}$), that means inter-sector pairs are affected by the mixing variable M, whereas intra-sector pairs are affected by the mixing variable $\Lambda_M^{(j)}$. Lemma 2.3 implies that $C_{\varphi_0\circ\Psi_j}\geq C_{\varphi_0}$, and, hence, also $UTD_{C_{\varphi_0\circ\Psi_j}}\geq UTD_{C_{\varphi_0}}$. This means that intra-sector tail dependence is always greater than or equal to inter-sector tail dependence. Going one step further, it is interesting to investigate how the parametric models for M and the Lévy subordinator translate into properties of the implied upper-tail dependence coefficients. If either $\mathbb{E}[M]$ or $\mathbb{E}[\Lambda_1^{(j)}]$ is finite, it is possible to draw conclusions about the implied upper-tail dependence coefficients. In particular, it follows from Lemma 2.1 that finite expectation of M implies zero upper-tail dependence of C_{φ_0}. Moreover, it is necessary to have $\mathbb{E}[\Lambda_1^{(j)}]=\infty$ to obtain an intra-sector upper-tail dependence which is strictly larger than the inter-sector one. The relations in Table 2.1 are verified by taking the respective limits and constructing suitable examples and counterexamples.

Table 2.1 Upper-tail dependence parameters within a group and between groups, depending on the first moment of M and $\Lambda_1^{(j)}$, respectively.

	$\mathbb{E}[\Lambda_1^{(j)}] < \infty$	$\mathbb{E}[\Lambda_1^{(j)}] = \infty$
$\mathbb{E}[M] < \infty$	$0 = UTD_{C_{\varphi_0}} = UTD_{C_{\varphi_0 \circ \Psi_j}}$	$0 = UTD_{C_{\varphi_0}} \leq UTD_{C_{\varphi_0 \circ \Psi_j}}$
$\mathbb{E}[M] = \infty$	$UTD_{C_{\varphi_0}} = UTD_{C_{\varphi_0 \circ \Psi_j}}$	$UTD_{C_{\varphi_0}} \leq UTD_{C_{\varphi_0 \circ \Psi_j}}$

2.4.4 *Examples*

The first example shows how to construct h-extendible copulas with respect to the Gumbel family, i.e. with respect to the family $\{C_{\varphi_\vartheta}\}_{\vartheta \in [1,\infty)}$, where $\varphi_\vartheta(x) = \exp(-x^{1/\vartheta})$ as given in (2.12). This example is taken from Mai and Scherer (2011b).

Example 2.6 (H-Extendible Gumbel Copulas)
Let $M \sim \mathcal{S}(1/\vartheta_0)$, $\vartheta_0 > 1$, and all involved Lévy subordinators are specified as stable Lévy subordinators, i.e. $\Psi_j(x) = x^{1/\vartheta_j}$, $\vartheta_j > 1$ for $j = 1, \ldots, J$ (see Section A.2.4). Then one readily verifies that the inner generators are given by $\varphi_j(x) = \varphi_0 \circ \Psi_j(x) = \exp(-x^{1/(\vartheta_j \vartheta_0)})$, i.e. they remain within the class of Gumbel generators. Hence, all exchangeable margins of the constructed h-extendible vector have a Gumbel copula as the dependence structure.

Let us now present two examples of hierarchical Archimedean copulas with different group-specific dependence to illustrate the flexibility of this approach.

Example 2.7 (A Geo $\circ$ (Γ, α-stable)-Archimedean Copula)
In this example, we construct a four-dimensional hierarchical Archimedean copula with two groups, each having two members. We let $M \sim Geo(1-\vartheta)$, $\vartheta \in [0,1)$. For the first sector, we take a Gamma subordinator $\Lambda^{(1)}$ whose Laplace exponent is given by $\Psi_1(x) = \beta \log(1+x)$, $\beta > 0$ (see Section A.2.2). For the second sector, an α-stable subordinator $\Lambda^{(2)}$ with Laplace exponent $\Psi_2(x) = x^\alpha$, $\alpha \in (0,1)$, is chosen. Recall from Section 2.2.4 that φ_0, the Laplace transform of M, is the generator of the Ali–Mikhail–Haq family, given by $\varphi_0(x) = (1-\vartheta)/(\exp(x)-\vartheta)$, $\vartheta \in [0,1)$. The inner generators are given by $\varphi_1(x) = (1-\vartheta)/((1+x)^\beta - \vartheta)$ and $\varphi_2(x) = (1-\vartheta)/(\exp(x^\alpha)-\vartheta)$. For the first sector, step (3) of Algorithm 2.4 involves sampling a $\Gamma(\beta\, M, 1)$-distribution. For the second sector it involves sampling a distribution with Laplace transform $\exp(-M\, x^\alpha)$, which can be sampled as $M^{1/\alpha} S$, where $S \sim \mathcal{S}(\alpha)$ (see Hofert (2008)). Concerning

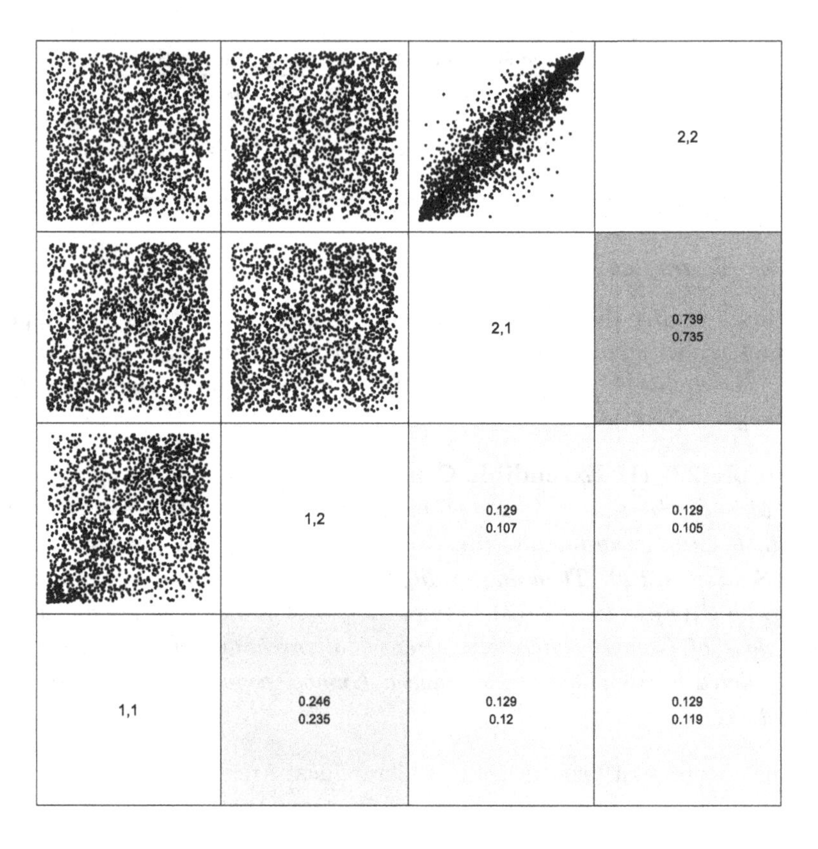

Fig. 2.10 2 500 vectors of random variates from the hierarchical Archimedean copula of Example 2.7, constructed by a Geometric random time M and the Laplace exponents of a Gamma and an α-stable subordinator, respectively. The parameters are $(\vartheta, \beta, \alpha) = (0.5, 2, 0.3)$. Theoretical Kendall's taus (upper entry) are compared to empirical Kendall's taus (lower entry) for each pair of columns. Shades of grey indicate the degree of dependence.

the dependence properties of this copula, Kendall's tau corresponding to the outer copula is given by $\tau_{C_{\varphi_0}} = 1-2/(3\,\vartheta^2)((1-\vartheta)^2 \log(1-\vartheta)+\vartheta)$. *Although there is no closed form for* $\tau_{C_{\varphi_0 \circ \Psi_1}}$ *and* $\tau_{C_{\varphi_0 \circ \Psi_2}}$ *available, numerical evaluation of Equation (2.5) suggests that for any choice of* ϑ, *values for* β *and* α *can be chosen such that the sector copulas have desired values for Kendall's tau. It is therefore possible to choose the parameters such that for any possible* $\tau_{C_{\varphi_0}}$, *desired values for* $\tau_{C_{\varphi_0 \circ \Psi_1}}$ *and* $\tau_{C_{\varphi_0 \circ \Psi_2}}$ *greater than or equal to* $\tau_{C_{\varphi_0}}$ *are obtained. Moreover, the involved upper-tail dependence parameters are given by* $UTD_{C_{\varphi_0}} = UTD_{C_{\varphi_0 \circ \Psi_1}} = 0$ *and* $UTD_{C_{\varphi_0 \circ \Psi_2}} = 2 - 2^{\alpha}$. *The lower-tail dependence parameters are given by* $LTD_{C_{\varphi_0}} = LTD_{C_{\varphi_0 \circ \Psi_2}} = 0$ *and* $LTD_{C_{\varphi_0 \circ \Psi_1}} = 2^{-\beta}$. *Figure 2.10 shows* 2 500 *random variates drawn from the constructed copula.*

Example 2.8 (An $IG \circ (\Gamma$, **cPP)-Archimedean Copula)**
This example also involves two groups with two members. The random time is specified as $M \sim IG(1, 1/\vartheta)$, $\vartheta > 0$, *i.e. an inverse Gaussian distribution with Laplace transform* $\varphi_0(x) = \exp\left((1 - \sqrt{1 + 2\vartheta^2 x})/\vartheta\right)$, $\vartheta \in (0, \infty)$ *(see Section 2.2.4.6). For the first sector, a Gamma subordinator* $\Lambda^{(1)}$ *with* 0 *drift is chosen. Its Laplace exponent is* $\Psi_1(x) = \beta \log(1+x)$ *for an intensity parameter* $\beta > 0$. *For the second sector, the Laplace exponent is determined as* $\Psi_2(x) = x+1-\exp(-x^{\alpha})$, $\alpha \in (0,1)$, *corresponding to a compound Poisson subordinator* $\Lambda^{(2)}$ *with drift* 1, *jump intensity* 1, *and jumps following an* α*-stable distribution (see Section A.2.1). For the resulting hierarchical Archimedean copula, this yields the inner generators* $\varphi_0 \circ \Psi_1$ *and* $\varphi_0 \circ \Psi_2$. *The upper-tail dependence parameters are given by* $UTD_{C_{\varphi_0}} = 0$ *for the outer copula, and by* $UTD_{C_{\varphi_0 \circ \Psi_1}} = 0$ *and* $UTD_{C_{\varphi_0 \circ \Psi_2}} = 2 - 2^{\alpha}$ *for the inner copulas. Figure 2.11 shows* 2 500 *random variates drawn from this copula.*

2.5 Other Topics Related to Archimedean Copulas

2.5.1 *Simulating from the Generator*

This section shows how to sample bivariate Archimedean copulas only from their analytic expressions. This is the case, for example, when we know that a function φ is a two-monotone generator, but cannot determine the distribution of the random variable R of the corresponding stochastic model (2.14). The following algorithm for the simulation of a bivariate Archimedean copula was introduced by Genest and MacKay (1986). It allows us to simulate the respective copula from φ and φ^{-1}. Required, however, are $\left((\varphi^{-1})'\right)^{-1}$ and $(\varphi^{-1})'$; both need to be derived for the respective generator.

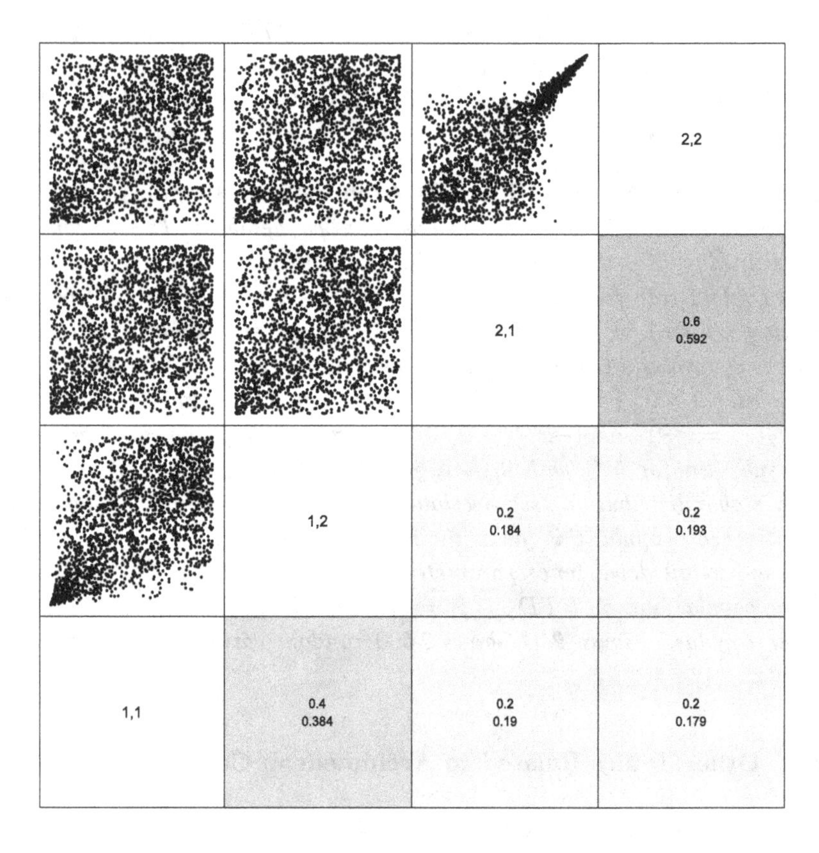

Fig. 2.11 2 500 vectors of random variates from the hierarchical Archimedean copula of Example 2.8, constructed by an Inverse Gaussian random time M and the Laplace exponents of a Gamma subordinator and a compound Poisson subordinator, respectively. Shades of grey indicate the degree of dependence. The parameters are specified to obtain Kendall's taus within the groups of 0.4 and 0.6, respectively, and in between the groups of 0.2. This is again compared to the respective empirical Kendall's taus.

Algorithm 2.5 (Sampling Bivariate Archimedean Copulas)

(1) Sample U_1 and T independently with $\mathcal{U}[0,1]$-distribution.
(2) Define $W := \big((\varphi^{-1})'\big)^{-1}((\varphi^{-1})'(U_1)/T)$.
(3) Define $U_2 := \varphi\big(\varphi^{-1}(W) - \varphi^{-1}(U_1)\big)$.
(4) Return $(U_1, U_2) \sim C_\varphi$.

Proof. See Genest and MacKay (1986). □

Example 2.9 (The Bivariate Frank Copula)
For the Frank copula, the required quantities to apply Algorithm 2.5 are given below. A scatterplot of the bivariate Frank copula, generated with Algorithm 2.5, is given in Figure 2.3.

$$\big((\varphi^{-1})'\big)^{-1}(x) = -\frac{1}{\vartheta}\log\Big(\frac{x}{x-\vartheta}\Big), \quad (\varphi^{-1})'(x) = \frac{\vartheta\, e^{-\vartheta x}}{e^{-\vartheta x}-1}.$$

Closely related to Algorithm 2.5 is Algorithm 2.6.

Algorithm 2.6 (Sampling Bivariate Archimedean Copulas 2)

(1) Sample U_2 and V independently with $U[0,1]$-distribution.
(2) Define $U_1 := \varphi\big((\varphi')^{-1}(V\,\varphi'(\varphi^{-1}(U_2))) - \varphi^{-1}(U_2)\big)$.
(3) Return $(U_1, U_2) \sim C_\varphi$.

Proof. Algorithm 2.6 is an application of Algorithm 1.2, i.e. the conditional sampling method for bivariate copulas. Hence, we first need to compute

$$F_{U_1|U_2}(u_1) = \frac{\partial}{\partial u_2} C_\varphi(u_1,u_2) = \varphi'\big(\varphi^{-1}(u_1)+\varphi^{-1}(u_2)\big)\,(\varphi^{-1})'(U_2).$$

To finally apply Algorithm 1.2, only the inverse $F^{-1}_{U_1|U_2}(x)$ is required. It is given by

$$F^{-1}_{U_1|U_2}(x) = \varphi\big((\varphi')^{-1}(x\,\varphi'(\varphi^{-1}(U_2))) - \varphi^{-1}(U_2)\big).$$ □

2.5.2 *Asymmetrizing Archimedean Copulas*

This section shows a simple methodology, known as the Khoudraji transformation (see Khoudraji (1995)), for how Archimedean copulas (and other copulas as well) can be asymmetrized (see also McNeil et al. (2005, p. 224)). More on asymmetrizing copulas is presented in Liebscher (2008) and Durante (2009). Applied to Archimedean copulas, the motivation is to overcome exchangeability. Consider independent random vectors $(V_1,\dots,V_d) \sim C_\varphi$ and $(\tilde U_1,\dots,\tilde U_d) \sim \Pi$ and parameters $(\alpha_1,\dots,\alpha_d) \in [0,1]^d$. Define a new random vector $(U_1,\dots,U_d)$ by

$$U_k := \max\big\{V_k^{1/\alpha_k}, \tilde U_k^{1/(1-\alpha_k)}\big\}, \quad k = 1,\dots,d.$$

It easily follows that each margin U_k is $U[0,1]$-distributed and

$$\mathbb{P}\Big(\bigcap_{k=1}^d \{U_k \le u_k\}\Big) = \mathbb{P}\Big(\bigcap_{k=1}^d \{V_k \le u_k^{\alpha_k}\}\Big)\,\mathbb{P}\Big(\bigcap_{k=1}^d \{\tilde U_k \le u_k^{1-\alpha_k}\}\Big)$$
$$= C_\varphi\big(u_1^{\alpha_1},\dots,u_d^{\alpha_d}\big)\,\Pi\big(u_1^{1-\alpha_1},\dots,u_d^{1-\alpha_d}\big).$$

One easily recognizes that other copulas (instead of Π) might also be included and how related combinations with more than two copulas are constructed. With a probabilistic model at hand, it is straightforward to formulate the following sampling algorithm.

Algorithm 2.7 (Sampling Asymmetric Archimedean Copulas)

(1) Sample $(V_1, \ldots, V_d) \sim C_\varphi$.
(2) Sample $(\tilde{U}_1, \ldots, \tilde{U}_d) \sim \Pi$, *independent of* $(V_1, \ldots, V_d)$.
(3) Return $(U_1, \ldots, U_d)$, *where* $U_k := \max\{V_k^{1/\alpha_k}, \tilde{U}_k^{1/(1-\alpha_k)}\}$.

Examples for scatterplots, generated with Algorithm 2.7, are presented in Figure 2.12.

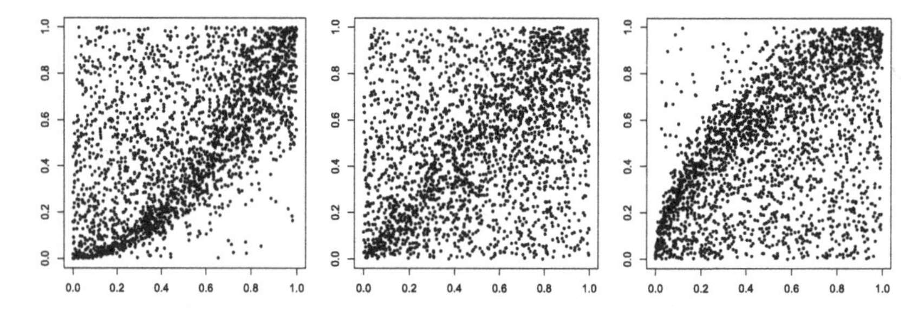

Fig. 2.12 2 500 samples from the bivariate asymmetrized Clayton copula with parameters $\vartheta = 10$, $(\alpha_1, \alpha_2) = (0.95, 0.5)$ (left), $(\alpha_1, \alpha_2) = (0.5, 0.5)$ (middle), and $(\alpha_1, \alpha_2) = (0.5, 0.95)$ (right).

Chapter 3

Marshall–Olkin Copulas

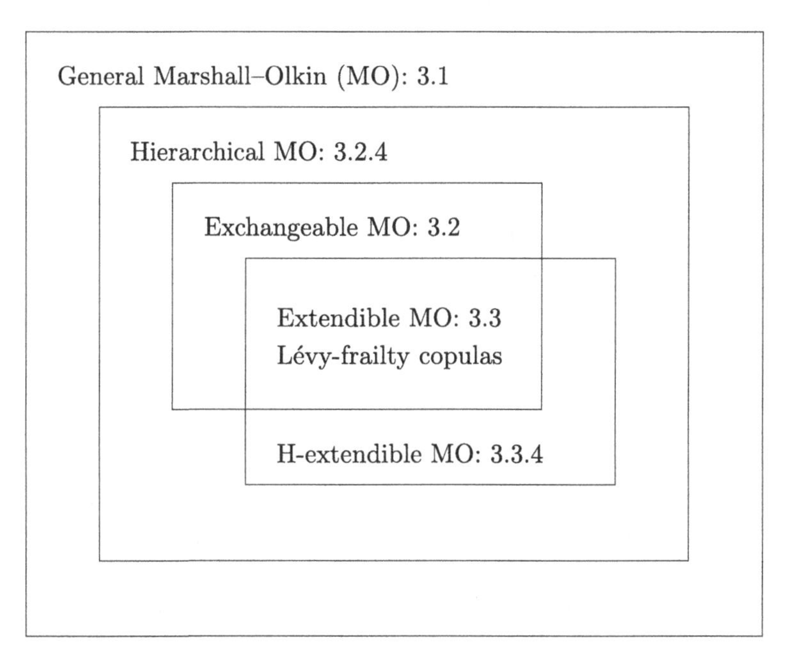

Fig. 3.1 Classification of subfamilies of Marshall–Olkin copulas, including the sections where these are discussed in this chapter.

This chapter on Marshall–Olkin copulas is organized as follows (see Figure 3.1). First, Section 3.1 introduces the most general form of Marshall–Olkin copulas. Second, Section 3.2 explicitly treats the subfamily of exchangeable Marshall–Olkin copulas. These have the advantage that the functional form and the respective sampling algorithms are much simpler and thus more efficient compared to the general case. Section 3.2.4 treats hierarchical versions of Marshall–Olkin copulas, constructed from exchangeable building blocks. These lie between the most general and the exchangeable Marshall–Olkin family. Finally, Section 3.3 studies the subfamily of (exchangeable and) extendible Marshall–Olkin copulas, which are also called Lévy-frailty copulas. These are even more efficient to simulate due to their latent factor structure. Furthermore, they give rise to quite convenient parameterizations of Marshall–Olkin copulas via so-called Laplace exponents of Lévy subordinators. Moreover, it is indicated in Section 3.3.4 how h-extendible Marshall–Olkin copulas can be constructed from extendible building blocks. This technique allows us to design parametric subfamilies of Marshall–Olkin copulas that are convenient to parameterize, flexible, and efficient to simulate.

3.1 The General Marshall–Olkin Copula

It is a well-known fact that a random variable X with support $[0,\infty)$, defined on a probability space $(\Omega,\mathcal{F},\mathbb{P})$, is exponentially[1] distributed if and only if for all $x, y \geq 0$ it holds that

$$\mathbb{P}(X > x + y \,|\, X > y) = \mathbb{P}(X > x) \tag{3.1}$$

(see, e.g., Billingsley (1995, p. 190)). Condition (3.1) is typically called the *lack of memory property.* Interpreting X as a lifetime, it intuitively means that the residual lifetime is independent of age. In fact, it is precisely this property which renders the exponential law one of the most popular and most tractable probability distributions. For instance, it is noted in David and Nagaraja (1970, p. 121) that "the exponential [distribution] occupies as commanding a position in life testing as does the normal [distribution] elsewhere in parametric theory". The article by Marshall and Olkin (1967) is concerned with a multivariate analog of the lack of memory property. Lifting (3.1) to larger dimensions, a probability space $(\Omega,\mathcal{F},\mathbb{P})$ is considered on which a random vector $(X_1,\ldots,X_d)$ with support $[0,\infty)^d$ is defined, which satisfies the following property: for all $x_1,\ldots,x_d,\, y \geq 0$, it holds

[1]Recall that an exponentially distributed random variable X with intensity parameter λ has survival function $\bar{F}(x) = \exp(-\lambda x)$ for $x \geq 0$. An important property of the exponential distribution is the min-stability: given independent, exponentially distributed random variables X and Y with parameter λ and η, respectively, the minimum $Z := \min\{X, Y\}$ is also exponentially distributed with intensity parameter $\lambda + \eta$.

that

$$\mathbb{P}\big(X_1 > x_1 + y, \ldots, X_d > x_d + y \,\big|\, X_1 > y, \ldots, X_d > y\big) = \\ \mathbb{P}\big(X_1 > x_1, \ldots, X_d > x_d\big). \tag{3.2}$$

Interpreting $X_1, \ldots, X_d$ as lifetimes, Condition (3.2) intuitively means that the residual lifetimes of all components are independent of the age of the whole system, in analogy with the one-dimensional case. In further analogy with the univariate case, Marshall and Olkin (1967) show that there is precisely one parametric family of multivariate distributions satisfying the lack of memory property (3.2). More clearly, (3.2) has to be read iteratively as follows: in dimension $d = 1$ we already know that (3.2) implies that X_1 is exponentially distributed. In dimension $d = 2$ one postulates that X_1, X_2 both are exponentially distributed and (3.2) holds. It is shown in Marshall and Olkin (1967, Lemma 2.2) that this implies the existence of parameters $\lambda_{\{1\}}, \lambda_{\{2\}}, \lambda_{\{1,2\}} \geq 0$, with $\lambda_{\{k\}} + \lambda_{\{1,2\}} > 0$, $k = 1, 2$, such that for $x_1, x_2 \geq 0$ one has

$$\begin{aligned} \bar{F}(x_1, x_2) :&= \mathbb{P}(X_1 > x_1, X_2 > x_2) \\ &= \exp\big(- \lambda_{\{1\}}\, x_1 - \lambda_{\{2\}}\, x_2 - \lambda_{\{1,2\}} \max\{x_1, x_2\}\big). \end{aligned}$$

This distribution is called the bivariate *Marshall–Olkin distribution*. Inductively, in dimension $d \geq 2$ one can show that if all $(d-1)$-dimensional subvectors of $(X_1, \ldots, X_d)$ have a Marshall–Olkin distribution and if (3.2) is satisfied, then it follows that there exist parameters $\lambda_I \geq 0$, $\emptyset \neq I \subset \{1, \ldots, d\}$, with $\sum_{I:k \in I} \lambda_I > 0$, $k = 1, \ldots, d$, such that for all $x_1, \ldots, x_d \geq 0$ one has

$$\begin{aligned} \bar{F}(x_1, \ldots, x_d) :&= \mathbb{P}(X_1 > x_1, \ldots, X_d > x_d) \\ &= \exp\Big(- \sum_{\emptyset \neq I \subset \{1,\ldots,d\}} \lambda_I \max_{i \in I}\{x_i\}\Big) \end{aligned} \tag{3.3}$$

(see Marshall and Olkin (1967, p. 39)). This distribution is called the (d-dimensional) *Marshall–Olkin distribution*. In total $2^d - 1$ parameters are involved, since $\{1, \ldots, d\}$ has 2^d subsets (including the empty set, which is not needed). In large dimensions, the Marshall–Olkin distribution is very flexible but also difficult to work with, both due to the huge number of parameters.

Other multivariate exponential distributions are proposed in the literature (see, e.g., Gumbel (1960b)), which do not share the lack of memory property. In fact, plugging exponential marginal laws into an arbitrary copula yields a multivariate distribution by virtue of Sklar's theorem, which one might call "exponential". However, the previous motivation by means of the lack of memory property suggests that not every distribution with exponential margins truly deserves the name "multivariate exponential distribution". From this perspective, the Marshall–Olkin exponential distribution is the "right" multivariate exponential distribution. For a nice treatment of the characterization of the Marshall–Olkin distribution by the lack of memory property, the interested reader is referred to the work by Galambos and Kotz (1978, p. 103–132).

So far, the Marshall–Olkin distribution has only been introduced analytically. However, Marshall and Olkin (1967) also provide a canonical construction of this distribution. The intuition behind it is a system of initially fully functional components which are affected by exogenous shocks destroying them. The random vector of extinction times of the components exhibits the Marshall–Olkin distribution. A shock can hit one or more components at the same time, rendering the extinction times dependent. In particular, when a shock hits, e.g., five components at a time, then all corresponding extinction times have the same value, i.e. the distribution has a singular component. This property together with the intuitive interpretation makes this kind of distributions interesting in financial applications such as risk management, insurance, and credit risk modeling (see, e.g., Embrechts et al. (2003), Giesecke (2003), Lindskog and McNeil (2003), and Mai (2010)).

3.1.1 *Canonical Construction of the MO Distribution*

In order to outline the construction of Marshall and Olkin (1967) we consider a probability space $(\Omega, \mathcal{F}, \mathbb{P})$. For each non-empty subset $\emptyset \neq I \subset \{1, \ldots, d\}$ let E_I be an exponentially distributed random variable with mean $1/\lambda_I > 0$ and assume that these $2^d - 1$ random variables are independent. Some λ_I are allowed to be 0, in which case we mean that $E_I \equiv \infty$ with probability 1. However, we must guarantee that $\sum_{I:k \in I} \lambda_I > 0$ for all $k = 1, \ldots, d$. This means that for each $k = 1, \ldots, d$ there is at least one subset $I \subset \{1, \ldots, d\}$ containing the index k such that λ_I is strictly positive. In this case, the following random variables are almost surely well defined in $[0, \infty)$:

$$X_k := \min\big\{E_I \,\big|\, I \subset \{1, \ldots, d\},\, k \in I\big\}, \quad k = 1, \ldots, d. \tag{3.4}$$

Lemma 3.1 (Canonical Construction of the MO Distribution)
The random vector $\big(X_1, \ldots, X_d\big)$ defined by (3.4) has the Marshall–Olkin

distribution with the survival function given by (3.3).

Proof. This can be seen from the following computation with $x_1, \ldots, x_d \geq 0$:

$$\begin{aligned}\bar{F}(x_1, \ldots, x_d) &:= \mathbb{P}(X_1 > x_1, \ldots, X_d > x_d) \\ &= \mathbb{P}\big(E_I > \max_{i \in I}\{x_i\}, \; \forall \emptyset \neq I \subset \{1, \ldots, d\}\big) \\ &= \prod_{\emptyset \neq I \subset \{1,\ldots,d\}} e^{-\lambda_I \max_{i \in I}\{x_i\}} \\ &= \exp\Big(- \sum_{\emptyset \neq I \subset \{1,\ldots,d\}} \lambda_I \max_{i \in I}\{x_i\}\Big).\end{aligned}$$

□

For each $k = 1, \ldots, d$ the distribution of X_k is exponential with parameter $\sum_{I:k \in I} \lambda_I > 0$. This follows from the fact that the minimum of independent exponential random variables is again exponentially distributed and the parameters are simply added up (see also the proof of Lemma 3.2). Intuitively, the random variable E_I is interpreted as the arrival time of an exogenous shock affecting all those components of $(X_1, \ldots, X_d)$ which are indexed by a number in I. Accordingly, the kth component is destroyed when hit by the first shock E_I with $k \in I$, motivating the definition in Equation (3.4). The survival copula $\hat{C}$ of $(X_1, \ldots, X_d)$ is computed in Li (2008, Proposition 1). For our purpose, however, a slightly different expression is more appropriate, which can be deduced from (3.3).

Lemma 3.2 (Survival Copula of the MO Distribution)
The survival copula $\hat{C}$ of the random vector $(X_1, \ldots, X_d)$ as defined in (3.4) is given by

$$\hat{C}(u_1, \ldots, u_d) = \prod_{\emptyset \neq I \subset \{1,\ldots,d\}} \min_{k \in I}\Big\{u_k^{\frac{\lambda_I}{\sum_{J:k \in J} \lambda_J}}\Big\}, \quad u_1, \ldots, u_d \in [0,1].$$

Proof. In a first step one verifies that the marginal laws are exponential: for the kth marginal survival function $\bar{F}_k$ of X_k, one has

$$\begin{aligned}\bar{F}_k(x) &= \mathbb{P}(X_k > x) = \mathbb{P}\big(E_I > x, \; \forall I \subset \{1, \ldots, d\} \, : \, k \in I\big) \\ &= \exp\Big(- x \sum_{I:k \in I} \lambda_I\Big), \quad x \geq 0.\end{aligned}$$

Hence, it follows from (3.3) that

$$\bar{F}(x_1, \ldots, x_d) = \prod_{\emptyset \neq I \subset \{1,\ldots,d\}} \min_{k \in I}\Big\{\bar{F}_k(x_k)^{\frac{\lambda_I}{\sum_{J:k \in J} \lambda_J}}\Big\}, \quad x_1, \ldots, x_d \geq 0.$$

By an application of Theorem 1.3 the claim is thus established. □

The survival copula of the Marshall–Olkin distribution is called the *Marshall–Olkin copula* in the sequel, and we denote it by C instead of $\hat{C}$ to simplify notation. It has been studied extensively in the literature (see, e.g., Embrechts et al. (2003), Li (2008), and Mai and Scherer (2010)). The best-studied example of Marshall–Olkin copulas is the bivariate Marshall–Olkin copula, which by virtue of Lemma 3.2 is given by

$$C(u_1, u_2) = u_1^{\frac{\lambda_{\{1\}}}{\lambda_{\{1\}}+\lambda_{\{1,2\}}}} u_2^{\frac{\lambda_{\{2\}}}{\lambda_{\{2\}}+\lambda_{\{1,2\}}}} \min\left\{u_1^{\frac{\lambda_{\{1,2\}}}{\lambda_{\{1\}}+\lambda_{\{1,2\}}}}, u_2^{\frac{\lambda_{\{1,2\}}}{\lambda_{\{2\}}+\lambda_{\{1,2\}}}}\right\}$$

$$= \min\left\{u_1^{1-\frac{\lambda_{\{1,2\}}}{\lambda_{\{1\}}+\lambda_{\{1,2\}}}} u_2, u_1\, u_2^{1-\frac{\lambda_{\{1,2\}}}{\lambda_{\{2\}}+\lambda_{\{1,2\}}}}\right\}. \tag{3.5}$$

In some textbooks (e.g., Embrechts et al. (2003) and Nelsen (2006)) the bivariate Marshall–Olkin copula (3.5) is more conveniently parameterized by the two parameters

$$\alpha := \frac{\lambda_{\{1,2\}}}{\lambda_{\{1\}} + \lambda_{\{1,2\}}} \in [0,1], \quad \beta := \frac{\lambda_{\{1,2\}}}{\lambda_{\{2\}} + \lambda_{\{1,2\}}} \in [0,1].$$

Figure 3.2 shows scatterplots of several bivariate Marshall–Olkin copulas. One can clearly see a singular component of the copula on the line

$$\left\{(u_1, u_2) \in [0,1]^2 \,:\, u_1^{\lambda_{\{2\}}+\lambda_{\{1,2\}}} = u_2^{\lambda_{\{1\}}+\lambda_{\{1,2\}}}\right\}.$$

Furthermore, one can observe in the plot on the right of Figure 3.2 that if $\lambda_{\{2\}} = 0$, then all realizations lie below this line, i.e. the support of the copula is not all of $[0,1]^2$. This can easily be explained from the canonical construction (3.4), since $\lambda_{\{2\}} = 0$ implies $E_{\{2\}} \equiv \infty$ and therefore $X_1 = \min\{E_{\{1\}}, E_{\{1,2\}}\} \leq E_{\{1,2\}} = X_2$ almost surely in this case.

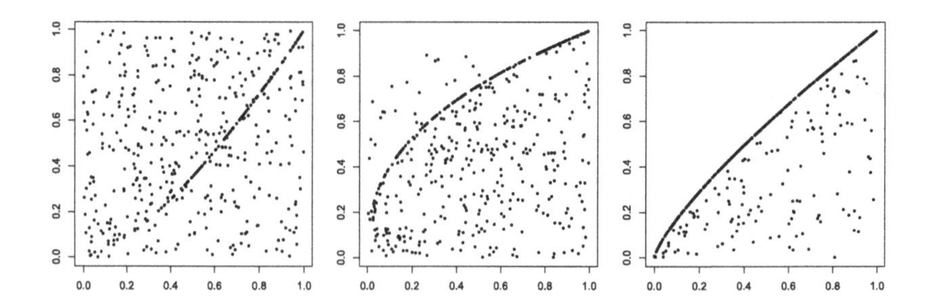

Fig. 3.2 Scatterplots of 500 samples from a bivariate Marshall–Olkin copula, corresponding to the parameters $(\lambda_{\{1\}}, \lambda_{\{2\}}, \lambda_{\{1,2\}}) = (1, 2, 1)$ (left), $(\lambda_{\{1\}}, \lambda_{\{2\}}, \lambda_{\{1,2\}}) = (10, 1, 5)$ (middle), and $(\lambda_{\{1\}}, \lambda_{\{2\}}, \lambda_{\{1,2\}}) = (3, 0, 10)$ (right).

Regarding implementation, one difficulty when working with the Marshall–Olkin distribution is the fact that its parameters are indexed by subsets $I \subset \{1, \ldots, d\}$. To tackle this problem, we propose associating any of the $2^d - 1$ subsets $\emptyset \neq I \subset \{1, \ldots, d\}$ (uniquely) with a number in the set $\{1, 2, \ldots, 2^d - 1\}$. To this end, recall that each number i in the set $\{1, 2, \ldots, 2^d - 1\}$ has a unique binary representation of the form[2]

$$i = \sum_{j=0}^{d-1} c_j \, 2^j, \quad c_0, \ldots, c_{d-1} \in \{0, 1\}.$$

Denoting the power set of $\{1, \ldots, d\}$ by $\mathcal{P}_d$, we can therefore define the following bijection:

$$\mathcal{P}_d \setminus \{\emptyset\} \to \{1, \ldots, 2^d - 1\}, \quad I \mapsto \sum_{j=0}^{d-1} \mathbb{1}_{\{j+1 \in I\}} \, 2^j, \tag{3.6}$$

$$\{1, \ldots, 2^d - 1\} \to \mathcal{P}_d \setminus \{\emptyset\}, \quad \sum_{j=0}^{d-1} c_j \, 2^j = i \mapsto \big\{ j \in \{1, \ldots, d\} \, : \, c_{j-1} = 1 \big\}.$$

Here is an example in the case $d = 3$.

Example 3.1 (The Binary Representation of $\mathcal{P}_3 \setminus \{\emptyset\}$)
For $d = 3$, the Marshall–Olkin distribution has parameters λ_I, for

$$I \in \big\{ \{1\}, \{2\}, \{3\}, \{1, 2\}, \{1, 3\}, \{2, 3\}, \{1, 2, 3\} \big\}.$$

All $2^3 - 1 = 7$ non-empty subsets of $\{1, 2, 3\}$ are associated with elements of the set $\{1, \ldots, 7\}$ via bijection (3.6), as follows:

$$1 = 2^0 \leftrightarrow \{1\},\ 2 = 2^1 \leftrightarrow \{2\},\ 3 = 2^0 + 2^1 \leftrightarrow \{1, 2\},\ 4 = 2^2 \leftrightarrow \{3\},$$
$$5 = 2^0 + 2^2 \leftrightarrow \{1, 3\},\ 6 = 2^1 + 2^2 \leftrightarrow \{2, 3\},\ 7 = 2^0 + 2^1 + 2^2 \leftrightarrow \{1, 2, 3\}.$$

A sampling algorithm for general Marshall–Olkin copulas, based on the canonical construction from Marshall and Olkin (1967) outlined earlier, is given as follows.

Algorithm 3.1 (Sampling Marshall–Olkin Copulas)
The inputs for the algorithm are the parameters $\{\lambda_I\}_{\emptyset \neq I \subset \{1, \ldots, d\}}$. To this end, one associates uniquely each subset $\emptyset \neq I \subset \{1, \ldots, d\}$ with a number $i \in \{1, \ldots, 2^d - 1\}$ via the binary representation (3.6). Thus, the input

[2] For example, the number 13 corresponds to $(c_0, c_1, c_2, c_3) = (1, 0, 1, 1)$, since $13 = 1 \cdot 2^0 + 0 \cdot 2^1 + 1 \cdot 2^2 + 1 \cdot 2^3$.

parameters are passed to the algorithm in a vector of length $2^d - 1$*, denoted* $\boldsymbol{\lambda}$.

```
FUNCTION sample_MO (vector: λ)
    E := vector(1 : 2^d − 1)                          (1)
    Max := 0                                          (2)
    FOR i = 1, ..., 2^d − 1                           (3)
        E[i] := sample_EXP(λ[i])                      (3a)
        Max := max{Max, E[i]}                         (3b)
    END FOR
    X := vector(1 : d)                                (4)
    U := vector(1 : d)                                (4)
    FOR k = 1, ..., d                                 (5)
        X[k] := Max                                   (5a)
    END FOR
    FOR k = 1, ..., d                                 (6)
        rate := 0                                     (6a)
        FOR i = 1, ..., 2^d − 1                       (6b)
            IF (within(k, i) = TRUE)                  (6c)
                rate := rate + λ[i]                   (6d)
                X[k] := min{X[k], E[i]}               (6e)
            END IF
        END FOR
        U[k] := exp(−rate · X[k])                     (6f)
    END FOR
    RETURN U

FUNCTION within (integers: k, i)
    Represent i as binary number: i ≅ (c_0, c_1, ..., c_{d−1}) ∈ {0, 1}^d   (7)
    IF (c_{k−1} = 1)                                                      (8)
        RETURN TRUE
    ELSE
        RETURN FALSE
```

The individual steps of Algorithm 3.1 are explained in the sequel:

(1) The arrival times of all exogenous shocks $\{E_I\}_{\emptyset \neq I \subset \{1,\ldots,d\}}$ are stored in the vector $\boldsymbol{E}$ of length $2^d - 1$, which is initialized here.
(2) The variable Max is used to denote the maximum of all E_I later on, i.e. $Max := \max\{E_I : \emptyset \neq I \subset \{1,\ldots,d\}\}$.
(3) The independent random variables $\{E_I\}_{\emptyset \neq I \subset \{1,\ldots,d\}}$ are simulated in step (3a), and the maximum is stored as variable Max in step (3b).
(4) The vectors $\boldsymbol{X}, \boldsymbol{U}$ of length d are defined. $\boldsymbol{U}$ is the vector we are later going to return, which contains a sample of the Marshall–Olkin copula in question. $\boldsymbol{X}$ is related to $\boldsymbol{U}$ via the componentwise exponential transformation in step (6f). This means that $\boldsymbol{X}$ is a sample from the Marshall–Olkin distribution (not the copula) in question.
(5) The initial value for all components of $\boldsymbol{X}$ is set to Max in step (5a).
(6) This step computes the extinction times of the components, i.e. $X[k] := \min\{E_I : k \in I\}$, $k = 1,\ldots,d$. For each component k, the FOR loop in step (6b) walks through all non-empty subsets $I \subset \{1,\ldots,d\}$. Step (6c) checks if $k \in I$. If so, step (6e) takes E_I into account in the minimum of the definition of $X[k]$. Additionally, for each $k = 1,\ldots,d$, step (6a) defines the variable $rate$, which equals the exponential rate of the kth margin $X[k]$, i.e. $rate = \sum_{I\,:\,k \in I} \lambda_I$. To this end, step (6d) accumulates the required λ_I's. This rate is used to transform the univariate margins to $U[0,1]$-distributions in step (6f).
(7) The conversion of an integer to a binary number is easily accomplished in most programming languages, since integers are typically stored as binaries in the computer's memory. Alternatively, the following "division-by-2" algorithm represents a given integer i (corresponding to a subset $I \subset \{1,\ldots,d\}$) in binary code. In Algorithm 3.2 below, the number $i \bmod 2$ is in $\{0,1\}$, depending on whether i is even or odd, respectively. The number $\lfloor i/2 \rfloor$ denotes the greatest integer less than or equal to $i/2$.
(8) The $(k-1)$st position in the binary code of i is 1 if and only if $k \in I$.

Algorithm 3.2 (Computing the Binary Representation of i)
The input for the algorithm is an integer $i \in \mathbb{N}$, as well as an empty list object l. The algorithm recursively appends 0's or 1's to the list l, until it contains the complete binary representation.

```
FUNCTION compute_binary(integer : i, list : l)
    IF (i = 1)
        l := append(l, 1)
        RETURN(l)
    ELSE
        l := append(l, i mod 2)
        compute_binary(⌊i/2⌋, l)
    END IF
```

In small dimensions, say $d \leq 15$, the runtime of Algorithm 3.1 is moderate. Generally speaking, the runtime (and memory requirement) for the simulation of one d-dimensional random vector is of order $\mathcal{O}(2^d)$. For large dimensions $d \gg 2$, this implies that it is practically impossible to simulate the Marshall–Olkin distribution with acceptable computation time. The main reason for this is that one has to consider all possible shocks when computing the components as minima taken over them. One way to circumvent this problem is to consider substructures of the general case, which is done in Sections 3.2 and 3.3.

3.1.2 *Alternative Construction of the MO Distribution*

There is an alternative construction of the Marshall–Olkin distribution via multivariate geometric compounding (see Arnold (1975)). This alternative approach implies a different sampling algorithm and allows us to speed up the simulation in the exchangeable case later on. Therefore we present it here. As a first step, the following lemma is proved.

Lemma 3.3 (Multivariate Geometric Compounding of Exponentials)
Consider a probability space $(\Omega, \mathcal{F}, \mathbb{P})$ supporting the following independent objects:

(1) A sequence $\{\epsilon_j\}_{j \in \mathbb{N}}$ of i.i.d. random variables with $\epsilon_1 \sim Exp(1/\lambda)$ for $\lambda > 0$.

(2) A sequence $\{Y_j\}_{j \in \mathbb{N}}$ of i.i.d. random variables taking values in the finite set $\{1, 2, \ldots, m\}$. Denote $p_k := \mathbb{P}(Y_1 = k)$, $k = 1, \ldots, m$. Without loss of generality, we assume $p_k > 0$ for all $k = 1, \ldots, m$.

Define the random variables $N_k := \min\{j \in \mathbb{N} : Y_j = k\}$ *and* $E_k := \epsilon_1 + \ldots + \epsilon_{N_k}$ *for* $k = 1, \ldots, m$. *Then:*

(1) For each $k = 1, \ldots, m$, *we have* $N_k \sim Geo(p_k)$ *and* $E_k \sim Exp(p_k/\lambda)$.
(2) $E_1, \ldots, E_m$ *are stochastically independent.*

Proof. This proof goes back to Arnold (1975).

(1) Fix $k \in \{1, \ldots, m\}$. $N_k \sim Geo(p_k)$ is clear by definition. To see that $E_k \sim Exp(p_k/\lambda)$, recall that the Laplace transform of an $Exp(1/c)$-distributed random variable E is given by

$$u \mapsto \mathbb{E}\Big[e^{-u\,E}\Big] = \int_0^\infty e^{-u\,x}\,\frac{1}{c}\,e^{-\frac{1}{c}\,x}\,dx = \frac{1}{1+u\,c}, \quad u \geq 0.$$

It suffices to show that the Laplace transform of E_k is of this form with $c = \lambda/p_k$. So consider $u \geq 0$ and check that

$$\begin{aligned}
\mathbb{E}\Big[e^{-u\,E_k}\Big] &= \mathbb{E}\Big[e^{-u\,\sum_{j=1}^{N_k}\epsilon_j}\Big] = \mathbb{E}\Big[\mathbb{E}\Big[e^{-u\,\sum_{j=1}^{N_k}\epsilon_j}\,\Big|\,N_k\Big]\Big] \\
&= \mathbb{E}\Big[\prod_{j=1}^{N_k}\underbrace{\mathbb{E}\Big[e^{-u\,\epsilon_j}\,\Big|\,N_k\Big]}_{=\mathbb{E}\left[e^{-u\,\epsilon_j}\right]=\frac{1}{1+u\,\lambda}}\Big] = \mathbb{E}\Big[\Big(\frac{1}{1+u\,\lambda}\Big)^{N_k}\Big] \\
&= \sum_{i=1}^{\infty}\Big(\frac{1}{1+u\,\lambda}\Big)^i p_k\,(1-p_k)^{i-1} = \frac{p_k}{1+u\,\lambda}\sum_{i=0}^{\infty}\Big(\underbrace{\frac{1-p_k}{1+u\,\lambda}}_{<1}\Big)^i \\
&= \frac{p_k}{1+u\,\lambda}\,\frac{1}{1-\frac{1-p_k}{1+u\,\lambda}} = \frac{1}{1+u\,\frac{\lambda}{p_k}}.
\end{aligned}$$

This verifies $E_k \sim Exp(p_k/\lambda)$.

(2) Having proved part (1) already, to establish independence it suffices to prove for $u_1, \ldots, u_m \geq 0$ that the joint Laplace transform factorizes, i.e.

$$\mathbb{E}\Big[e^{-\sum_{l=1}^m u_l\,E_l}\Big] = \prod_{l=1}^m\Big(\frac{1}{1+u_l\,\frac{\lambda}{p_l}}\Big). \tag{3.7}$$

The proof is carried out by induction over $m \in \mathbb{N}$. For $m = 1$ the statement is trivial. Now suppose the statement is true for some $m-1 \in \mathbb{N}$. Fix some $k \in \{1, \ldots, m\}$. Conditioned on the event $\{Y_1 = k\}$, one has $N_l - 1 \sim Geo(p_l)$ for all $l \neq k$, and also the $m-1$ random variables

$\sum_{j=1}^{N_l-1} \epsilon_{j+1}$ for $l = 1, \ldots, m$, $l \neq k$, are independent by induction hypothesis (IH). This implies that

$$\begin{aligned}
&\mathbb{E}\Big[e^{-\sum_{l=1}^m u_l E_l}\Big] \\
&= \sum_{k=1}^m p_k \, \mathbb{E}\Big[e^{-\sum_{l=1}^m u_l \sum_{j=1}^{N_l} \epsilon_j} \,\Big|\, Y_1 = k\Big] \\
&= \sum_{k=1}^m p_k \, \mathbb{E}\Big[e^{-\epsilon_1 \sum_{l=1}^m u_l - \sum_{l=1, l\neq k}^m u_l \sum_{j=1}^{N_l-1} \epsilon_{j+1}} \,\Big|\, Y_1 = k\Big] \\
&= \sum_{k=1}^m p_k \, \mathbb{E}\Big[e^{-\epsilon_1 \sum_{l=1}^m u_l}\Big] \, \mathbb{E}\Big[e^{-\sum_{l=1, l\neq k}^m u_l \sum_{j=1}^{N_l-1} \epsilon_{j+1}} \,\Big|\, Y_1 = k\Big] \\
&\overset{(IH)}{=} \Big(\frac{1}{1 + \sum_{l=1}^m u_l \, \lambda}\Big) \sum_{k=1}^m p_k \prod_{\substack{l=1 \\ l\neq k}}^m \Big(\frac{1}{1 + u_l \, \frac{\lambda}{p_l}}\Big).
\end{aligned}$$

Hence, it is left to show that

$$\prod_{l=1}^m \Big(\frac{1}{1 + u_l \, \frac{\lambda}{p_l}}\Big) = \Big(\frac{1}{1 + \sum_{l=1}^m u_l \, \lambda}\Big) \sum_{k=1}^m p_k \prod_{\substack{l=1 \\ l\neq k}}^m \Big(\frac{1}{1 + u_l \, \frac{\lambda}{p_l}}\Big).$$

By multiplying both sides of the last equality by

$$1 + \sum_{k=1}^m u_k \, \lambda = \sum_{k=1}^m p_k \, \Big(1 + u_k \, \frac{\lambda}{p_k}\Big),$$

we can easily verify (3.7). □

Recall that the canonical construction of the Marshall–Olkin distribution is based on the arrival times of $2^d - 1$ external shocks, one for each subset of $\{1, \ldots, d\}$. On a slightly different note, equipped with Lemma 3.3, let us now model the shocks one after another. The first shock can hit any subset of components of $\{1, \ldots, d\}$. This random subset Y_1 is modeled as the outcome of an experiment with values in the power set (excluding the empty set) $\mathcal{P}_d \setminus \{\emptyset\}$ of $\{1, \ldots, d\}$. Imagine we have been given an i.i.d. sequence of this experiment, say $\{Y_j\}_{j\in\mathbb{N}}$. Then the subset Y_2 tells us which components are killed by the second shock. Namely, the second shock kills all components in $Y_2 \setminus Y_1$, unless this set is empty because all components have already been killed by Y_1. If it is empty, then we ignore the outcome of Y_2 and proceed with Y_3. The interarrival times between the shocks are modeled by exponential random variables. The following lemma makes this alternative construction more precise.

Lemma 3.4 (Alternative Construction of the MO Distribution)
Given parameters λ_I for $\emptyset \neq I \subset \{1,\ldots,d\}$ with $\sum_{I\,:\,k\in I}\lambda_I > 0$ for each $k = 1,\ldots,d$, consider a probability space $(\Omega,\mathcal{F},\mathbb{P})$ supporting the following independent objects:

(1) A sequence $\{Y_i\}_{i\in\mathbb{N}}$ of i.i.d. $\mathcal{P}_d\backslash\{\emptyset\}$-valued random variables with distribution[3]

$$\mathbb{P}(Y_1 = I) = \frac{\lambda_I}{\sum\limits_{\emptyset\neq J\subset\{1,\ldots,d\}}\lambda_J},\qquad I\in\mathcal{P}_d\backslash\{\emptyset\}.$$

(2) A sequence $\{\epsilon_i\}_{i\in\mathbb{N}}$ of i.i.d. exponential random variables with parameter[4]

$$\sum_{\emptyset\neq J\subset\{1,\ldots,d\}}\lambda_J.$$

Then $(X_1,\ldots,X_d)$ has the Marshall–Olkin distribution with parameters $\{\lambda_I\}$, where

$$X_k := \epsilon_1+\epsilon_2+\ldots+\epsilon_{\min\{i\in\mathbb{N}\,:\,k\in Y_i\}},\qquad k=1,\ldots,d.$$

Proof. The proof is due to Arnold (1975) and consists of two steps:

(a) We define the random variables

$$N_I := \min\{i\in\mathbb{N}\,:\,Y_i = I\},\qquad \emptyset\neq I\subset\{1,\ldots,d\}. \tag{3.8}$$

Then, it is obvious that

$$X_k = \min\Big\{\sum_{j=1}^{N_I}\epsilon_j\,:\,\emptyset\neq I\subset\{1,\ldots,d\},\,k\in I\Big\},\qquad k=1,\ldots,d.$$

The last equality resembles the definition of the canonical construction (3.4), if we denote $E_I := \sum_{j=1}^{N_I}\epsilon_j$ for $\emptyset\neq I\subset\{1,\ldots,d\}$. Given this, the proof boils down to step (b).

(b) We have to show that $\{E_I\}$ is a collection of independent random variables with $E_I\sim Exp(\lambda_I)$. But this follows directly from Lemma 3.3, if we associate the power set $\mathcal{P}_d\setminus\{\emptyset\}$ with the set $\{1,2,\ldots,2^d-1\}$ via the bijection (3.6), i.e. by setting $m = 2^d-1$ in Lemma 3.3. □

Lemma 3.4 shows that a Marshall–Olkin copula can also be simulated by the following alternative algorithm:

[3]These probabilities give the distribution of the subset of components killed by the first shock.

[4]This is precisely the rate of the first shock to come.

Algorithm 3.3 (Sampling Marshall–Olkin Copulas II)
The inputs for the algorithm are the parameters $\{\lambda_I\}_{\emptyset \neq I \subset \{1,\ldots,d\}}$. *To this end, one associates uniquely each subset* $\emptyset \neq I \subset \{1,\ldots,d\}$ *with a number* $i \in \{1,\ldots,2^d-1\}$ *via the binary representation (3.6). Thus, the input parameters are passed to the algorithm in the vector* $\boldsymbol{\lambda}$ *of length* 2^d-1.

FUNCTION sample_MO (vector: $\boldsymbol{\lambda}$)

- $\lambda_{sum} := sum(\boldsymbol{\lambda})$ (1a)
- $\boldsymbol{rate} := vector(1:d)$ (1b)
- FOR $k=1,\ldots,d$ (1c)
  - $rate[k] := 0$
  - FOR $i=1,\ldots,2^d-1$
    - IF $(within(k,i) = TRUE)$
      - $rate[k] := rate[k] + \lambda[i]$
    - END IF
  - END FOR
- END FOR
- $\boldsymbol{y} := vector(1:2^d-1)$ (2a)
- $\boldsymbol{p} := vector(1:2^d-1)$ (2b)
- FOR $i=1,\ldots,2^d-1$
  - $y[i] := i$ (2c)
  - $p[i] := \lambda[i]/\lambda_{sum}$ (2d)
- END FOR
- $destroyed := 0$ (3a)
- $\boldsymbol{X} := (0,\ldots,0)$ (3b)
- $\boldsymbol{U} := (0,\ldots,0)$ (3b)
- $\epsilon := 0$ (3c)
- WHILE $(destroyed < d)$ (4)
  - $Y := sample\_discrete(\boldsymbol{y},\boldsymbol{p})$ (4a)
  - $\epsilon := \epsilon + sample\_EXP(\lambda_{sum})$ (4b)
  - FOR $k=1,\ldots,d$
    - IF $\big((within(k,Y) = TRUE)$ AND $(X[k]=0)\big)$ (4c)
      - $destroyed := destroyed + 1$ (4d)
      - $X[k] := \epsilon$ (4e)
      - $U[k] := \exp(-rate[k]\cdot X[k])$ (4f)
    - END IF
  - END FOR
- END WHILE
- RETURN $(U[1],\ldots,U[d])$

The individual steps of Algorithm 3.3 are explained as follows:

(1) This is an initializing step. Step (1a) defines λ_{sum} as the sum of all parameters, i.e. $\sum_{\emptyset \neq I \subset \{1,\ldots,d\}} \lambda_I$. This is required later on to compute the probability law of the Y_i's. Step (1b) defines the d-dimensional vector $\boldsymbol{rate}$ whose kth component $rate[k]$ is supposed to be the exponential rate $\sum_{I\,:\,k \in I} \lambda_I$ of the kth margin of the Marshall–Olkin distribution in question. The FOR loop in step (1c) accomplishes this computation, where the function *within* is the same as in Algorithm 3.1. The vector $\boldsymbol{rate}$ is later required to transform the univariate margins to $U[0,1]$-distributions in (4f).

(2) The $(2^d - 1)$-dimensional vectors $\boldsymbol{y}$ and $\boldsymbol{p}$ are defined in steps (2a) and (2b). The vector $\boldsymbol{y}$ is simply $(1, 2, \ldots, 2^d - 1)$, where each component represents a subset $\emptyset \neq I \subset \{1, \ldots, d\}$ via the binary correspondence (see step (2c)). The vector $\boldsymbol{p}$ contains the probabilities $\mathbb{P}(Y_1 = I)$, again using the binary representation of a set as a number (see step (2d)).

(3) The upcoming WHILE loop is prepared. The variable *destroyed* denotes the number of components of the Marshall–Olkin vector $\boldsymbol{X}$ (defined in step (3b)) that are already destroyed. Initially, *destroyed* $:= 0$, since all components are "alive", see step (3a). The variable ϵ denotes the sum of accumulated ϵ_j from Lemma 3.4, hence it is set to 0 at the beginning (see step (3c)). The vector $\boldsymbol{U}$ in step (3b) has the same meaning as in Algorithm 3.1, i.e. it is the transformation of the vector $\boldsymbol{X}$ to uniform margins.

(4) While not all components are destroyed yet, i.e. WHILE (*destroyed* $< d$), the following steps are repeated:

    (4a) The next set Y of affected components is simulated. The respective sampling algorithm *sample_discrete* is given in Algorithm 3.4.

    (4b) The sum over the ϵ_j is increased by one more sample of an $Exp(\lambda_{sum})$-distribution.

    (4c) For each component $k = 1, \ldots, d$, it is checked whether $k \in Y$ and whether component k is still "alive", i.e. whether $X[k] = 0$.

    (4d) If so, then the kth component is destroyed, i.e. the number of destroyed components is increased by one.

    (4e) Moreover, $X[k]$ is set to the accumulated sum over the ϵ_j's.

    (4f) This step simply transforms the univariate exponential margin of the destroyed components to the $U[0,1]$-law, i.e. the vector $\boldsymbol{X}$ (a sample from the Marshall–Olkin distribution) is transformed to the vector $\boldsymbol{U}$ (a sample from the Marshall–Olkin copula).

Notice that the simulation of the set-valued random variables Y_i requires a similar correspondence between $\mathcal{P}_d \setminus \{\emptyset\}$ and $\{1, \ldots, 2^d - 1\}$ as Algorithm 3.1. This correspondence renders the Y_i discrete random variables taking values in the set $\{1, \ldots, 2^d - 1\}$, which is huge for large d. Using a bisection routine one may sample these discrete random variables Y_i by the following sampling algorithm (one has to call it with $n = 2^d - 1$).

Algorithm 3.4 (Sampling RVs with Finitely Many Values)
The input for the algorithm is the value set $\boldsymbol{y} = \{y_1, \ldots, y_n\} \subset \mathbb{R}$ *and the corresponding probabilities* $\{p_1, \ldots, p_n\} \subset (0, 1]$ *with* $p_1 + \ldots + p_n = 1$. *The function* $x \mapsto \lfloor x \rfloor$ *returns the greatest integer less than or equal to the real number* x.

```
FUNCTION sample_discrete (vector: y, p)
    n := length(y)
    cum := vector(0 : n);    cum[0] := 0
    FOR i = 1, ..., n
        cum[i] := cum[i − 1] + p[i]
    END FOR
    U := sample_U[0, 1]
    upper := n;    lower := 0
    current := ⌊(upper + lower)/2⌋
    WHILE ((U > cum[current + 1]) OR (U < cum[current]))
        IF (U > cum[current + 1])
            lower := current
        ELSE
            upper := current
        END IF
        current := ⌊(upper + lower)/2⌋
    END WHILE
    RETURN y[current + 1]
```

The computation of $\boldsymbol{cum}$ requires $\mathcal{O}(n) = \mathcal{O}(2^d)$ operations. But this step has to be done only once if many samples are simulated. If we neglect this computation, being a typical "divide-and-conquer" algorithm,[5] the expected runtime of Algorithm 3.4 is of order $\mathcal{O}(\log n) = \mathcal{O}(d)$. Therefore, sampling Y_i is possible with an expected runtime of $\mathcal{O}(d)$ (since $n = 2^d - 1$ in this case). The total runtime of Algorithm 3.3 is of order $\mathcal{O}(2^d)$, since the computation of the sum over all parameters in step (1a) requires $2^d - 1$ steps in general. Similarly, the computation of the exponential rates of the margins in step (1c) is tedious.

Assume for a minute that $\mathbf{p}, \lambda_{sum}$, as well as all marginal exponential rates and the vector $\boldsymbol{cum}$ in Algorithm 3.4, are known beforehand and passed as additional arguments to Algorithm 3.3, for example due to a special parameterization, which is chosen for simplicity. Then the remaining runtime of the sampling scheme mainly depends on the number of required WHILE loops in Algorithm 3.3. Let N denote this random number, i.e. N equals the smallest integer j such that $Y_1 \cup \ldots \cup Y_j = \{1, \ldots, d\}$. It is intuitively clear that there might be a huge number of repetitions $i < N$ with $Y_i \subset (Y_1 \cup Y_2 \cup \ldots \cup Y_{i-1})$. This means that we have to simulate many of the Y_i's to no avail. The expected value $\mathbb{E}[N]$ of required repetitions can be estimated by the following argument. Fix one component $k \in \{1, \ldots, d\}$. The probability that component k is destroyed in one repetition is given by

$$p_k := \mathbb{P}(k \in Y_i) = \mathbb{P}\big(Y_i \in \{I \,:\, k \in I\}\big) = \frac{\sum_{I\,:\,k \in I} \lambda_I}{\sum_{\emptyset \neq I \subset \{1,\ldots,d\}} \lambda_I}.$$

Denote by $N_k := \min\{i \in \mathbb{N} \,:\, k \in Y_i\}$ the WHILE loop destroying component k. Then $N_k \sim Geo(p_k)$, since all Y_i, $i \in \mathbb{N}$, are i.i.d., and hence

$$\mathbb{E}[N_k] = \sum_{i=1}^{\infty} i\, \mathbb{P}(N_k = i) = \sum_{i=1}^{\infty} i\, p_k\, (1 - p_k)^{i-1} = \frac{1}{p_k}.$$

Therefore, we obtain for each $k = 1, \ldots, d$ the estimate

$$\begin{aligned} \frac{1}{p_k} &= \mathbb{E}[N_k] \leq \mathbb{E}[N] = \mathbb{E}[\max\{N_1, \ldots, N_d\}] \\ &\leq \mathbb{E}[N_1 + \ldots + N_d] = \frac{1}{p_1} + \ldots + \frac{1}{p_d}. \end{aligned}$$

If the parameters $\{\lambda_I\}$ are chosen such that

$$0 \approx p_k \ll \min\{p_1, \ldots, p_{k-1}, p_{k+1}, \ldots, p_d\} \approx 1/(d-1)$$

for one component k, i.e. if this component is very unlikely to fail compared to the other components, then $1/p_k \approx 1/p_1 + \ldots + 1/p_d \gg 1$. This example shows that the (expected) overall runtime of the algorithm strongly

[5] In each step the remaining problem is divided in two parts from which one can be removed.

depends on the parameters. In contrast, in the most favorable case when all components are equally likely to fail, i.e. if $p_1 = \ldots = p_d = 1/d$, then we obtain $\mathbb{E}[N] \leq d^2$. Taking into account the fact that steps (3d) respectively (4f) are of order $\mathcal{O}(d)$, in this most favorable case we hence obtain at most the order $\mathcal{O}(d^3)$ for the expected overall runtime of Algorithm 3.3 (still neglecting the initial computations of step (1), and ***cum*** in Algorithm 3.4).

Notice furthermore that both sampling algorithms for a general Marshall–Olkin copula, Algorithm 3.1 and Algorithm 3.3, require a storage capacity of order $\mathcal{O}(2^d)$, since for example the parameters $\{\lambda_I\}$ have to be stored. This may cause serious problems for $d \gg 2$.

So far, we have mainly discussed the computational effort for the simulation of one d-dimensional random vector in dependence of the dimension d. For many applications it is important to be able to simulate a huge number $n \in \mathbb{N}$ of i.i.d. random vectors. In this case, one simply has to run Algorithms 3.1 or 3.3 n times. If doing so, for Algorithm 3.3 one only has to perform the initializing step (1) as well as the computation of ***cum*** in Algorithm 3.4 once and store the values λ_{sum}, ***rate***, and ***cum*** for all later runs. Hence, for the simulation of n i.i.d. d-dimensional Marshall–Olkin vectors we obtain the runtime order $\mathcal{O}(n\,2^d)$ for Algorithm 3.1, and the expected runtime order $\mathcal{O}(2^d + n\,d^3)$ for Algorithm 3.3, which may be considerably smaller for $n \gg 2$.

Figure 3.3 illustrates scatterplots of samples from trivariate Marshall–Olkin copulas. It can be observed that the measure dC induced by C assigns positive mass to the "twisted" diagonal

$$\begin{aligned}
&\big\{(u_1,u_2,u_3) \in [0,1]^3 : u_1^{r_2\,r_3} = u_2^{r_1\,r_3} = u_3^{r_1\,r_2}\big\},\\
&r_k := \sum_{I \subset \{1,2,3\}\,:\,k \in I} \lambda_I, \quad k = 1,2,3.
\end{aligned}$$

More difficult to recognize in the plots is that dC also assigns positive mass to the planes $\{(u_1,u_2,u_3) \in [0,1]^3 : u_1^{r_2} = u_2^{r_1}\}$, $\{(u_1,u_2,u_3) \in [0,1]^3 : u_1^{r_3} = u_3^{r_1}\}$, and $\{(u_1,u_2,u_3) \in [0,1]^3 : u_2^{r_3} = u_3^{r_2}\}$.

3.1.3 *Properties of Marshall–Olkin Copulas*

There are multiple reasons why Marshall–Olkin copulas are interesting for applications. We list some of their most important properties in the sequel.

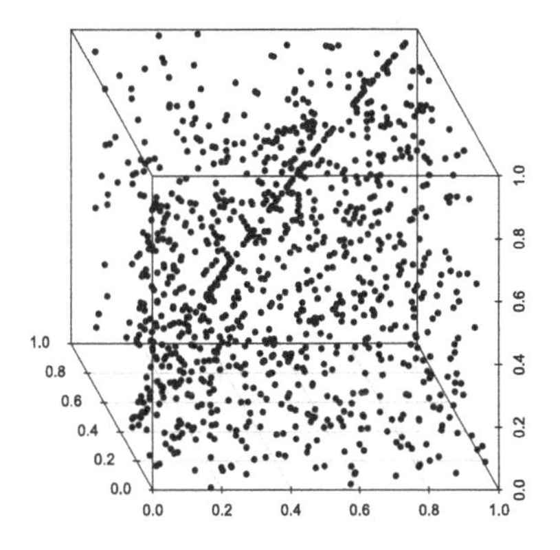
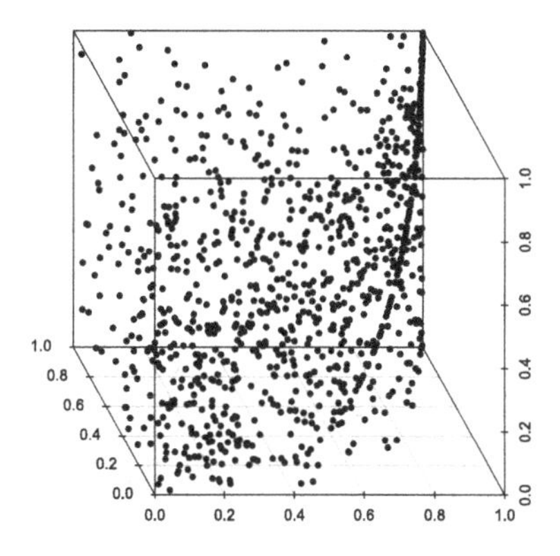

Fig. 3.3 Scatterplots of 1 000 samples from a Marshall–Olkin copula, corresponding to the parameters $(\lambda_{\{1\}}, \lambda_{\{2\}}, \lambda_{\{3\}}, \lambda_{\{1,2\}}, \lambda_{\{1,3\}}, \lambda_{\{2,3\}}, \lambda_{\{1,2,3\}}) = (10, 1, 10, 1, 1, 1, 2)$ (left) and $(\lambda_{\{1\}}, \lambda_{\{2\}}, \lambda_{\{3\}}, \lambda_{\{1,2\}}, \lambda_{\{1,3\}}, \lambda_{\{2,3\}}, \lambda_{\{1,2,3\}}) = (0, 8, 5, 1, 0, 1, 2)$ (right).

(1) **Lack of memory property:** As already indicated, the Marshall–Olkin distribution has a natural motivation via the multivariate extension of the univariate lack of memory property. Therefore, it is also sometimes called *the* "multivariate exponential distribution", even though this notion is nowadays typically used otherwise. This intuitive property renders the Marshall–Olkin distribution a standard model in the field of reliability theory (see, e.g., Barlow and Proschan (1975)).
(2) **Extreme-value copula:** The Marshall–Olkin copula is a so-called *extreme-value copula* (see Definition 1.12), even though the Marshall–Olkin distribution is not a multivariate extreme-value distribution. Such copulas have important applications in the field of multivariate extreme-value theory (see, e.g., Joe (1997), Beirlant et al. (2004), and Mai and Scherer (2010)).
(3) **Singular component:** The Marshall–Olkin distribution is one of the few prominent examples of multivariate distributions that are not absolutely continuous, but still simple enough to be of practical value. Unlike, for instance, the multivariate normal distribution or Archimedean copulas, in the case of a Marshall–Olkin distribution there is a positive probability that several components take the same value. In some applications, this property is highly desirable. Indeed, let $(X_1, \ldots, X_d)$ have as the joint distribution function a Marshall–Olkin distribution with parameters $\{\lambda_I\}$. We introduce the notation

$$r_k := \sum_{I \subset \{1,\ldots,d\} \,:\, k \in I} \lambda_I, \quad k = 1, \ldots, d,$$

where the number $r_k > 0$ is precisely the exponential rate of component X_k. This means that the random vector

$$(U_1, \ldots, U_d) := \big(e^{-r_1 X_1}, \ldots, e^{-r_d X_d}\big)$$

has the associated Marshall–Olkin copula. Then it holds true that

$$\mathbb{P}\Big(U_1^{\frac{1}{r_1}} = U_2^{\frac{1}{r_2}} = \ldots = U_d^{\frac{1}{r_d}}\Big) = \mathbb{P}(X_1 = \ldots = X_d) = \frac{\lambda_{\{1,\ldots,d\}}}{\sum_{\emptyset \neq I \subset \{1,\ldots,d\}} \lambda_I}. \tag{3.9}$$

To verify the last equality, we use the notation from the canonical construction (3.4) and see that

$$\mathbb{P}(X_1 = \ldots = X_d) = \mathbb{P}\Big(E_{\{1,\ldots,d\}} < \underbrace{\min\big\{E_I \,:\, \emptyset \neq I \subset \{1, \ldots, d\},\, |I| < d\big\}}_{=:\tilde{E}}\Big).$$

Now $E_{\{1,\ldots,d\}}$ and $\tilde{E}$ are independent and both exponentially distributed with rates $\lambda_{\{1,\ldots,d\}}$ and $\tilde{\lambda} := \sum_{\emptyset \neq I \subset \{1,\ldots,d\},\, |I|<d} \lambda_I$, respectively. This implies that

$$\mathbb{P}(E_{\{1,\ldots,d\}} < \tilde{E}) = \int_0^\infty \int_x^\infty \lambda_{\{1,\ldots,d\}}\, e^{-\lambda_{\{1,\ldots,d\}}\, x}\, \tilde{\lambda}\, e^{-\tilde{\lambda}\, y}\, dy\, dx = \frac{\lambda_{\{1,\ldots,d\}}}{\lambda_{\{1,\ldots,d\}} + \tilde{\lambda}},$$

which implies (3.9).

(4) **Exogenous shock model interpretation:** The canonical probability space, which constructs a Marshall–Olkin distribution by (3.4), has an intuitive interpretation. The random variables E_I correspond to the arrival times of exogenous shocks, and the X_k equal the first time a shock hits the respective component. Such an intuitive interpretation of a multivariate distribution is useful for applications, since it allows for a good understanding of (the sensitivity of the model with respect to) its parameters. Applications in insurance and credit risk modeling can be found in Lindskog and McNeil (2003) and Giesecke (2003).

(5) **Positive upper-tail dependence:** A careful and precise investigation of extremal dependence coefficients of the Marshall–Olkin distribution can be found in Li (2008). Loosely speaking, Marshall–Olkin copulas exhibit strong dependence in the "upper extremes", but independence in the "lower extremes". Such an asymmetry of extremal dependence can be a convincing argument in favor of a stochastic model, particularly in the field of credit risk modeling. For example, the upper-tail dependence coefficient of the bivariate Marshall–Olkin copula (3.5) is given by

$$\begin{aligned} UTD_C &= \lim_{u \uparrow 1} \frac{u^{2 - \frac{\lambda_{\{1,2\}}}{\lambda_{\{1,2\}} + \max\{\lambda_{\{1\}}, \lambda_{\{2\}}\}}} - 2\,u + 1}{1 - u} \\ &= \frac{\lambda_{\{1,2\}}}{\lambda_{\{1,2\}} + \max\{\lambda_{\{1\}}, \lambda_{\{2\}}\}}, \end{aligned}$$

where the last equality follows from de l'Hospital's rule. For exchangeable Marshall–Olkin copulas, even the UEDC is known (see Mai (2010, p. 108)).

(6) **Concordance measures:** As a subclass of extreme-value copulas, Marshall–Olkin copulas are positive orthant dependent (POD) (see, e.g., Joe (1997, Theorem 6.7, p. 177)). In the bivariate case, this readily implies that concordance measures such as Spearman's rho and Kendall's tau are non-negative. More precisely, for the bivariate Marshall–Olkin copula (3.5) one can show that

$$\begin{aligned} \rho_C &= \frac{3\,\lambda_{\{1,2\}}}{3\,\lambda_{\{1,2\}} + 2\,\lambda_{\{1\}} + 2\,\lambda_{\{2\}}} \in [0,1], \\ \tau_C &= \frac{\lambda_{\{1,2\}}}{\lambda_{\{1,2\}} + \lambda_{\{1\}} + \lambda_{\{2\}}} \in [0,1] \end{aligned}$$

(see, e.g., Embrechts et al. (2003) for a derivation with alternative parameterization). In particular, Marshall–Olkin copulas parameterize a rich spectrum of dependence structures containing independence and complete monotonicity as marginal special cases. For exchangeable Marshall–Olkin copulas, even the multivariate Spearman's rho is known (see Mai (2010, p. 109)).

3.2 The Exchangeable Case

Even though Marshall–Olkin copulas have desirable properties for many practical applications (particularly in large dimensions), we have seen that a simulation using the original shock model is quite inefficient, namely of the exponential order $\mathcal{O}(2^d)$. This is unacceptable if the dimension d is larger than, say, $d \approx 15$. Especially for some applications in portfolio credit risk modeling, dimensions such as $d = 125$ or even larger are considered. In this section the exchangeable subfamily of Marshall–Olkin copulas is determined. Recall from Chapter 1 that exchangeable distribution functions are the ones that are invariant with respect to permutations of their arguments. It turns out that in this special case the sampling algorithm can be sped up enormously to obtain computational efficiency of polynomial order in d. Moreover, the serious problem of overparameterization can be tackled by considering the exchangeable subfamily. Instead of the huge number of $2^d - 1$ parameters one is left with only d parameters. Lemmas 3.5 and 3.6 prove this by clarifying which Marshall–Olkin distributions are exchangeable. It is established that the characterizing condition is as follows: the random vector $(X_1, \ldots, X_d)$, distributed according to a Marshall–Olkin distribution with parameters $\lambda_I \geq 0$, $\emptyset \neq I \subset \{1, \ldots, d\}$, such that $\sum_{I:k \in I} \lambda_I > 0$, $k = 1, \ldots, d$, is exchangeable if and only if its parameters satisfy the exchangeability condition

$$|I| = |J| \;\Rightarrow\; \lambda_I = \lambda_J. \tag{3.10}$$

To establish this result, we first assume that condition (3.10) holds for some random vector $(X_1, \ldots, X_d)$. The survival function and copula then simplify considerably (see Lemma 3.5 below), and we observe that both are invariant with respect to permutations of their arguments.

Lemma 3.5 (Exchangeable Marshall–Olkin Survival Copula) *The survival copula of the random vector $(X_1, \ldots, X_d)$, as defined in Equation (3.4), with parameters satisfying (3.10), is given by*

$$C(u_1, \ldots, u_d) = \prod_{k=1}^{d} u_{(k)}^{\frac{\sum_{i=0}^{d-k} \binom{d-k}{i} \lambda_{i+1}}{\sum_{i=0}^{d-1} \binom{d-1}{i} \lambda_{i+1}}}, \tag{3.11}$$

where $u_{(1)} \leq \ldots \leq u_{(d)}$ denotes the ordered list of $u_1, \ldots, u_d \in [0,1]$ and λ_i is defined via

$$\{\lambda_i\} := \{\lambda_I : |I| = i\}, \quad i = 1, \ldots, d.$$

Equivalently, the survival function of $(X_1, \ldots, X_d)$ *is given by*

$$\bar{F}(x_1, \ldots, x_d) = \exp\Big(-\sum_{k=1}^{d} x_{(d+1-k)} \sum_{i=0}^{d-k} \binom{d-k}{i} \lambda_{i+1} \Big),$$

where $x_{(1)} \leq x_{(2)} \leq \ldots \leq x_{(d)}$ *denotes the ordered list of* $x_1, \ldots, x_d \geq 0$.

Clearly, both C *and* $\bar{F}$ *are invariant with respect to permutations of their arguments.*

Proof. The statement is obtained by applying Lemma 3.2 in the case where the parameters satisfy (3.10). It is observed in this case that $\sum_{I:k\in I} \lambda_I$ is independent of k:

$$\sum_{I:k\in I} \lambda_I = \sum_{i=0}^{d-1} \binom{d-1}{i} \lambda_{i+1} =: c.$$

This is due to the fact that for each index k there are precisely $(d-1)$ choose i subsets I of $\{1, \ldots, d\}$ with $(i+1)$ elements containing k, $i = 0, \ldots, d-1$. Hence, an application of Lemma 3.2 implies

$$\begin{aligned} C(u_1, \ldots, u_d) &= \prod_{k=1}^{d} \prod_{1\leq i_1 < \ldots < i_k \leq d} \Big(\min\{u_{i_1}, \ldots, u_{i_k}\} \Big)^{\frac{\lambda_k}{c}} \\ &= u_{(d)}^{\frac{\lambda_1}{c}} \, u_{(d-1)}^{\frac{\lambda_1}{c} + \frac{\lambda_2}{c}} \, u_{(d-2)}^{\frac{\lambda_1}{c} + 2\frac{\lambda_2}{c} + \frac{\lambda_3}{c}} \cdots u_{(1)}^{\sum_{i=0}^{d-1} \binom{d-1}{i} \frac{\lambda_{i+1}}{c}} \\ &= \prod_{k=1}^{d} u_{(k)}^{\frac{1}{c} \sum_{i=0}^{d-k} \binom{d-k}{i} \lambda_{i+1}}. \end{aligned}$$

The second equation illustrates the required combinatorial observation: the kth largest element $u_{(k)}$ of $u_1, \ldots, u_d$ is once the minimum of a set with one element (namely of $\{u_{(k)}\}$), $(d-k)$ times the minimum of a set with two elements (namely of $\{u_{(i)}, u_{(k)}\}$ for $i > k$), $(d-k)$ choose two times the minimum of a set with three elements, and so on. The claimed survival function follows immediately from (the survival analog of) Sklar's theorem, since the margins are $Exp(c)$-distributed. □

It is now shown that condition (3.10) is necessary and sufficient to obtain an exchangeable Marshall–Olkin distribution and copula, respectively.

Lemma 3.6 (Exchangeable Marshall–Olkin Distribution)
On a probability space $(\Omega, \mathcal{F}, \mathbb{P})$ *let* $(X_1, \ldots, X_d)$ *be a random vector with a Marshall–Olkin distribution, i.e. with survival function (3.3) for parameters* $\lambda_I \geq 0$, $\emptyset \neq I \subset \{1, \ldots, d\}$, *such that* $\sum_{I:k\in I} \lambda_I > 0$, $k = 1, \ldots, d$. *Then* $(X_1, \ldots, X_d)$ *is exchangeable if and only if its parameters satisfy the exchangeability condition (3.10).*

Proof. First, suppose that (3.10) is valid. Lemma 3.5 establishes via the survival function (and copula, respectively) that the random vector $(X_1, \ldots, X_d)$ (and its associated copula, respectively) is exchangeable.

Conversely, assume that $(X_1, \ldots, X_d)$ is exchangeable. This means that the survival function (3.3) is invariant with respect to permutations of its arguments. In order to simplify the notation, we write $\bar{F}(\boldsymbol{x})$ instead of $\bar{F}(x_1, \ldots, x_d)$, where $\boldsymbol{x} := (x_1, \ldots, x_d)$. Moreover, the ith unit vector in $\mathbb{R}^d$ is denoted by $\boldsymbol{e}_i$. We prove (3.10) by induction over the cardinality of subsets of $\{1, \ldots, d\}$. To begin with, we verify $\lambda_{\{1\}} = \lambda_{\{2\}} = \ldots = \lambda_{\{d\}}$: for each $k = 2, \ldots, d$, exchangeability implies that

$$\sum_{\substack{\emptyset \neq I \subset \{1,\ldots,d\} \\ I \neq \{1\}}} \lambda_I = -\log \bar{F}\Big(\sum_{i=2}^{d} \boldsymbol{e}_i\Big) = -\log \bar{F}\Big(\sum_{\substack{i=1 \\ i \neq k}}^{d} \boldsymbol{e}_i\Big) = \sum_{\substack{\emptyset \neq I \subset \{1,\ldots,d\} \\ I \neq \{k\}}} \lambda_I.$$

When we subtract the sum of all parameters on both sides, this in turn verifies $\lambda_{\{1\}} = \lambda_{\{2\}} = \ldots = \lambda_{\{d\}}$. Now by induction hypothesis we assume that all parameters λ_I corresponding to subsets $I \subset \{1, \ldots, d\}$ of cardinality $|I| \leq k$ are identical. We now prove that all parameters λ_I corresponding to subsets $I \subset \{1, \ldots, d\}$ of cardinality $|I| = k+1 \leq d$ are identical. To this end, let I_0 be an arbitrary subset of $\{1, \ldots, d\}$ of cardinality $|I_0| = k + 1$. Then

$$\sum_{\substack{\emptyset \neq I \subset \{1,\ldots,d\} \\ I \not\subset I_0}} \lambda_I = -\log \bar{F}\Big(\sum_{\substack{i=1 \\ i \notin I_0}}^{d} \boldsymbol{e}_i\Big) = -\log \bar{F}\Big(\sum_{i=k+2}^{d} \boldsymbol{e}_i\Big) = \sum_{\substack{\emptyset \neq I \subset \{1,\ldots,d\} \\ I \not\subset \{1,\ldots,k+1\}}} \lambda_I.$$

When we subtract the sum of all parameters on both sides, this implies

$$\lambda_{I_0} + \sum_{\substack{\emptyset \neq I \subset I_0 \\ |I| \leq k}} \lambda_I = \lambda_{\{1,\ldots,k+1\}} + \sum_{\substack{\emptyset \neq I \subset \{1,\ldots,k+1\} \\ |I| \leq k}} \lambda_I.$$

Using the induction hypothesis, this verifies that $\lambda_{\{1,\ldots,k+1\}} = \lambda_{I_0}$. Since I_0 was an arbitrary subset with cardinality $k + 1$, we may conjecture that all parameters λ_I with $|I| = k + 1$ are identical. □

For the exchangeable subfamily of Marshall–Olkin distributions, Lemma 3.6 shows that the sets $\{\lambda_I : |I| = k\}$ are singletons for $k = 1, \ldots, d$. This means that instead of $2^d - 1$ parameters λ_I, $\emptyset \neq I \subset \{1, \ldots, d\}$, an exchangeable Marshall–Olkin distribution is parameterized by only d parameters $\lambda_1, \ldots, \lambda_d \geq 0$, where $\lambda_k := \lambda_{\{1,\ldots,k\}}$, $k = 1, \ldots, d$. It is important to stress that the case $\lambda_1 = \ldots = \lambda_d = 0$ is excluded by the earlier assumptions on the λ_I's in order for construction (3.4) to be well defined.

The parametric family of copulas of form (3.11) is denoted by eMO (standing for *exchangeable Marshall–Olkin*) in the sequel. For the sake of clarity, we explicitly define the class of exchangeable Marshall–Olkin survival copulas by

$$eMO := \left\{ \prod_{k=1}^{d} u_{(k)}^{\frac{\sum_{i=0}^{d-k} \binom{d-k}{i} \lambda_{i+1}}{\sum_{i=0}^{d-1} \binom{d-1}{i} \lambda_{i+1}}} \,\middle|\, (0,\ldots,0) \neq (\lambda_1,\ldots,\lambda_d) \in [0,\infty)^d \right\}.$$

Any copula $C \in eMO$ is invariant under permutations of its arguments. This implies that for $2 \leq i \leq d$, all i-margins of C are of the same structural kind. In particular, two-margins are bivariate Cuadras–Augé copulas with parameter[6]

$$\alpha := 1 - \frac{\sum_{i=0}^{d-2} \binom{d-2}{i} \lambda_{i+1}}{\sum_{i=0}^{d-1} \binom{d-1}{i} \lambda_{i+1}} \tag{3.12}$$

(compare Example 1.4). Hence, the class eMO is a multivariate extension of bivariate Cuadras–Augé copulas. In fact, this is an example showing that extensions to higher dimensions need not be unique: for example, the two three-dimensional eMO-copulas with parameters $\boldsymbol{\lambda} := (\lambda_1, \lambda_2, \lambda_3) = (1,1,1)$, respectively $\tilde{\boldsymbol{\lambda}} := (\tilde{\lambda}_1, \tilde{\lambda}_2, \tilde{\lambda}_3) = (3,2,3)$, have the same bivariate margins, since formula (3.12) implies $\alpha = 1/2$ for both parameter choices. But the two associated trivariate eMO-copulas are not the same, since they are given by

$$C_{\boldsymbol{\lambda}}(u_1,u_2,u_3) = u_{(1)}\, u_{(2)}^{\frac{1}{2}}\, u_{(3)}^{\frac{1}{4}}, \quad C_{\tilde{\boldsymbol{\lambda}}}(u_1,u_2,u_3) = u_{(1)}\, u_{(2)}^{\frac{1}{2}}\, u_{(3)}^{\frac{3}{10}},$$

respectively, i.e. the exponent of $u_{(3)}$ differs. This has no effect on bivariate margins, but it does on trivariate events such as $\{U_1 = U_2 = U_3\}$. For instance, when we use formula (3.9) it holds that

$$\mathbb{P}_{\boldsymbol{\lambda}}(U_1 = U_2 = U_3) = \frac{1}{7} \neq \frac{1}{6} = \mathbb{P}_{\tilde{\boldsymbol{\lambda}}}(U_1 = U_2 = U_3).$$

For a deeper discussion of this issue and its implications for parameter estimation, the interested reader is referred to Hering and Mai (2012).

Figure 3.4 illustrates scatterplots from trivariate eMO-copulas. It is observed that the measure dC induced by C assigns positive mass to the diagonal of the unit cube. More difficult to recognize in the plots is that dC also assigns positive mass to the planes $\{(u_1,u_2,u_3) \in [0,1]^3 : u_1 = u_2\}$,

[6]The first author wants to use this opportunity to mention that the expression for the Cuadras–Augé parameter in his dissertation, Mai (2010, p. 65, l. 3), is a misprint. Formula (3.12) is the correct expression.

$\{(u_1, u_2, u_3) \in [0,1]^3 : u_1 = u_3\}$, and $\{(u_1, u_2, u_3) \in [0,1]^3 : u_2 = u_3\}$. Comparing the plots with those in Figure 3.3 for general Marshall–Olkin copulas, we see that the most obvious difference is that the diagonal of the unit cube is always a part of the singular component, rather than a "twisted" diagonal in the general case.

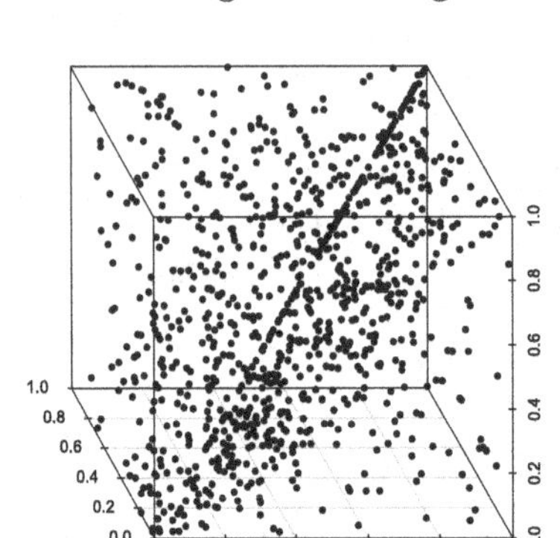

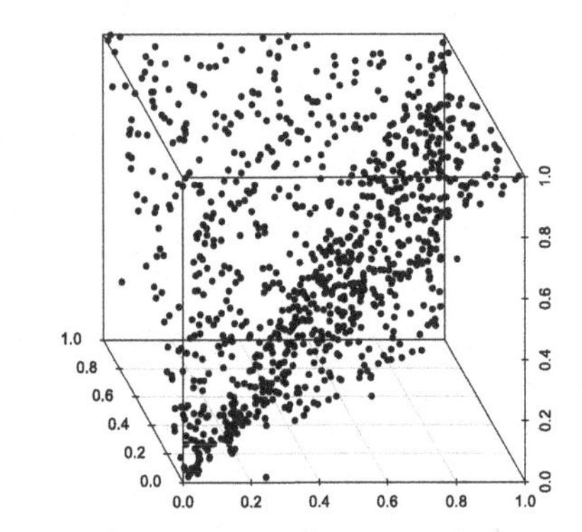

Fig. 3.4 Scatterplots of 1 000 samples from the copula $C(u_1, u_2, u_3) \in eMO$, corresponding to the parameters $(\lambda_1, \lambda_2, \lambda_3) = (1.2, 1.5, 1)$ (left) and $(\lambda_1, \lambda_2, \lambda_3) = (0.1, 5, 0.1)$ (right). In the second case, the probability of having exactly two identical components is quite high, which can be observed in the scatterplot.

3.2.1 *Reparameterizing Marshall–Olkin Copulas*

Although the analytic expression of eMO-copulas is already quite handy, the exponents still appear somewhat complicated. Therefore, we propose a reparameterization of eMO-copulas to write them in a more compact form

$$C(u_1, \ldots, u_d) = \prod_{k=1}^{d} u_{(k)}^{a_{k-1}}, \quad u_1, \ldots, u_d \in [0,1],$$

by introducing the notation

$$a_k := \frac{\sum_{i=0}^{d-k-1} \binom{d-k-1}{i} \lambda_{i+1}}{\sum_{i=0}^{d-1} \binom{d-1}{i} \lambda_{i+1}}, \quad k = 0, \ldots, d-1.$$

The advantage of this reparameterization is two-fold. First, the analytic expression and therefore the notations in the sequel are simplified. Second, it is a first and important step in determining the extendible subclass of eMO-copulas (see Section 3.3). Unfortunately, there is also a disadvantage of this reparameterization, of which one should be aware: the intuitive interpretation of the parameters $\lambda_1, \ldots, \lambda_d$ as intensities of exogenous shocks is lost. Moreover, whereas the parameter space in terms of the original parameters is simply $[0, \infty)^d \setminus \{(0, \ldots, 0)\}$, it is not immediately clear how the new parameters $(a_0, \ldots, a_{d-1})$ can be chosen. Lemma 3.7 clarifies this issue. The following notation is required.

Definition 3.1 (Difference Operator)
For a given sequence of real numbers $a_0, a_1, \ldots$, *the* difference operator Δ *is defined by* $\Delta a_k := a_{k+1} - a_k$, $k \in \mathbb{N}_0$.

The difference operator Δ is a linear operator in the sense that for sequences $a_0, a_1, \ldots$, $b_0, b_1, \ldots$, and $\beta \in \mathbb{R}$, one has

$$\begin{aligned}\Delta(\beta\, a_k + b_k) &= (\beta\, a_{k+1} + b_{k+1}) - (\beta\, a_k + b_k) \\ &= \beta\,(a_{k+1} - a_k) + (b_{k+1} - b_k) = \beta\, \Delta a_k + \Delta b_k.\end{aligned}$$

In the sequel, Δ is often applied iteratively. For example, we write $\Delta^2 a_k$ for the expression $\Delta(\Delta a_k) = \Delta(a_{k+1} - a_k) = \Delta a_{k+1} - \Delta a_k = a_{k+2} - 2\,a_{k+1} + a_k$. More generally, we write $\Delta^j a_k$ when Δ is applied j times to a_k. Notice in particular that the expression $\Delta^j a_k$ involves the $j+1$ numbers[7] $a_k, \ldots, a_{k+j}$. Moreover, it is convenient to introduce the notation $\Delta^0 a_k := a_k$. The following definition is absolutely crucial for the study of eMO-copulas.

Definition 3.2 (d-Monotonicity of Sequences)
A finite sequence $(a_0, a_1, \ldots, a_{d-1})$ *of real numbers is called* d-monotone *if*

$$(-1)^j\, \Delta^j a_k \geq 0, \quad k = 0, 1, \ldots, d-1,\ j = 0, 1, \ldots, d-k-1.$$

Intuitively, the difference operator can be considered a derivative for sequences, if the latter are interpreted as discrete functions $\mathbb{N}_0 \to \mathbb{R}$. For example, the fact that all terms $\Delta^0 a_k = a_k$ are non-negative implies that the sequence is non-negative. The fact that all terms $-\Delta a_k = a_k - a_{k+1}$ are non-negative implies that the sequence is non-increasing (heuristically, this corresponds to a non-positive first derivative). The fact that all terms $\Delta^2 a_k = \Delta a_{k+1} - \Delta a_k = a_{k+2} - 2\,a_{k+1} + a_k$ are non-negative implies that the sequence is "convex" in the sense that $a_{k+1} \leq (a_{k+2} + a_k)/2$ (heuristically, this corresponds to a non-negative second derivative). The following theorem clarifies the role of this notion in the context of eMO-copulas.

[7]One has $\Delta^j a_k = \sum_{i=0}^{j} (-1)^{i+j} \binom{j}{i} a_{k+i}$ (see Mai (2010, p. 54)).

Lemma 3.7 (Reparameterization of eMO-Copulas)
The family eMO can alternatively be written as

$$eMO = \Big\{ \prod_{k=1}^{d} u_{(k)}^{a_{k-1}} \,\Big|\, (a_0, \ldots, a_{d-1}) \text{ d-monotone, } a_0 = 1 \Big\}.$$

Moreover, the original parameters $(\lambda_1, \ldots, \lambda_d) \in [0,\infty)^d \setminus \{(0,\ldots,0)\}$ *and the new parameters* $(a_0, \ldots, a_{d-1})$ *are related via*

$$a_k = \frac{\sum_{i=0}^{d-k-1} \binom{d-k-1}{i} \lambda_{i+1}}{\sum_{i=0}^{d-1} \binom{d-1}{i} \lambda_{i+1}}, \quad k = 0, \ldots, d-1,$$
$$\lambda_k = c\,(-1)^{k-1}\,\Delta^{k-1} a_{d-k}, \quad k = 1, \ldots, d,$$

where $c > 0$ *is an arbitrary positive constant which has no effect on the copula (but on the marginal law of the associated Marshall–Olkin distribution).*

Furthermore, one can show that the function $\prod_{k=1}^{d} u_{(k)}^{a_{k-1}}$ *is a copula if and only if* $a_0 = 1$ *and* $(a_0, \ldots, a_{d-1})$ *is d-monotone. This means that there is a one-to-one correspondence between eMO-copulas and d-monotone sequences starting from* $a_0 = 1$.

Proof. The proof is technical and can be found in Mai and Scherer (2011a, 2009a) and Mai (2010, Chapter 3). However, the crucial idea of the reparameterization as well as the probabilistic meaning of the new parameters are explained in Section 3.2.2. □

Recall that a random vector $\boldsymbol{X}$ with an exchangeable Marshall–Olkin distribution is parameterized by $(\lambda_1, \ldots, \lambda_d)$. The univariate margins are exponential distributions with parameter $c := \sum_{i=0}^{d-1} \binom{d-1}{i} \lambda_{i+1} > 0$, identical for all components by exchangeability. Hence, the distribution of $\boldsymbol{X}$ is completely determined by its survival copula, which is an eMO-copula, and the marginal rate c. If we use the reparameterization discussed earlier, this distribution is now determined by the parameters $(c, a_0, a_1, \ldots, a_{d-1})$. Notice that the dimension of this new parameter vector only appears to be $d+1$ (i.e. increased by 1 compared with the original parameter vector $(\lambda_1, \ldots, \lambda_d)$) but in fact is not, since $a_0 = 1$. Given these new parameters $(c, a_0, a_1, \ldots, a_{d-1})$, the original parameters $(\lambda_1, \ldots, \lambda_d)$ are obtained precisely as stated in Lemma 3.7, where now c is not arbitrary but the given marginal rate.

3.2.2 *The Inverse Pascal Triangle*

There is a clever way of computing the new parameters $(a_0, \ldots, a_{d-1})$ from the original ones $(\lambda_1, \ldots, \lambda_d)$ by recursion (see Mai and Scherer (2011c)). Moreover, this computation sheds some light on the probabilistic meaning of the new parameters. To this end, consider a probability space $(\Omega, \mathcal{F}, \mathbb{P})$ supporting a random vector $(X_1, \ldots, X_d)$, which follows an exchangeable Marshall–Olkin distribution with parameters $(\lambda_1, \ldots, \lambda_d) \in [0, \infty)^d \setminus \{(0, \ldots, 0)\}$. Without loss of generality we let $(\Omega, \mathcal{F}, \mathbb{P})$ be the probability space from the original construction (3.4) and recall that λ_k equals the intensity of exogenous shocks affecting k-dimensional subvectors of $(X_1, \ldots, X_d)$. Now consider a $(d-1)$-dimensional subvector, w.l.o.g. $(X_1, \ldots, X_{d-1})$. Rewriting the canonical construction (3.4) of the Marshall–Olkin distribution, this implies for $k = 1, \ldots, d-1$ that

$$\begin{aligned} X_k &= \min\big\{E_I \,\big|\, k \in I\big\} \\ &= \min\big\{E_I \,\big|\, k \in I,\, d \notin I\big\} \cup \big\{E_I \,\big|\, \{k, d\} \subset I\big\} \\ &= \min\Big\{\min\big\{E_I \,\big|\, k \in I,\, d \notin I\big\},\, \min\big\{E_I \,\big|\, \{k, d\} \subset I\big\}\Big\}. \end{aligned}$$

The first minimum corresponds to the definition of a $(d-1)$-dimensional exchangeable Marshall–Olkin distribution with parameters $(\lambda_1, \ldots, \lambda_{d-1})$. However, the appearance of the second minimum suggests that exogenous shocks affecting the last component, which we have now eliminated, have to be taken into account as well. More precisely, the components indexed by $I \subset \{1, \ldots, d-1\}$ are affected by E_I and $E_{I \cup \{d\}}$. Since $\min\{E_I, E_{I \cup \{d\}}\} \sim Exp(\lambda_{|I|} + \lambda_{|I|+1})$, this observation implies that all $(d-1)$-dimensional subvectors of $(X_1, \ldots, X_d)$ follow a $(d-1)$-dimensional exchangeable Marshall–Olkin distribution corresponding to the parameters $(\lambda_1 + \lambda_2, \lambda_2 + \lambda_3, \ldots, \lambda_{d-1} + \lambda_d) \in [0, \infty)^{d-1} \setminus \{(0, \ldots, 0)\}$. Consequently, one can compute the parameters of the $(d-1)$-dimensional subvectors from $\lambda_1, \ldots, \lambda_d$ via the following geometric scheme:

$$\begin{array}{ccccccccc} \lambda_1 + \lambda_2 & & \lambda_2 + \lambda_3 & & & & \lambda_{d-1} + \lambda_d & & \\ \uparrow & \nwarrow & \uparrow & \nwarrow & \cdots & \nwarrow & \uparrow & \nwarrow & \\ \lambda_1 & & \lambda_2 & & & & \lambda_{d-1} & & \lambda_d \end{array}$$

One can now proceed iteratively and compute the following triangular scheme, the kth row of which corresponds to the parameters of all k-dimensional subvectors of $(X_1, \ldots, X_d)$:

$$\sum_{i=1}^{d} \binom{d-1}{i-1} \lambda_i$$

$$\sum_{i=1}^{d-1} \binom{d-2}{i-1} \lambda_i \quad \sum_{i=2}^{d} \binom{d-2}{i-2} \lambda_i$$

$$\vdots \qquad \ddots$$

$$\lambda_1 + 2\lambda_2 + \lambda_3 \quad \lambda_2 + 2\lambda_3 + \lambda_4 \quad \ldots$$

$$\lambda_1 + \lambda_2 \qquad \lambda_2 + \lambda_3 \qquad \ldots \quad \lambda_{d-1} + \lambda_d$$

$$\lambda_1 \qquad \lambda_2 \qquad \ldots \qquad \lambda_{d-1} \qquad \lambda_d$$

We call this triangular scheme the *inverse Pascal triangle*, since it resembles the classical Pascal triangle for the computation of binomial coefficients. In particular, the value at the tip of the triangle[8] $c := \sum_{i=1}^{d} \binom{d-1}{i-1} \lambda_i$ corresponds to the exponential rate of the univariate margins, i.e. of the random variables X_k, $k = 1, \ldots, d$. Dividing the numbers in the whole triangular scheme by c, one obtains the values $a_0, a_1, \ldots, a_{d-1}$ in the left column, i.e.

$$a_0 = \frac{1}{c} \sum_{i=1}^{d} \binom{d-1}{i-1} \lambda_i, \quad a_1 = \frac{1}{c} \sum_{i=1}^{d-1} \binom{d-2}{i-1} \lambda_i, \ldots,$$

$$a_{d-2} = \frac{1}{c} (\lambda_1 + \lambda_2), \quad a_{d-1} = \lambda_1 / c.$$

Clearly, this triangular scheme of non-negative numbers determines the distribution of $(X_1, \ldots, X_d)$, since already the bottom row does that. If we start from the new parameters, instead of computing the triangle from the bottom row we can alternatively start from the left column and compute the whole triangular scheme as follows:

$$\begin{array}{ccccccc}
c\,a_0 & & & & & & \\
 & \searrow & & & & & \\
c\,a_1 & \rightarrow & \underbrace{c\,(a_0 - a_1)}_{-c\,\Delta a_0} & & & & \\
 & \searrow & & \searrow & & & \\
c\,a_2 & \rightarrow & \underbrace{c\,(a_1 - a_2)}_{-c\,\Delta a_1} & \rightarrow & \underbrace{-c\,\Delta a_0 + c\,\Delta a_1}_{c\,\Delta^2 a_0} & & \\
 & \searrow & & \searrow & & \searrow & \\
\vdots & & \vdots & & \vdots & & \ddots
\end{array}$$

---

[8]This value was already denoted c in the proof of Lemma 3.5.

Hence, the survival copula of $(X_1, \ldots, X_d)$ (i.e. the distribution up to the constant c) is alternatively determined by the first column of the triangular scheme: the parameters $(c, a_0, \ldots, a_{d-1})$. This is basically the idea of the proof of Lemma 3.7 and shows how the difference operator Δ naturally comes into play. Notice that we can also extract from this argument a probabilistic meaning of the new parameters $(c, a_0, \ldots, a_{d-1})$: the value $c\,a_k$ equals the exponential rate at which a single component in a $(d-k)$-dimensional subvector of $(X_1, \ldots, X_d)$ is destroyed, $k = 0, \ldots, d-1$.

3.2.3 *Efficiently Sampling eMO*

Exchangeability allows us to massively improve Algorithms 3.1 and 3.3 with respect to runtime and memory requirements. The required modification is based on a quite intuitive idea, presented in Mai and Scherer (2011d): the original frailty model involves $2^d - 1$ shocks, but only at most d of these shocks are relevant for determining the components' lifetimes. These are precisely the shocks that represent the minimum of a set of shocks related to some component. Considering this, it is more efficient to only sample shocks that are truly required and to avoid simulating shocks that are not important. This is indeed possible by pursuing the following recursive strategy: (1) Simulate the time until the first shock appears. Being the minimum of all (exponentially distributed) shocks, this is again an exponential distribution.[9] (2) Simulate the number of components destroyed by the first shock ($|Y_1|$ in the notation of Lemma 3.4). The crucial point here is that it is enough to know the cardinality $|Y_1|$ of Y_1, due to the exchangeability of the distribution. (3) Set the lifetime of all components that have been destroyed by the shock. (4) Update the parameters of the remaining components. The remaining vector of lifetimes of the survived components has again a Marshall–Olkin distribution, due to the lack of memory property. (5) Continue (with the remaining model and updated parameters) with step (1), until all components are destroyed.[10] To formally derive this algorithm we require the following technical result. Denote by $\mathcal{S}_d$ the set of all permutation maps, i.e. bijections, on $\{1, \ldots, d\}$.

---

[9] ϵ_1 using the notation of Lemma 3.4.

[10] In principle, a related treatment is also possible in the general Marshall–Olkin case. Updating and organizing the parameters requires in the general case computational effort of order $\mathcal{O}(2^d)$, so we can only improve the efficiency by exploiting some knowledge about the parameters, e.g. as in the exchangeable case.

Lemma 3.8 (Technical Lemma)
Let $(X_1, \ldots, X_d)$ be an exchangeable random vector on a probability space $(\Omega, \mathcal{F}, \mathbb{P})$. Furthermore, let $\mathfrak{S}_1, \mathfrak{S}_2 : \Omega \to \mathcal{S}_d$ be two random permutations such that $\mathfrak{S}_1$ is independent of $(X_1, \ldots, X_d)$ and $\mathfrak{S}_2$. Moreover, assume that the distribution of $\mathfrak{S}_1$ is uniform on $\mathcal{S}_d$, i.e. given by $\mathbb{P}(\mathfrak{S}_1 = \sigma) = 1/d!$, $\sigma \in \mathcal{S}_d$. Then it follows that

$$(X_1, \ldots, X_d) \stackrel{d}{=} \Big(X_{\mathfrak{S}_1(\mathfrak{S}_2(1))}, \ldots, X_{\mathfrak{S}_1(\mathfrak{S}_2(d))}\Big),$$

where $\stackrel{d}{=}$ denotes equality in distribution. In words, this means that no matter how $\mathfrak{S}_2$ is defined, even if it contains some information about $(X_1, \ldots, X_d)$, it is totally blurred by $\mathfrak{S}_1$.

Proof. By exchangeability, it follows for every (deterministic) $\sigma \in \mathcal{S}_d$ that

$$(X_1, \ldots, X_d) \stackrel{d}{=} (X_{\sigma(1)}, \ldots, X_{\sigma(d)}).$$

Since $\mathfrak{S}_1$ is independent of $(X_1, \ldots, X_d)$, this readily implies that

$$(X_1, \ldots, X_d) \stackrel{d}{=} (X_{\mathfrak{S}_1(1)}, \ldots, X_{\mathfrak{S}_1(d)}). \tag{3.13}$$

Even though $\mathfrak{S}_2$ may not be, $\mathfrak{S}_1 \circ \mathfrak{S}_2$ is independent of $(X_1, \ldots, X_d)$, since for any Borel set $B \subset \mathbb{R}^d$ and $\sigma \in \mathcal{S}_d$ one has

$$\begin{aligned}
\mathbb{P}\big(\mathfrak{S}_1 \circ \mathfrak{S}_2 = \sigma, (X_1, \ldots, X_d) \in B\big) &= \mathbb{P}\big(\mathfrak{S}_1 = \sigma \circ \mathfrak{S}_2^{-1}, (X_1, \ldots, X_d) \in B\big) \\
&= \mathbb{E}\big[\mathbb{P}\big(\mathfrak{S}_1 = \sigma \circ \mathfrak{S}_2^{-1}, (X_1, \ldots, X_d) \in B \,\big|\, \mathfrak{S}_2, (X_1, \ldots, X_d)\big)\big] \\
&= \mathbb{E}\big[\mathbb{1}_{\{(X_1, \ldots, X_d) \in B\}}\, \mathbb{P}\big(\mathfrak{S}_1 = \sigma \circ \mathfrak{S}_2^{-1} \,\big|\, \mathfrak{S}_2\big)\big] \\
&= \mathbb{E}\big[\mathbb{1}_{\{(X_1, \ldots, X_d) \in B\}}\, 1/d!\big] \\
&= \mathbb{P}\big((X_1, \ldots, X_d) \in B\big)\, \mathbb{P}(\mathfrak{S}_1 \circ \mathfrak{S}_2 = \sigma).
\end{aligned}$$

The last equality uses the fact that $\mathbb{P}(\mathfrak{S}_1 \circ \mathfrak{S}_2 = \sigma) = 1/d!$, $\sigma \in \mathcal{S}_d$, which is true by the independence of $\mathfrak{S}_1$ and $\mathfrak{S}_2$. Therefore,

$$\big(\mathfrak{S}_1, (X_1, \ldots, X_d)\big) \stackrel{d}{=} \big(\mathfrak{S}_1 \circ \mathfrak{S}_2, (X_1, \ldots, X_d)\big).$$

This, together with (3.13), implies the claim. □

Now consider a random vector $(X_1, \ldots, X_d)$ with eMO-distribution with parameters $(\lambda_1, \ldots, \lambda_d)$ and denote by $(\tilde{X}_1, \ldots, \tilde{X}_d)$ an independent copy of $(X_1, \ldots, X_d)$. For later reference, we denote by $(c, a_0, \ldots, a_{d-1})$ the new parameters of this eMO-distribution by virtue of Lemma 3.7. Without loss of generality, assume that both vectors are defined on $(\Omega, \mathcal{F}, \mathbb{P})$ and $(X_1, \ldots, X_d)$ is constructed as in Lemma 3.4 from $\big(\{Y_i\}_{i \in \mathbb{N}}, \{\epsilon_i\}_{i \in \mathbb{N}}\big)$

and $(\tilde{X}_1, \ldots, \tilde{X}_d)$ is constructed as in Lemma 3.4 from $\big(\{\tilde{Y}_i\}_{i\in\mathbb{N}}, \{\tilde{\epsilon}_i\}_{i\in\mathbb{N}}\big)$. Lemma 3.4 implies that

$$(X_1, \ldots, X_d) = (\epsilon_1, \ldots, \epsilon_1) + (\eta_1, \ldots, \eta_d),$$

where

$$\eta_k := \left\{ \begin{matrix} 0 & , k \in Y_1 \\ \epsilon_2 + \ldots + \epsilon_{\min\{i \geq 2 \,:\, k \in Y_i\}} & , k \notin Y_1 \end{matrix} \right\}, \quad k = 1, \ldots, d.$$

Given $Y_1 =: \{y_1, \ldots, y_{|Y_1|}\}$ and $\{1, \ldots, d\} \setminus Y_1 =: \{x_1, \ldots, x_{d-|Y_1|}\}$, define a random permutation $\mathfrak{S}_2$ on $\{1, \ldots, d\}$ by

$$\mathfrak{S}_2^{-1}(y_i) := d - |Y_1| + i, \quad i = 1, \ldots, |Y_1|,$$
$$\mathfrak{S}_2^{-1}(x_i) := i, \quad i = 1, \ldots, d - |Y_1|.$$

In words, $\mathfrak{S}_2^{-1}$ sorts the components of $(\eta_1, \ldots, \eta_d)$ in such a way that the $|Y_1|$ zeros are at the end of the vector. In particular, $\mathfrak{S}_2$ depends on Y_1 and therefore on $(\eta_1, \ldots, \eta_d)$. Conditioned on Y_1, the random subvector $\big(\eta_{\mathfrak{S}_2(1)}, \ldots, \eta_{\mathfrak{S}_2(d-|Y_1|)}\big)$ obviously satisfies

$$\begin{aligned} &\big(\eta_{\mathfrak{S}_2(1)}, \ldots, \eta_{\mathfrak{S}_2(d-|Y_1|)}\big) \\ &= \Big(\epsilon_2 + \ldots + \epsilon_{\min\{i\geq 2 \,:\, \mathfrak{S}_2(1) \in Y_i\}}, \ldots, \epsilon_2 + \ldots + \epsilon_{\min\{i \geq 2 \,:\, \mathfrak{S}_2(d-|Y_1|) \in Y_i\}}\Big) \\ &\stackrel{d}{=} \Big(\tilde{\epsilon}_1 + \ldots + \tilde{\epsilon}_{\min\{i\geq 1 \,:\, 1 \in \tilde{Y}_i\}}, \ldots, \tilde{\epsilon}_1 + \ldots + \tilde{\epsilon}_{\min\{i\geq 1 \,:\, d-|Y_1| \in \tilde{Y}_i\}}\Big) \\ &= (\tilde{X}_1, \ldots, \tilde{X}_{d-|Y_1|}). \end{aligned}$$

A $(d-k)$-dimensional subvector of $(\tilde{X}_1, \ldots, \tilde{X}_d)$ follows a $(d-k)$-dimensional eMO-distribution with parameters $(c, a_0, \ldots, a_{d-k-1})$. Hence, conditioned on Y_1, the random vector $\big(\eta_{\mathfrak{S}_2(1)}, \ldots, \eta_{\mathfrak{S}_2(d-|Y_1|)}\big)$ has a $(d - |Y_1|)$-dimensional eMO-distribution with parameters $(c, a_0, \ldots, a_{d-|Y_1|-1})$.

Denote by $\mathfrak{S}_1$ a random permutation on $(\Omega, \mathcal{F}, \mathbb{P})$, which is independent of all other random objects and satisfies $\mathbb{P}(\mathfrak{S}_1 = \sigma) = 1/d!$, $\sigma \in \mathcal{S}_d$. By Lemma 3.8 it follows that

$$(\eta_1, \ldots, \eta_d) \stackrel{d}{=} \big(\eta_{\mathfrak{S}_1 \circ \mathfrak{S}_2(1)}, \ldots, \eta_{\mathfrak{S}_1 \circ \mathfrak{S}_2(d)}\big).$$

It is thus established that

$$(X_1, \ldots, X_d) \stackrel{d}{=} (\epsilon_1, \ldots, \epsilon_1) + \big(\eta_{\mathfrak{S}_1 \circ \mathfrak{S}_2(1)}, \ldots, \eta_{\mathfrak{S}_1 \circ \mathfrak{S}_2(d)}\big).$$

This implies that we can simulate $(X_1, \ldots, X_d)$ as follows:

(1) Simulate ϵ_1.
(2) Simulate $|Y_1|$ and then $(\tilde{X}_1, \ldots, \tilde{X}_{d-|Y_1|}) \sim eMO$, a $(d-|Y_1|)$-dimensional vector with parameters $(c, a_0, \ldots, a_{d-|Y_1|-1})$. The latter simulation recursively calls the sampling algorithm again.
(3) Set $\big(\eta_{\mathfrak{S}_2(1)}, \ldots, \eta_{\mathfrak{S}_2(d)}\big) := (\tilde{X}_1, \ldots, \tilde{X}_{d-|Y_1|}, 0, \ldots, 0)$.
(4) Simulate $\mathfrak{S}_1$.
(5) Return $(X_1, \ldots, X_d) := (\epsilon_1, \ldots, \epsilon_1) + \big(\eta_{\mathfrak{S}_1 \circ \mathfrak{S}_2(1)}, \ldots, \eta_{\mathfrak{S}_1 \circ \mathfrak{S}_2(d)}\big)$.

In particular, the simulation steps (1), (2), and (4) of the above strategy are independent of each other. This verifies Algorithm 3.5, which is implemented precisely along these lines. The transformation $(U_1, \ldots, U_d) := \big(\exp(-c\,X_1), \ldots, \exp(-c\,X_d)\big)$ again provides a sample from the respective eMO-copula, rather than from the Marshall–Olkin distribution, i.e. the margins are standardized to the $U[0,1]$-law.

A pseudo-code for this sampling strategy is provided with Algorithm 3.5. Before the algorithm is provided, let us make a few remarks for the sake of clarity:

(1) The scheme is implemented recursively; it is called at most d times for the simulation of a d-dimensional vector. The runtime for the simulation is random itself, since it depends on how many components are destroyed in each step. A worst-case estimate for the runtime is $\mathcal{O}(d^2 \log d)$ (see the following investigation).
(2) The constant c is set to $c = 1$ w.l.o.g. This is possible because Algorithm 3.5 samples an eMO-copula and not an eMO-distribution. The latter is easily obtained by transforming the margins to an $Exp(c)$-distribution afterwards.
(3) Since the algorithm is based on the probabilistic construction of Lemma 3.4, it is helpful to recall the distributions of the involved random variables $\{\epsilon_i\}_{i\in\mathbb{N}}$ and $\{|Y_i|\}_{i\in\mathbb{N}}$. The distribution of the interarrival times $\{\epsilon_i\}_{i\in\mathbb{N}}$ in Lemma 3.4 in the exchangeable special case is exponential with parameter

$$\sum_{l=1}^{d} \binom{d}{l} \lambda_l = \sum_{l=1}^{d} \binom{d}{l} (-1)^{l-1}\, \Delta^{l-1} a_{d-l} = \sum_{j=0}^{d-1} a_j,$$

where the first equation applies the reparameterization of Theorem 3.7 and the second uses the summation identity of Mai (2010, Lemma 4.4.6, p. 112). Similarly, recalling the distribution of Y_1 in Lemma 3.4 in the exchangeable special case, we obtain

$$\begin{aligned}\mathbb{P}(|Y_1| = k) &= \frac{\binom{d}{k}\lambda_k}{\sum_{l=1}^{d}\binom{d}{l}\lambda_l} \\ &= \frac{\binom{d}{k}(-1)^{k-1}\,\Delta^{k-1}a_{d-k}}{\sum_{l=1}^{d}\binom{d}{l}(-1)^{l-1}\,\Delta^{l-1}a_{d-l}} \\ &= \frac{\binom{d}{k}\sum_{j=0}^{k-1}(-1)^j\binom{k-1}{j}\,a_{d-k+j}}{\sum_{j=0}^{d-1}a_j},\end{aligned}$$

where again the second equality uses the reparameterization of Theorem 3.7 and the third equality uses the summation identities of Mai (2010, Lemma 2.5.2, p. 54, and Lemma 4.4.6, p. 112).

Algorithm 3.5 (Sampling eMO-Copulas)

FUNCTION sample_eMO (integer: d, vector: $\boldsymbol{a}, \boldsymbol{X}$)

Set alive $:= length(\boldsymbol{a})$ (1a)

Set destroyed $:= d - alive$ (1b)

IF $(destroyed = 0)$ (2)

  Set $t_0 := 0$ (2a)

ELSE

  Set $t_0 := \max_{i=1,\ldots,destroyed}\{X[i]\}$ (2b)

END IF

Set $\lambda_{next} := \sum_{j=0}^{alive-1} a[j]$ (3)

Set $t_{next} := t_0 + sample\_EXP(\lambda_{next})$ (4)

Set $h := sample\_type\_of\_shock(\boldsymbol{a})$ (5)

FOR $j = 1, \ldots, h$ (6)

  Set $X[j + destroyed] := t_{next}$ (6a)

END FOR

IF $(alive > h)$ (7)

$\tilde{\boldsymbol{a}} := \boldsymbol{a}[0 : alive - h - 1]$

RETURN $sample\_eMO\,(d, \tilde{\boldsymbol{a}}, \boldsymbol{X})$ (7a)

ELSE

RETURN $sample\_permutate\,\big(\exp(-\boldsymbol{X})\big)$ (7b)

END IF

Note that the algorithm requires the following functions:

FUNCTION $sample\_type\_of\_shock\,(vector\text{: } \boldsymbol{a})$

$alive := length(\boldsymbol{a})$

RETURN a sample of $|Y|$, *where* $|Y| \sim \mathbb{P}(|Y| = k) =$

$$\frac{\binom{alive}{k} \sum_{j=0}^{k-1} (-1)^j \binom{k-1}{j} a[alive - k + j]}{\sum_{j=0}^{alive-1} a[j]}$$

FUNCTION $sample\_permutate\,(vector\text{: } \boldsymbol{X})$

RETURN a random permutation of $\boldsymbol{X}$

drawn uniform from $S_{length(\boldsymbol{X})}$

Descriptions of individual steps, as well as runtime estimations for Algorithm 3.5, are provided in the sequel. The function *sample_eMO* is called with argument d for the dimension, and the vectors $\boldsymbol{a}$ and $\boldsymbol{X}$. The vector $\boldsymbol{X}$ is the zero vector when the function is called for the first time. The components are then recursively simulated, so that $\boldsymbol{X}$ contains the previously sampled extinction times (if any). The vector $\boldsymbol{a}$ contains the d-monotone parameters of the *eMO*-copula of the remaining components, where $a_0 = 1$. Here one advantage of the reparameterization is observed: a k-margin of a (d-dimensional) *eMO*-copula with parameters $(a_0, \ldots, a_{d-1})$ is again a (k-dimensional) *eMO*-copula with truncated parameter vector $(a_0, \ldots, a_{k-1})$. Therefore, no time-consuming parameter updates are needed, which would be the case when working with the original parameterization.

(1a) The variable *alive* denotes the dimension of the remaining problem, i.e. the number of components that are still "alive". When the function is called for the first time, $alive = d$.

(1b) The variable *destroyed* denotes the number of already destroyed components, i.e. $destroyed + alive = d$ always holds.

(2) The position of the largest extinction time (so far) is stored in t_0. This may also be interpreted as the time that all remaining components

have survived so far. If the vector $\boldsymbol{X}$ is still empty, i.e. *destroyed* $= 0$, then $t_0 := 0$. Otherwise, $t_0 := \max_{i=1,\ldots,destroyed}\{X[i]\}$. Finding the maximum of a set with cardinality smaller or equal to d requires at most $d-1$ comparisons, so it belongs to $\mathcal{O}(d)$ (see Schöning (2001, p. 130)).

(3) The intensity of the next extinction time is computed. The minimum of all upcoming extinction times (one has *alive* shocks affecting precisely one component, *alive* (*alive* − 1)/2 shocks affecting precisely two components, ..., one shock affecting all *alive* components) is again an exponential distribution. Its intensity is the sum of the shocks' intensities, which in the new parameterization equals $\sum_{j=1}^{alive} a[j-1]$. Computing this intensity requires at most $\mathcal{O}(d)$ computation steps, since $alive \leq d$.

(4) The time until the next shock is simulated.

(5) The type of shock is simulated. The random variable $h \in \{1, \ldots, alive\}$ is the number of components destroyed in this step. This simulation of a discrete random variable with $alive \leq d$ possible outcomes requires at most $\mathcal{O}(d)$ steps (see Algorithm 3.4).

(6) The vector of extinction times is enlarged by the h components destroyed in this step.

(7) If at least one component is still alive, the function *sample_eMO* is recursively called again; now with truncated parameter vector $\tilde{\boldsymbol{a}}$. Otherwise, a random permutation is applied to overcome the increasing ordering that is convenient for the implementation. The transformation of $\boldsymbol{X}$ by the mapping $x \mapsto \exp(-x)$ corresponds to the normalization to uniform margins. Note that we chose $c = 1$ as the marginal intensity of the corresponding Marshall–Olkin distribution. Sampling the required random permutation is possible with effort $\mathcal{O}(d \log d)$. For this, one might simply simulate d independent $U[0,1]$-distributed random variables and use their order statistics to permutate the vector $\boldsymbol{X}$. Sorting a vector of dimension d requires at most $\mathcal{O}(d \log d)$ computation steps using, e.g., *HeapSort* or *MergeSort* (see Schöning (2001, p. 130)).

It is guaranteed that in each call of *sample_eMO* at least one component is destroyed. Hence, for sampling a d-dimensional exchangeable Marshall–Olkin distribution, the function *sample_eMO* is at most called d times (recursively). Therefore, the overall effort is at most $\mathcal{O}(d^2 \log d)$, i.e. polynomial in the dimension of the problem. Moreover, this is a worst-case estimate. The expected runtime is much shorter, since often multiple components are destroyed in one step.

Assuming that we start with the original parameters $(\lambda_1, \ldots, \lambda_d)$, Algorithm 3.5 requires computation of the new parameters. If this computation is performed using the inverse Pascal triangle and a regular Pascal triangle (containing the binomial coefficients), it requires (only once) $\mathcal{O}(d^2)$ steps. To summarize, the simulation of n i.i.d. samples of a d-dimensional eMO-random vector in this case is at most of computational order $\mathcal{O}(d^2 + n\, d^2 \log d)$. This is a considerable improvement compared with the general case of Algorithms 3.1 and 3.3.

3.2.4 *Hierarchical Extensions*

The simulation of exchangeable Marshall–Olkin distributions (and copulas) was shown to be quite efficient. Given this, Lemma 3.9 provides a tool to design and sample hierarchical and other flexible (non-exchangeable) Marshall–Olkin distributions as well. Recall that for two independent exponential random variables $E_1 \sim Exp(\lambda_1)$ and $E_2 \sim Exp(\lambda_2)$, one has $\min\{E_1, E_2\} \sim Exp(\lambda_1 + \lambda_2)$. This so-called *min-stability* of the exponential distribution can be lifted to the multivariate case as follows.

Lemma 3.9 (Min-Stability of the Marshall–Olkin Distribution)
On a probability space $(\Omega, \mathcal{F}, \mathbb{P})$ let $(X_1, \ldots, X_d)$ have a Marshall–Olkin distribution with parameters λ_I, $\emptyset \neq I \subset \{1, \ldots, d\}$, and let $(\tilde{X}_1, \ldots, \tilde{X}_d)$ have a Marshall–Olkin distribution with parameters $\tilde{\lambda}_I$, $\emptyset \neq I \subset \{1, \ldots, d\}$, independent of $(X_1, \ldots, X_d)$. It follows that

$$\big(\min\{X_1, \tilde{X}_1\}, \ldots, \min\{X_d, \tilde{X}_d\}\big)$$

has a Marshall–Olkin distribution with parameters $\lambda_I + \tilde{\lambda}_I$, $\emptyset \neq I \subset \{1, \ldots, d\}$.

Proof. Referring to the canonical construction (3.4), assume w.l.o.g. that $(X_1, \ldots, X_d)$ is constructed from the collection $\{E_I \,|\, \emptyset \neq I \subset \{1, \ldots, d\}\}$ and $(\tilde{X}_1, \ldots, \tilde{X}_d)$ is constructed from the collection $\{\tilde{E}_I \,|\, \emptyset \neq I \subset \{1, \ldots, d\}\}$. Furthermore, all random variables $E_I, \tilde{E}_I$, $\emptyset \neq I \subset \{1, \ldots, d\}$, are assumed to be independent, corresponding to the independence of $(X_1, \ldots, X_d)$ and $(\tilde{X}_1, \ldots, \tilde{X}_d)$. Then, it follows that

$$\min\{X_k, \tilde{X}_k\} = \min\big\{ \min\{E_I, \tilde{E}_I\} \,\big|\, k \in I\big\}, \quad k = 1, \ldots, d.$$

Since $\min\{E_I, \tilde{E}_I\}$ is exponentially distributed with parameter $\lambda_I + \tilde{\lambda}_I$, the claim follows immediately. □

Lemma 3.9 can be used to design, e.g., hierarchical (non-exchangeable) Marshall–Olkin distributions. Assuming a given partition $d_1 + \ldots + d_J = d$ of the dimension d, one can design a random vector $(X_{1,1}, \ldots, X_{1,d_1}, \ldots, X_{J,1}, \ldots, X_{J,d_J}) \in (0, \infty)^d$ with a certain hierarchical Marshall–Olkin distribution. To provide a motivation, assume $X_{j,i}$ equals the bankruptcy time of one of d companies. One might think of the groups as being J industrial branches. Moreover, within a specific industrial branch j it might be intuitive to assume that $(X_{j,1}, \ldots, X_{j,d_j})$ are equally affected by branch-specific as well as global economic shocks. A reasonable model for the branch-specific shocks might be an exchangeable Marshall–Olkin distribution, say, the vector[11]

$$(X'_{j,1}, \ldots, X'_{j,d_j}) \sim eMO(\lambda_1^{(j)}, \ldots, \lambda_{d_j}^{(j)}), \quad j = 1, \ldots, J,$$

denotes the extinction times within sector j, if no global shocks were present, i.e. as if the groups $j = 1, \ldots, J$ were mutually independent. Additionally, it might be reasonable to model macroeconomic shocks affecting all industry sectors, e.g. stemming from events such as natural catastrophes that can hit companies of all branches in the same (exchangeable) way. The extinction times of the companies stemming only from such macroeconomic shocks (as if no group-specific shocks were present) might be modeled by a random vector

$$(X''_{1,1}, \ldots, X''_{1,d_1}, \ldots, X''_{J,1}, \ldots, X''_{J,d_J}) \sim eMO(\lambda''_1, \ldots, \lambda''_d).$$

Finally, the lifetimes of the companies are defined as

$$X_{j,i} := \min\{X'_{j,i}, X''_{j,i}\}, \quad j = 1, \ldots, J,\, i = 1, \ldots, d_j. \tag{3.14}$$

Lemma 3.9 implies that this stochastic model exhibits again a Marshall–Olkin distribution, which is no longer exchangeable, however. The simulation of such hierarchical structures is straightforward: one simulates $J + 1$ independent exchangeable distributions using Algorithm 3.5, and then computes the minima required in (3.14). Note that each layer of hierarchy added in this way increases the order of the overall effort (at most) by the factor d. If a sample from the corresponding (survival) copula is required, the group-specific rates of the marginal laws must be computed and the univariate marginals must be standardized to the required $U[0, 1]$-distribution.

[11] We denote by $eMO(\lambda_1, \ldots, \lambda_d)$ the d-dimensional eMO-distribution with parameters $(\lambda_1, \ldots, \lambda_d)$.

3.3 The Extendible Case

The goal of this section is to determine the subclass of extendible Marshall–Olkin copulas, i.e. copulas that stem from a stochastic model with conditionally independent and identically distributed components. The previous section showed that exchangeable Marshall–Olkin copulas can be simulated with computational effort of the order of at most $\mathcal{O}(d^2 \log d)$. In the extendible special case, this can even be improved to obtain the effort $\mathcal{O}(d \log d)$. Moreover, a completely new access to Marshall–Olkin distributions is provided, which allows us to construct non-trivial, low-parametric subfamilies of the general class. However, this requires quite a bit of theoretical work. As a starting point it is convenient to reconsider the concept of d-monotone parameter sequences. The following notion is useful in this regard.

Definition 3.3 (Complete Monotonicity of Sequences)
An infinite sequence of real numbers $\{a_k\}_{k \in \mathbb{N}_0}$ is called completely monotone *(c.m.) if $(a_0, a_1, \ldots, a_{d-1})$ is d-monotone for each $d \geq 2$, i.e. if*

$$(-1)^j \, \Delta^j a_k \geq 0, \quad k \in \mathbb{N}_0, \, j \in \mathbb{N}_0.$$

We have seen in the previous section that an eMO-copula is parameterized by a finite, d-monotone sequence $(a_0, a_1, \ldots, a_{d-1})$ of parameters with $a_0 = 1$. There are now two possibilities: either this sequence can be extended to an infinite completely monotone sequence, or not. If it can, then it is possible to derive an alternative stochastic representation of the associated Marshall–Olkin distribution. This alternative approach is based on the notion of *Lévy subordinators.* This section is structured as follows:

(1) This section is heavily based on the notion of Lévy subordinators. Readers that are unfamiliar with this subject are strongly encouraged to start with the respective summary in the Appendix.
(2) The following theorem is proved in Section 3.3.1.

    Theorem 3.1 (C.M. Sequences and Lévy Subordinators)
    Any pair $(c, \{a_k\}_{k \in \mathbb{N}_0})$ of a number $c > 0$ and a completely monotone sequence $\{a_k\}_{k \in \mathbb{N}_0}$ with $a_0 = 1$ is associated with a Lévy subordinator Λ, which is unique in distribution.

(3) Using (2), the following result is established in Section 3.3.2.

Theorem 3.2 (Extendibility of eMO-Copulas)
Let $(\Omega, \mathcal{F}, \mathbb{P})$ be a probability space supporting an i.i.d. sequence of unit exponential random variables $\{E_k\}_{k \in \mathbb{N}}$ and an independent non-zero Lévy subordinator Λ. Denote by $(c, \{a_k\}_{k \in \mathbb{N}_0})$ the pair associated with Λ by Theorem 3.1.
Then for each $d \geq 2$ the survival copula of the vector $(X_1, \ldots, X_d)$ is the eMO-copula associated with $(a_0, \ldots, a_{d-1})$, where $X_k := \inf\{t > 0 : \Lambda_t > E_k\}$, $k = 1, \ldots, d$, is the first passage time of Λ across the level E_k. Moreover, the random variables $X_1, \ldots, X_d$ are $Exp(c)$-distributed.

Theorem 3.2 reveals that a d-dimensional eMO-copula is extendible if its associated sequence $(a_0, \ldots, a_{d-1})$ can be extended to an infinite completely monotone sequence and can thus be associated with some Lévy subordinator. It can further be shown that these are all extendible eMO-copulas, i.e. extendibility of the copula corresponds precisely to extendibility of the parameter sequence $(a_0, \ldots, a_{d-1})$. Because of Theorem 3.2, extendible eMO-copulas are called *Lévy-frailty copulas.*

(4) In Section 3.3.3, Theorem 3.2 is applied to construct an unbiased and quick simulation algorithm for eMO-copulas derived from completely monotone sequences associated with compound Poisson subordinators.
(5) Hierarchical (h-extendible) versions of Lévy-frailty copulas are discussed in Section 3.3.4.

3.3.1 *Precise Formulation and Proof of Theorem 3.1*

Completely monotone sequences arise naturally in the context of probability theory. A well-known theorem by Hausdorff (1921, 1923) is stated as follows.

Theorem 3.3 (Hausdorff's Moment Problem)
A sequence $\{a_k\}_{k \in \mathbb{N}_0}$ with $a_0 = 1$ is completely monotone if and only if there exists a probability space $(\Omega, \mathcal{F}, \mathbb{P})$ supporting a random variable $X \in [0, 1]$ such that $a_k = \mathbb{E}[X^k]$, $k \in \mathbb{N}_0$. Moreover, the law of this random variable X is uniquely determined by its moments.

Proof. Originally due to Hausdorff (1921, 1923). Sufficiency is a one-liner, since for a given random variable X defining $a_k := \mathbb{E}[X^k]$, $k \in \mathbb{N}_0$, we see by the binomial formula that

$$(-1)^j \, \Delta^j a_k = \sum_{i=0}^{j} \binom{j}{i} (-1)^i \, \mathbb{E}[X^{k+i}] = \mathbb{E}[X^k \, (1-X)^j] \geq 0.$$

The converse implication is more difficult to derive and we refer to Feller (1966, Theorem 1, p. 225) for a full proof. The uniqueness statement relies on the boundedness of the interval $[0, 1]$ and follows from a classical result of Müntz (1914) and Szász (1916). □

To conclude, Theorem 3.3 states that there is a bijection between the set of all completely monotone sequences $\{a_k\}_{k \in \mathbb{N}_0}$ with $a_0 = 1$ and the set of all probability measures on $[0, 1]$. This will be useful later on.

The notion of complete monotonicity means that the "derivatives" of the sequence alternate in sign. More clearly, a sequence $\{a_k\}_{k \in \mathbb{N}_0}$ is completely monotone, if it is non-negative, non-increasing (non-positive first derivative), "convex" (non-negative second derivative), and so on. By definition, the first d members of a completely monotone sequence are d-monotone. The converse, however, is not true. An example of a three-monotone sequence which is not a subsequence of a completely monotone sequence is provided in Example 3.2.

Example 3.2 (Proper d-Monotone Sequence)
Consider the sequence $(a_0, a_1, a_2) := (1, 1/2, \epsilon)$ *with* $0 \leq \epsilon < 1/4$. *Then it is not difficult to check that* $(1, 1/2, \epsilon)$ *is three-monotone. However, there exists no (infinite) completely monotone sequence* $\{b_k\}_{k \in \mathbb{N}_0}$ *such that* $(b_0, b_1, b_2) = (1, 1/2, \epsilon)$. *If there were such a sequence, then by Theorem 3.3 there would be a probability space* $(\Omega, \mathcal{F}, \mathbb{P})$ *supporting a random variable* X *with values in* $[0, 1]$ *such that* $b_k = \mathbb{E}[X^k]$, $k \in \mathbb{N}_0$. *But Jensen's inequality would then imply that*

$$\frac{1}{4} > \epsilon = b_2 = \mathbb{E}[X^2] \geq \mathbb{E}[X]^2 = b_1^2 = \frac{1}{4},$$

which is a contradiction. Hence, the finite sequence $(1, 1/2, \epsilon)$ *is a proper three-monotone sequence in the sense that it cannot be extended to a completely monotone sequence.*

It is obvious that every two-monotone sequence $(1, a_1)$ can be extended to a completely monotone sequence, e.g. by the sequence $\{a_1^k\}_{k \in \mathbb{N}_0}$, which equals the sequence of moments of a constant random variable $X \equiv a_1$. However, Example 3.2 shows that for $d \geq 3$ this is no longer true in general. It follows from[12] Dette and Studden (1997, Theorem 1.4.3, p. 20)

[12]This result is originally derived in the monograph by Karlin and Shapley (1953).

that a sequence $(a_0, a_1, \ldots, a_{d-1})$ can be extended to a completely monotone sequence $\{a_k\}_{k \in \mathbb{N}_0}$ if and only if the so-called *Hankel determinants* $\hat{H}_1, \check{H}_1, \hat{H}_2, \check{H}_2, \ldots, \hat{H}_{d-1}, \check{H}_{d-1}$ are all non-negative, where for all $l \in \mathbb{N}$ with $2\,l \leq d-1$ and for all $k \in \mathbb{N}_0$ with $2\,k+1 \leq d-1$ one has

$$\hat{H}_{2\,l} := \det \begin{pmatrix} a_0 & \ldots & a_l \\ \vdots & & \vdots \\ a_l & \ldots & a_{2\,l} \end{pmatrix}, \quad \check{H}_{2\,l} := \det \begin{pmatrix} -\Delta a_1 & \ldots & -\Delta a_l \\ \vdots & & \vdots \\ -\Delta a_l & \ldots & -\Delta a_{2\,l-1} \end{pmatrix},$$

$$\hat{H}_{2\,k+1} := \det \begin{pmatrix} a_1 & \ldots & a_{k+1} \\ \vdots & & \vdots \\ a_{k+1} & \ldots & a_{2\,k+1} \end{pmatrix}, \quad \check{H}_{2\,k+1} := \det \begin{pmatrix} -\Delta a_0 & \ldots & -\Delta a_k \\ \vdots & & \vdots \\ -\Delta a_k & \ldots & -\Delta a_{2\,k} \end{pmatrix}.$$

For example, a sequence $(1, a_1, a_2)$ is extendible to a completely monotone sequence if and only if $1 \geq a_1 \geq a_2 \geq a_1^2$. Figure 3.5 illustrates the set of all three-monotone sequences.

The following theorem, together with Theorem 3.3, is a more precise formulation of Theorem 3.1.

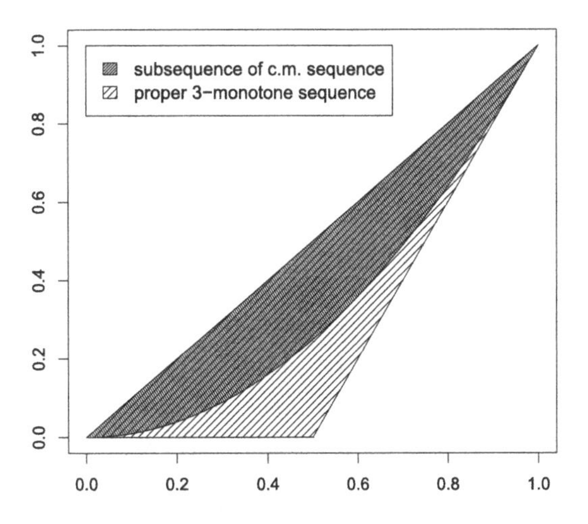

Fig. 3.5 Illustration of all three-monotone sequences $(1, a_1, a_2)$ (a_1 on the x-axis and a_2 on the y-axis). The set is subdivided into sequences that can be obtained as moments of a random variable and proper three-monotone sequences. Computing the areas in the plot, one may conclude that the share of sequences, which can be extended to a completely monotone sequence, is $2/3$.

Theorem 3.4 (A Useful Bijection)
There is a bijection from the set of all probability measures on $[0,1]$ to the set of all characteristics (μ,ν) of Lévy subordinators with Laplace exponent Ψ satisfying $\Psi(1)=1$. Denoting a probability law on $[0,1]$ by $\mathbb{P}(X\in dx)$ with a generic random variable X on $[0,1]$, the bijection is given as follows:

(1) $\mathbb{P}(X\in dx)\mapsto(\mu,\nu)$, *where*

$$\mu := \mathbb{P}(X=1),$$
$$\nu(B) := \mathbb{E}\Big[\frac{1}{1-X}\,\mathbb{1}_{\{-\log X\in B\}}\Big],\quad B\in\mathcal{B}\big((0,\infty]\big). \tag{3.15}$$

(2) $(\mu,\nu)\mapsto\mathbb{P}(X\in dx)$, *where for* $B\in\mathcal{B}\big([0,1]\big)$

$$\mathbb{P}(X\in B) := \mu\,\mathbb{1}_{\{1\in B\}} + \int_{\{-\log b\,|\,b\in B\setminus\{1\}\}}\big(1-e^{-t}\big)\,\nu(dt). \tag{3.16}$$

Proof.

(a) The first step is to check that the measure ν defined in (3.15) actually is a Lévy measure, i.e. satisfies (A.4). To this end, consider a probability space $(\Omega,\mathcal{F},\mathbb{P})$ supporting a random variable $X\in[0,1]$ with the given distribution. With $\epsilon>0$ it follows that

$$\nu\big((\epsilon,\infty]\big) = \mathbb{E}\Big[\frac{1}{1-X}\,\mathbb{1}_{\{-\log X\in(\epsilon,\infty]\}}\Big]$$
$$= \mathbb{E}\Big[\frac{1}{1-X}\,\mathbb{1}_{\{X\in[0,e^{-\epsilon})\}}\Big] \le \frac{1}{1-e^{-\epsilon}} < \infty$$

and

$$\int_{(0,1]} t\,\nu(dt) = \int_{(0,1]} t\,\big(1-e^{-t}\big)^{-1}\,\mathbb{P}(-\log X\in dt)$$
$$= \mathbb{E}\Big[\frac{-\log X}{1-X}\,\mathbb{1}_{\{X\in[e^{-1},1)\}}\Big] \overset{(*)}{\le} e < \infty.$$

Equation $(*)$ holds, since using the series expansion $\log(1+y)=\sum_{k=1}^{\infty}\frac{(-1)^{k+1}}{k}\,y^k$ implies for $y\in[\exp(-1),1)$ that

$$\Big|\frac{\log y}{1-y}\Big| = \Big|\frac{\log\big(1+(-(1-y))\big)}{1-y}\Big| = \Big|\sum_{k=1}^{\infty}\frac{-1}{k}\,(1-y)^{k-1}\Big|$$
$$\le \Big(\sum_{k=0}^{\infty}(1-y)^k\Big) = \frac{1}{y} \le e.$$

Denoting by Ψ the Laplace exponent corresponding to the characteristics (μ,ν) defined in (3.15), it is verified that

$$\begin{aligned}\Psi(1) &= \mu + \int_{(0,\infty]} \left(1 - e^{-t}\right) \nu(dt)\\ &= \mathbb{P}(X = 1) + \int_{(0,\infty]} \left(1 - e^{-t}\right) \left(1 - e^{-t}\right)^{-1} \mathbb{P}(-\log X \in dt)\\ &= \mathbb{P}(X = 1) + \mathbb{P}(0 \leq X < 1) = \mathbb{P}(0 \leq X \leq 1) = 1,\end{aligned}$$

establishing that $\Psi(1) = 1$.

(b) Verify that

$$\mathbb{P}(X \in [0,1]) = \mu + \int_{(0,\infty]} \left(1 - e^{-t}\right) \nu(dt) = \Psi(1) = 1,$$

implying that the measure defined by the right-hand side of (3.16) is indeed a probability measure.

(c) It is left to check that the claimed mappings are the inverses of each other. A rigorous proof is provided in Mai (2010, Lemma 4.1.3, p. 91). The idea of this mapping is based on the concatenation of three bijections: (i) the bijection of Theorem 3.3 between probability measures on $[0,1]$ and completely monotone sequences $\{a_k\}_{k\in\mathbb{N}_0}$, (ii) a bijection between completely monotone sequences $\{a_k\}_{k\in\mathbb{N}_0}$ and Bernstein functions Ψ induced by the equality $a_k = \Psi(k+1) - \Psi(k)$, and (iii) a bijection between Bernstein functions Ψ and Lévy subordinators $\Lambda = \{\Lambda_t\}_{t\geq 0}$ induced via the Laplace transforms $\mathbb{E}[\exp(-x\,\Lambda_t)] = \exp(-t\,\Psi(x))$. The bijection (ii) follows from a result in Gnedin and Pitman (2008). The bijection (iii) is a classical result that can be found, for instance, in Applebaum (2004, Theorem 1.3.23, p. 52) or Schilling et al. (2010, Theorem 5.2, p. 35). □

By Theorem 3.3, we can find for each completely monotone sequence $\{a_k\}_{k\in\mathbb{N}_0}$ with $a_0 = 1$ a unique probability measure $\mathbb{P}(X \in dx)$ on $[0,1]$. In combination with Theorem 3.4 we can thus find for each completely monotone sequence $\{a_k\}_{k\in\mathbb{N}_0}$ with $a_0 = 1$ unique Lévy characteristics (μ, ν), whose associated Laplace exponent Ψ satisfies $\Psi(1) = 1$. This Laplace exponent additionally satisfies

$$\begin{aligned}\Psi(k+1) - \Psi(k) &= \mu + \int_{(0,\infty]} e^{-k\,t} \left(1 - e^{-t}\right) \nu(dt)\\ &= \mathbb{P}(X = 1) + \int_{(0,\infty]} e^{-k\,t} \left(1 - e^{-t}\right) \left(1 - e^{-t}\right)^{-1} \mathbb{P}(-\log X \in dt)\\ &= \mathbb{E}\big[X^k\, \mathbb{1}_{\{X=1\}}\big] + \mathbb{E}\big[X^k\, \mathbb{1}_{\{X\in[0,1)\}}\big] = \mathbb{E}[X^k] = a_k, \quad k \in \mathbb{N}_0.\end{aligned}$$

To understand the role of the constant $c > 0$ in Theorem 3.1, notice that if Ψ is the Laplace exponent of a Lévy subordinator with $\Psi(1) = 1$, then

$\tilde{\Psi} := c\,\Psi$ is the Laplace exponent of a Lévy subordinator with $\tilde{\Psi}(1) = c$. Multiplying the completely monotone sequence $\{a_k\}_{k\in\mathbb{N}_0}$ by c, we obtain a completely monotone sequence $\{c\,a_k\}_{k\in\mathbb{N}_0}$ associated with the Laplace exponent $\tilde{\Psi} = c\,\Psi$ satisfying $\tilde{\Psi}(1) = c$. Hence, we can extend the bijection induced by Theorems 3.3 and 3.4 to the one claimed in Theorem 3.1, getting rid of the constraint $\Psi(1) = 1$.

Formulated without the constant $c > 0$, for a given Laplace exponent Ψ of an arbitrary (non-zero) Lévy subordinator, the sequence $\{\Psi(k+1) - \Psi(k)\}_{k\in\mathbb{N}_0}$ is completely monotone (and non-zero). And conversely, for a given (non-zero) completely monotone sequence $\{a_k\}_{k\in\mathbb{N}_0}$ it is possible to find a (unique and non-zero) Laplace exponent Ψ satisfying $\Psi(k+1) - \Psi(k) = a_k$, $k \in \mathbb{N}_0$. This correspondence was originally determined in Gnedin and Pitman (2008). For our purpose, the formulation with the constant c, although appearing a little complicated, is appropriate, since it reflects the decomposition of the eMO-distribution into its exponential margins on the one hand and its eMO-survival copula on the other hand.

3.3.2 *Proof of Theorem 3.2*

The correspondence between completely monotone sequences and Lévy subordinators from Theorem 3.4 motivates the following definition.

Definition 3.4 (Lévy-Frailty Copula)
For a given Laplace exponent Ψ of a Lévy subordinator satisfying $\Psi(1) = 1$, we define the copula

$$C_{\Psi}(u_1, \ldots, u_d) := \prod_{i=1}^{d} u_{(i)}^{\Psi(i)-\Psi(i-1)}, \tag{3.17}$$

where, as before, $u_{(1)} \leq u_{(2)} \leq \ldots \leq u_{(d)}$ denotes the ordered list of $u_1, \ldots, u_d \in [0,1]$. C_{Ψ} is called a Lévy-frailty copula.

The term "Lévy-frailty copula" stems from Mai and Scherer (2009a) and is justified by Theorem 3.2, which is repeated and proved in the sequel.

Theorem 3.5 (Extendibility of eMO-Copulas)
Let $(\Omega, \mathcal{F}, \mathbb{P})$ be a probability space supporting an i.i.d. sequence of unit exponential random variables $\{E_k\}_{k\in\mathbb{N}}$ and an independent non-zero Lévy subordinator Λ. Denote by $(c, \{a_k\}_{k\in\mathbb{N}_0})$ the pair associated with Λ by Theorems 3.3 and 3.4.

Then for each $d \geq 2$ the survival copula of the vector $(X_1, \ldots, X_d)$ is the Lévy-frailty copula C_{Ψ} associated with $(a_0, \ldots, a_{d-1})$ via Ψ, where $X_k :=$

$\inf\{t > 0 : \Lambda_t > E_k\}$, $k = 1, \ldots, d$, is the first passage time of Λ across the random level E_k. Moreover, the random variables $X_1, \ldots, X_d$ are $Exp(c)$-distributed. It follows that $(U_1, \ldots, U_d) \sim C_\Psi$, where $U_k := \exp(-c\, X_k)$, $k = 1, \ldots, d$.

Proof. According to Definition A.2 of a (killed) Lévy subordinator we denote

$$\Lambda_t = \tilde{\Lambda}_t + \infty \cdot \mathbb{1}_{\{N_t \geq 1\}}, \quad t \geq 0,$$

where $\tilde{\Lambda}$ is a classical (real-valued) Lévy subordinator and N is an independent Poisson process. Recall from the remark after Theorem 3.4 that the Laplace exponent of Λ is of the form $c\,\Psi$, where $c > 0$ is the given constant and Ψ satisfies $\Psi(1) = 1$. Referring to the Lévy–Khinchin representation (see Theorem A.4), we split the Laplace exponent $c\,\Psi$ of Λ into two parts via

$$\begin{aligned} c\,\Psi(x) &= c\,\mu\,x + c \int_{(0,\infty)} \big(1 - e^{-t\,x}\big)\,\nu(dt) + c\,\nu(\{\infty\})\,\mathbb{1}_{\{x>0\}} \\ &=: c\,\tilde{\Psi}(x) + c\,\nu(\{\infty\})\,\mathbb{1}_{\{x>0\}},\ x \geq 0, \end{aligned}$$

where $c\,\tilde{\Psi}$ denotes the Laplace exponent of $\tilde{\Lambda}$ and $c\,\nu(\{\infty\})$ is the intensity of the Poisson process N. Recall that if $\nu(\{\infty\}) = 0$ this is conveniently interpreted as $\Lambda = \tilde{\Lambda}$, i.e. "$N$ never jumps". For arbitrary $t_1, t_2, \ldots, t_d \in [0, \infty)$ with ordered list $t_{(1)} \leq \ldots \leq t_{(d)}$ and $t_{(0)} := 0$, it is verified that

$$\begin{aligned} \sum_{i=1}^{d} (d+1-i)\,\big(\tilde{\Lambda}_{t_{(i)}} - \tilde{\Lambda}_{t_{(i-1)}}\big) &= \sum_{i=1}^{d} (d+1-i)\,\tilde{\Lambda}_{t_{(i)}} - \sum_{i=0}^{d-1} (d-i)\,\tilde{\Lambda}_{t_{(i)}} \\ &= \sum_{i=1}^{d} \tilde{\Lambda}_{t_i}. \end{aligned}$$

$\tilde{\Lambda}$ being a classical Lévy subordinator implies that the random vector of increments

$$\big(\tilde{\Lambda}_{t_{(d)}} - \tilde{\Lambda}_{t_{(d-1)}}, \ldots, \tilde{\Lambda}_{t_{(1)}} - \tilde{\Lambda}_{t_{(0)}}\big)$$

has independent components and the component $\tilde{\Lambda}_{t_{(i)}} - \tilde{\Lambda}_{t_{(i-1)}}$ has the same distribution as $\tilde{\Lambda}_{t_{(i)} - t_{(i-1)}}$. Hence, one obtains

$$\begin{aligned} \mathbb{E}\Big[e^{-\sum_{i=1}^{d} \tilde{\Lambda}_{t_i}}\Big] &= \prod_{i=1}^{d} \mathbb{E}\Big[e^{-(d+1-i)\,\tilde{\Lambda}_{(t_{(i)} - t_{(i-1)})}}\Big] \\ &= \prod_{i=1}^{d} \exp\Big(-\big(t_{(i)} - t_{(i-1)}\big)\,c\,\tilde{\Psi}(d+1-i)\Big). \end{aligned}$$

Furthermore, since N is a Poisson process with intensity $c\,\nu(\{\infty\})$, it follows with a telescope argument that

$$\mathbb{P}(N_{t_{(d)}} = 0) = e^{-c\,\nu(\{\infty\})\,t_{(d)}} = \prod_{i=1}^{d} \exp\Big(- (t_{(i)} - t_{(i-1)})\,c\,\nu(\{\infty\})\Big).$$

From this, using the conditional independence of events in the third equality (conditioned on the σ-algebra $\sigma(\Lambda_t \,:\, t \geq 0)$), and the convention $\exp(-\infty) = 0$ in the fourth, it is straightforward to compute

$$\begin{aligned}
\bar{F}(t_1, \ldots, t_d) &:= \mathbb{P}\big(X_1 > t_1, X_2 > t_2, \ldots, X_d > t_d\big) \\
&= \mathbb{P}\big(E_1 > \Lambda_{t_1}, E_2 > \Lambda_{t_2}, \ldots, E_d > \Lambda_{t_d}\big) \\
&= \mathbb{E}\left[\prod_{i=1}^{d} e^{-\Lambda_{t_i}}\right] = \mathbb{E}\left[\mathbb{1}_{\{N_{t_{(d)}}=0\}}\, e^{-\sum_{i=1}^{d} \tilde{\Lambda}_{t_i}}\right] + 0 \\
&= \mathbb{P}(N_{t_{(d)}} = 0)\, \mathbb{E}\left[e^{-\sum_{i=1}^{d} \tilde{\Lambda}_{t_i}}\right] \\
&= \prod_{i=1}^{d} \exp\Big(- (t_{(i)} - t_{(i-1)})\,\big(c\,\tilde{\Psi}(d+1-i) + c\,\nu(\{\infty\})\big)\Big) \\
&= \prod_{i=1}^{d} \exp\Big(- (t_{(i)} - t_{(i-1)})\,c\,\Psi(d+1-i)\Big).
\end{aligned}$$

In the univariate case, one obtains by the same argument for $i = 1, \ldots, d$ and $t \geq 0$ that

$$\bar{F}_i(t) := \mathbb{P}\big(X_i > t\big) = \mathbb{P}\big(E_i > \Lambda_t\big) = e^{-t\,c\,\Psi(1)} = e^{-c\,t}.$$

Thus, the X_i are *Exp*(c)-distributed. By the analog of Sklar's theorem for survival copulas (see Theorem 1.3), there exists a unique copula $\hat{C}$ which satisfies

$$\bar{F}(t_1, \ldots, t_d) = \hat{C}\big(e^{-c\,t_1}, \ldots, e^{-c\,t_d}\big).$$

On the other hand, since $t \mapsto \exp(-c\,t)$ is decreasing, Equation (3.17) in Definition 3.4 implies

$$\begin{aligned}
C_\Psi\big(e^{-c\,t_1}, \ldots, e^{-c\,t_d}\big) &= \prod_{i=1}^{d} e^{-c\,t_{(i)}\,\big(\Psi(d+1-i) - \Psi(d-i)\big)} \\
&= \prod_{i=1}^{d} e^{-t_{(i)}\,c\,\Psi(d+1-i)} \prod_{i=1}^{d-1} e^{t_{(i)}\,c\,\Psi(d-i)} \\
&= \prod_{i=1}^{d} e^{-t_{(i)}\,c\,\Psi(d+1-i)} \prod_{i=1}^{d} e^{t_{(i-1)}\,c\,\Psi(d+1-i)} \\
&= \bar{F}(t_1, \ldots, t_d).
\end{aligned}$$

Thus, by the uniqueness of $\hat{C}$, it holds that $\hat{C} = C_\Psi$. To finally see that the random vector $(U_1, \ldots, U_d)$ has joint distribution function C_Ψ, it suffices to observe that

$$\begin{aligned}\mathbb{P}\big(e^{-c X_1} &\leq u_1, \ldots, e^{-c X_d} \leq u_d\big) \\ &= \mathbb{P}\big(X_1 > -\log(u_1)/c, \ldots, X_d > -\log(u_d)/c\big) \\ &= C_\Psi(u_1, \ldots, u_d).\end{aligned}$$

Notice that the continuity of the exponential law allows one to replace "$\geq$" with "$>$" in the first equality. The claim is established. □

Example 3.3 (Exchangeable Multivariate Cuadras–Augé Copula) *Consider a parameter $\alpha \in (0,1)$. Then the sequence $\{a_k\}_{k\in\mathbb{N}_0} := \{(1-\alpha)^k\}_{k\in\mathbb{N}_0}$ is completely monotone, since*

$$(-1)^j\, \Delta^j a_k = \sum_{i=0}^{j} \binom{j}{i} (-1)^i\, (1-\alpha)^{k+i} = (1-\alpha)^k\, \alpha^j \geq 0, \quad j,k \in \mathbb{N}_0.$$

We now face the following problem: can we identify the Lévy subordinator that is associated with this sequence? As a first step, we have to find the unique random variable X on $[0,1]$ satisfying $(1-\alpha)^k = \mathbb{E}[X^k]$, $k \in \mathbb{N}_0$. It is obvious that the constant random variable $X \equiv 1-\alpha$ is the required one. Therefore, using Theorem 3.4 yields the associated Lévy characteristics (μ, ν), which are given by

$$\mu = \mathbb{P}(X = 1) = 0, \quad \nu(B) = \mathbb{1}_{\{-\log(1-\alpha)\in B\}}\, \frac{1}{\alpha}, \quad B \in \mathcal{B}\big((0,\infty]\big).$$

This corresponds precisely to a compound Poisson subordinator with intensity $1/\alpha$ and constant jump sizes of height $-\log(1-\alpha)$. The respective Lévy-frailty copula has the functional form

$$C_\alpha(u_1, \ldots, u_d) = \prod_{k=1}^{d} u_{(k)}^{(1-\alpha)^{k-1}}, \quad u_1, \ldots, u_d \in [0,1].$$

We have seen that Lévy-frailty copulas are conveniently parameterized by a Laplace exponent Ψ satisfying $\Psi(1) = 1$. Moreover, we have also seen that this parameterization is related to the parameterization in terms of the parameters $(a_0, \ldots, a_{d-1})$ via $a_k = \Psi(k+1) - \Psi(k)$, $k \in \mathbb{N}_0$. Combining this with Lemma 3.7, one obtains the following relation to the original parameters $(\lambda_1, \ldots, \lambda_d)$:

$$\lambda_k = \sum_{i=0}^{k-1} \binom{k-1}{i} (-1)^i \big(\Psi(d-k+i+1) - \Psi(d-k+i)\big), \quad k = 1, \ldots, d.$$

In terms of the Lévy characteristics, one can rewrite the last expression by an application of the binomial formula to obtain

$$\lambda_k = \mu\, \mathbb{1}_{\{k=1\}} + \int_{(0,\infty]} e^{-t\,(d-k)} \left(1 - e^{-t}\right)^k \nu(dt), \quad k = 1, \ldots, d.$$

Furthermore, notice that $\Psi(1) = 1$ implies $\mu = 1 - \int_{(0,\infty]}(1 - \exp(-t))\,\nu(dt) \in [0,1]$, which allows us to represent the parameters $(\lambda_1, \ldots, \lambda_d)$ as a function of the Lévy measure ν only.

3.3.3 Efficient Simulation of Lévy-Frailty Copulas

Theorem 3.2 (respectively Theorem 3.5) is the key to an efficient sampling algorithm, because it reveals Lévy-frailty copulas as a homogeneous mixture. To see this, one observes that a simulation of $(U_1, \ldots, U_d)$ from Theorem 3.5 is equivalent to the construction of Algorithm 1.3 with input $F = \{F_x\}_{x \in \mathbb{R}}$, where

$$F_x := \left(1 - e^{-\Lambda_{\max\{x,0\}}}\right) \mathbb{1}_{\{x>0\}}, \quad x \in \mathbb{R}. \tag{3.18}$$

In other words, conditioned on the σ-algebra $\sigma(\Lambda_t \, : \, t > 0)$, the random variables $X_1, \ldots, X_d$ from the proof of Theorem 3.5 are i.i.d. with distribution function F as given by (3.18). To summarize, this implies the following generic sampling algorithm for Lévy-frailty copulas.

Algorithm 3.6 (Generic Sampling of Lévy-Frailty Copulas)
To sample the copula $C_\Psi(u_1, \ldots, u_d)$ from Definition 3.4, the following steps may be exercised. The algorithm has as its input a Laplace exponent Ψ satisfying $\Psi(1) = 1$.

(1) *Generate $E_1, \ldots, E_d$ i.i.d. with $E_1 \sim Exp(1)$.*
(2) *Find the maximum $E_{(d)} := \max\{E_1, \ldots, E_d\}$.*
(3) *Independently of the random variables $E_1, \ldots, E_d$, simulate one path of a Lévy subordinator $\Lambda = \{\Lambda_t\}_{t \in [0,T]}$ with Laplace exponent Ψ, where T is chosen large enough[13] such that $\Lambda_T \geq E_{(d)}$.*
(4) *Determine the first passage times*

$$X_k := \inf\{t > 0 \, : \, \Lambda_t > E_k\}, \quad k = 1, \ldots, d.$$

(5) *Return $(U_1, \ldots, U_d)$, where $U_k := \exp(-X_k)$, $k = 1, \ldots, d$.*

Admittedly, the simulation of the path of a Lévy subordinator as well as the determination of the first passage times are itself not straightforward. But if the Lévy subordinator is of the compound Poisson type (with handy jump size distribution), then an unbiased and very quick sampling algorithm can be implemented. This algorithm is provided in the sequel. By virtue of Remark A.7, any Lévy subordinator can be approximated arbitrarily close by a compound Poisson subordinator. Hence, in theory, the following algorithm provides an approximate sampling scheme for arbitrary Lévy-frailty copulas.

[13]To find bounds for the expected time T until which the path of the Lévy subordinator must be sampled, note that $\mathbb{E}[E_{(d)}] \in \mathcal{O}(\log(d))$ (see Mai and Scherer (2009b, Lemma 3.1)). For most model specifications, $\mathbb{E}[\Lambda_t]$, $t \geq 0$, is known explicitly. Bounds, however, can additionally be found using $\Psi(1) = 1$ and Jensen's inequality, since $\exp(-\mathbb{E}[\Lambda_t]) \leq \mathbb{E}[\exp(-\Lambda_t)] = \exp(-t\,\Psi(1)) = \exp(-t)$, implying $\mathbb{E}[\Lambda_t] \geq t$.

Algorithm 3.7 (Sampling CPP Lévy-Frailty Copulas)
Input is the dimension $d \geq 2$, as well as the parameters of a compound Poisson subordinator, i.e. the drift constant $\mu \in [0,1]$, the intensity $\beta > 0$, and the parameter (vector) $\boldsymbol{\theta}$ of the jump size distribution. Note that $\mu \leq 1$ follows from $\Psi(1) = 1$.

```
FUNCTION sample_LFC (vector: (d, μ, β, θ))
    X := vector(1 : d)                                              (1a)
    U := vector(1 : d)                                              (1b)
    E := vector(1 : d)                                              (1c)
    FOR k = 1, ..., d
        E[k] := sample_EXP(1)                                       (1c)
    END FOR
    Max := max{E}                                                   (1d)
    Λ := list((0, 0))                                               (1e)
    maxvalue := 0                                                   (1f)
    WHILE (maxvalue < Max)                                          (2)
        nextjumptime := sample_EXP(β)                               (2a)
        nextjumpsize := sample_JUMP(θ)                              (2b)
        maxvalue := maxvalue + μ · nextjumptime + nextjumpsize      (2c)
        Λ := append(Λ, (nextjumptime, nextjumpsize))                (2d)
    END WHILE
    FOR k = 1, ..., d                                               (3)
        indi := 1                                                   (3a)
        maxvalue := 0                                               (3b)
        time := 0                                                   (3c)
        WHILE (maxvalue < E[k])                                     (4)
            indi := indi + 1                                        (4a)
            maxvalue := maxvalue + μ · Λ[indi][1] + Λ[indi][2]      (4b)
            time := time + Λ[indi][1]                               (4c)
        END WHILE
        IF (maxvalue − Λ[indi][2] > E[k])                           (5a)
            X[k] := time − (maxvalue − Λ[indi][2] − E[k])/μ
        ELSE                                                        (5b)
            X[k] := time
        END IF
        U[k] := exp(−X[k])                                          (6)
    END FOR
    RETURN U
```

The individual steps of Algorithm 3.7 are explained as follows:

(1) This is an initializing step. (1a) The d-dimensional vector $\boldsymbol{X}$ denotes the respective sample of the eMO-distribution in question, whereas (1b) the d-dimensional vector $\boldsymbol{U}$ denotes the sample from the corresponding Lévy-frailty copula. (1c) The d-dimensional vector $\boldsymbol{E}$ contains the exponential threshold levels $E_1, \ldots, E_d$, which are sampled and stored in this step. (1d) The largest threshold level is stored in the variable Max for later use. (1e) Λ is a list of two-dimensional vectors. It is indexed by two indices, i.e. $\Lambda[i]$ denotes the ith member of the list, which is a two-dimensional vector. Each list member corresponds to a jump of the compound Poisson subordinator. Hence $\Lambda[i][1]$, the first component of the ith list member, denotes the ith jump, and $\Lambda[i][2]$, the second component of the ith list member, denotes the jump size of the ith jump. The first list member is initialized to be the two-dimensional zero vector. (1f) The variable *maxvalue* is an auxiliary variable, which is used in the following loop.

(2) This *WHILE* loop computes and stores all required jump times and jump sizes of the compound Poisson subordinator in the list object Λ. These numbers completely determine the path of the subordinator until all threshold levels are crossed. The variable *maxvalue* denotes the current value of the subordinator. While it has not yet crossed the maximum Max of all threshold levels, (2a) the waiting time until the next jump is simulated,[14] (2b) the next jump size is simulated, and (2c) the value of the subordinator after the jump is adjusted accordingly. (2d) Finally, the jump times and sizes are stored in Λ. The command $append(\Lambda, (a, b))$ appends the vector (a, b) to the list object Λ and returns the extended list.

(3) This *FOR* loop computes all components of the random vector in question. (3a) The variable *indi* runs through all jump times of the subordinator, starting with the first. (3b) The variable *maxvalue* again denotes the current value of the subordinator. (3c) The variable *time* denotes the current jump time.

(4) While the current value of the subordinator is still less than the respective threshold level, (4a) the next jump time is considered by increasing *indi*, (4b) the current value is adjusted by taking into account the next jump, and (4c) the time until the first passage is accumulated.

(5) Two possibilities are distinguished: either (5a) the threshold level E_k

[14]It is exponentially distributed (see Equation (A.1)).

is hit between two jump times (because of the continuous linear growth with drift rate μ), or (5b) it is jumped over. In the first case, the exact first passage time has to be determined by an appropriate adjustment. In the second case, the jump time equals the first passage time.

(6) The random vector $\boldsymbol{X}$ has an eMO-distribution with $Exp(1)$-margins. It is transformed by $x \mapsto \exp(-x)$ to obtain the vector $\boldsymbol{U}$, which has the desired Lévy-frailty copula as joint distribution function.

3.3.4 *Hierarchical (H-Extendible) Lévy-Frailty Copulas*

Our next goal is to construct h-extendible copulas with respect to the family $\{C_\Psi\}$ of Lévy-frailty copulas. For an h-extendible structure with two levels, corresponding to a partition $d = d_1 + \ldots + d_J$, consider as ingredients $J+1$ independent Lévy subordinators $\Lambda^{(0)}, \Lambda^{(1)}, \ldots, \Lambda^{(J)}$ with associated Laplace exponents $\Psi_0, \Psi_1, \ldots, \Psi_J$, each having fixpoint 1. Moreover, let $\alpha \in (0,1)$ denote an additional free parameter. With independent i.i.d. unit exponential trigger variables $E_{1,1}, \ldots, E_{1,d_1}, \ldots, E_{J,1}, \ldots, E_{J,d_J}$, it is shown in Mai and Scherer (2011a) that the random vector $(U_{1,1}, \ldots, U_{1,d_1}, \ldots, U_{J,1}, \ldots, U_{J,d_J})$ has an h-extendible copula with respect to $\{C_\Psi\}$, where

$$U_{j,k} := \exp\Big(- \inf\big\{t > 0 \,:\, \Lambda^{(0)}_{\alpha\, t} + \Lambda^{(j)}_{(1-\alpha)\, t} > E_{j,k}\big\}\Big), \tag{3.19}$$

for $k = 1, \ldots, d_j$, $j = 1, \ldots, J$. The precise form of the copula is given in (3.20). This construction has an obvious factor structure and is easily verified to be h-extendible with two levels of hierarchy. However, sampling from the stochastic representation (3.19) is a challenging task (see the algorithm described in Mai and Scherer (2011a)). In contrast to the latter reference, we might as well construct the same distribution making use of the min-stability property in Lemma 3.9. Let $(U^{(0)}_{1,1}, \ldots, U^{(0)}_{1,d_1}, \ldots, U^{(0)}_{J,1}, \ldots, U^{(0)}_{J,d_J}) \sim C_{\Psi_0}$. Given a partition $d_1 + \ldots + d_J = d$, consider mutually independent random vectors

$$(U^{(1)}_1, \ldots, U^{(1)}_{d_1}) \sim C_{\Psi_1}, \ldots, (U^{(J)}_1, \ldots, U^{(J)}_{d_J}) \sim C_{\Psi_J},$$

all independent of $(U^{(0)}_{1,1}, \ldots, U^{(0)}_{1,d_1}, \ldots, U^{(0)}_{J,1}, \ldots, U^{(0)}_{J,d_J})$. Each of these $J+1$ independent Lévy-frailty vectors might be constructed like in Theorem 3.5. After the margin transformation $\tau^{(j)}_k := -\alpha \log(U^{(j)}_k)/(1-\alpha)$, $j = 1, \ldots, J$, $k = 1, \ldots, d_j$, with some $\alpha \in (0,1)$, the random vectors $(\tau^{(j)}_1, \ldots, \tau^{(j)}_{d_j})$ follow extendible Marshall–Olkin distributions, $j = 1, \ldots, J$.

The same holds for $\tau_{j,k}^{(0)} := -\log(U_{j,k}^{(0)})$, $k = 1, \ldots, d_j$, $j = 1, \ldots, J$. Using the min-stability property of Lemma 3.9, it follows that the random vector

$$(\tau_{1,1}, \ldots, \tau_{1,d_1}, \ldots, \tau_{J,1}, \ldots, \tau_{J,d_J}) := \\ \Big(\min\big\{\tau_{1,1}^{(0)}, \tau_1^{(1)}\big\}, \ldots, \min\big\{\tau_{J,d_J}^{(0)}, \tau_{d_J}^{(J)}\big\}\Big)$$

has a Marshall–Olkin distribution. Since each component is exponentially distributed with parameter $1/\alpha$, the distribution function of the random vector

$$(U_{1,1}, \ldots, U_{1,d_1}, \ldots, U_{J,1}, \ldots, U_{J,d_J}) := \\ \Big(\exp\big(-\frac{\tau_{1,1}}{\alpha}\big), \ldots, \exp\big(-\frac{\tau_{1,d_1}}{\alpha}\big), \ldots, \exp\big(-\frac{\tau_{J,1}}{\alpha}\big), \ldots, \exp\big(-\frac{\tau_{J,d_J}}{\alpha}\big)\Big)$$

is a copula. The following lemma shows that it is of the h-extendible Marshall–Olkin kind. To this end, let us introduce the notation $\Psi_\Pi(x) := x$, $x \geq 0$, i.e. the Laplace exponent inducing the independence copula $C_{\Psi_\Pi} = \Pi$.

Lemma 3.10 (H-Extendible Lévy-Frailty Copula)

The copula of the random vector $(U_{1,1}, \ldots, U_{1,d_1}, \ldots, U_{J,1}, \ldots, U_{J,d_J})$ *is h-extendible with respect to the family* $\{C_\Psi\}$ *of Lévy-frailty copulas. If* $1 \leq j_1 < \ldots < j_k \leq J$ *are selected from* k *distinct groups, it holds that* $(U_{j_1,1}, \ldots, U_{j_k,1}) \sim C_{\alpha\,\Psi_0 + (1-\alpha)\,\Psi_\Pi}$. *Moreover, the random subvector of the* j*th group satisfies*

$$(U_{j,1}, \ldots, U_{j,d_j}) \sim C_{\alpha\,\Psi_0 + (1-\alpha)\,\Psi_j}, \quad j = 1, \ldots, J.$$

Proof. We compute the survival function of the vector $(\tau_{1,1}, \ldots, \tau_{1,d_1}, \ldots, \tau_{J,1}, \ldots, \tau_{J,d_J})$ from above:

$$\begin{aligned} &\mathbb{P}(\tau_{1,1} > t_{1,1}, \ldots, \tau_{1,d_1} > t_{1,d_1}, \ldots, \tau_{J,1} > t_{J,1}, \ldots, \tau_{J,d_J} > t_{J,d_J}) \\ &\quad = \mathbb{P}\big(\tau_{j,k}^{(0)} > t_{j,k},\, k = 1, \ldots, d_j,\, j = 1, \ldots, J\big) \times \\ &\quad\quad \times \prod_{j=1}^{J} \mathbb{P}\big(\tau_k^{(j)} > t_{j,k},\, k = 1, \ldots, d_j\big) \\ &\quad = C_{\Psi_0}\Big(e^{-t_{1,1}}, \ldots, e^{-t_{J,d_J}}\Big) \prod_{j=1}^{J} C_{\Psi_j}\Big(e^{-\frac{1-\alpha}{\alpha}\,t_{j,1}}, \ldots, e^{-\frac{1-\alpha}{\alpha}\,t_{j,d_j}}\Big). \end{aligned}$$

Since $\tau_{j,k} \sim Exp(1/\alpha)$ for all j, k, the survival copula of $(\tau_{1,1}, \ldots, \tau_{J,d_J})$, which equals the distribution function of $(U_{1,1}, \ldots, U_{J,d_J})$, is given by

$$\begin{aligned}
&\mathbb{P}\big(\tau_{1,1} > -\alpha \log(u_{1,1}), \ldots, \tau_{J,d_J} > -\alpha \log(u_{J,d_J})\big) \\
&= C_{\Psi_0}\Big(u_{1,1}^{\alpha}, \ldots, u_{J,d_J}^{\alpha}\Big) \prod_{j=1}^{J} C_{\Psi_j}\Big(u_{j,1}^{1-\alpha}, \ldots, u_{j,d_j}^{1-\alpha}\Big) \\
&= \Big(C_{\Psi_0}(u_{1,1}, \ldots, u_{J,d_J})\Big)^{\alpha} \Big(\prod_{j=1}^{J} C_{\Psi_j}(u_{j,1}, \ldots, u_{j,d_j})\Big)^{1-\alpha}. \qquad (3.20)
\end{aligned}$$

The last equality holds, since Lévy-frailty copulas are extreme-value copulas (see Section 1.2.5), and therefore satisfy $C_{\Psi}(u_1^t, \ldots, u_d^t) = C_{\Psi}(u_1, \ldots, u_d)^t$, $t \geq 0$. Comparing the copula (3.20) with the one derived in Mai and Scherer (2011a, Theorem 3.4), the constructed model agrees in distribution with model (3.19). From this, the h-extendible structure is obvious. □

In contrast to construction (3.19), this alternative construction of h-extendible Marshall–Olkin copulas applying Lemma 3.9 implies a very convenient simulation algorithm, since it only requires us to simulate independent Lévy-frailty copulas. Simulation of the latter can be achieved, for example, by Algorithm 3.7. Notice, furthermore, that the convex combination of two Laplace exponents with fixpoint 1 is again of such kind. The parameter α interpolates between independent and fully dependent groups: the limiting case $\alpha = 0$ implies that the J groups are independent, the opposite case $\alpha = 1$ implies that the group-specific dependencies vanish, i.e. the overall copula is the extendible copula C_{Ψ_0}. Of course, the simulation of hierarchical Lévy-frailty copulas as in Lemma 3.10 is straightforward, as the following algorithm shows. Like in Algorithm 3.7 we assume that all involved Lévy subordinators are of the compound Poisson type. In this case, the sampling algorithm is unbiased.

Algorithm 3.8 (Sampling H-Extendible Lévy-Frailty Copulas)
The inputs for the algorithm are the parameter $\alpha \in (0,1)$, the number of groups J, the group sizes $d_1, \ldots, d_J$, as well as the parameters of all $J+1$ involved compound Poisson subordinators (corresponding to the Laplace exponents $\Psi_0, \Psi_1, \ldots, \Psi_J$). These comprise the intensities $\beta_0, \beta_1, \ldots, \beta_J$, the drifts $\mu_0, \mu_1, \ldots, \mu_J$, as well as parameters for the respective jump size distributions $\boldsymbol{\theta}_0, \boldsymbol{\theta}_1, \ldots, \boldsymbol{\theta}_J$.

(1) For each $j = 1, \ldots, J$, simulate (mutually independently) a d_j-dimensional random vector $(U_1^{(j)}, \ldots, U_{d_j}^{(j)})$ from the copula C_{Ψ_j} using Algorithm 3.7 with parameters $(\mu_j, \beta_j, \boldsymbol{\theta}_j)$.

(2) Independently of step (1), simulate a $(d_1+\ldots+d_J)$*-dimensional random vector* $(U_{1,1}^{(0)},\ldots,U_{1,d_1}^{(0)},\ldots,U_{J,1}^{(0)},\ldots,U_{J,d_J}^{(0)})$ *from the copula* C_{Ψ_0} *using Algorithm 3.7 with parameters* $(\mu_0,\beta_0,\boldsymbol{\theta}_0)$.

(3) Return the random vector

$$\Big(\max\big\{(U_{1,1}^{(0)})^{\frac{1}{\alpha}},(U_1^{(1)})^{\frac{1}{1-\alpha}}\big\},\ldots,\max\big\{(U_{1,d_1}^{(0)})^{\frac{1}{\alpha}},(U_{d_1}^{(1)})^{\frac{1}{1-\alpha}}\big\},\ldots$$
$$\ldots,\max\big\{(U_{J,1}^{(0)})^{\frac{1}{\alpha}},(U_1^{(J)})^{\frac{1}{1-\alpha}}\big\},\ldots,\max\big\{(U_{J,d_J}^{(0)})^{\frac{1}{\alpha}},(U_{d_J}^{(J)})^{\frac{1}{1-\alpha}}\big\}\Big). \tag{3.21}$$

Figure 3.6 illustrates a hierarchical Lévy-frailty copula by means of pairwise scatterplots in a five-dimensional example with two groups and $\alpha = 1/2$; Algorithm 3.8 was used to sample the respective copula. All three involved Lévy subordinators are of the compound Poisson type with exponential jump sizes. More precisely, their Laplace exponent is of the parametric form

$$\Psi(x) := \mu\,x + \beta\,\frac{x}{x+\eta}, \quad x \geq 0, \tag{3.22}$$
$$(\mu,\beta,\eta) \in \big\{(\mu,\beta,\eta) \in [0,1]\times(0,\infty)^2 \,:\, \beta \leq 1+\eta,\, \mu = 1-\beta/(1+\eta)\big\}.$$

This corresponds to a drift $\mu \in [0,1]$, an intensity β, and exponential jump sizes with mean $1/\eta$. The restrictions on the three parameters stem from the condition $\Psi(1) = 1$. For this specification, recall that the bivariate Lévy-frailty copula is given by

$$\begin{aligned} C_\Psi(u_1,u_2) &= \min\{u_1,u_2\}\,\max\{u_1,u_2\}^{\Psi(2)-1} \\ &= \min\{u_1,u_2\}\,\max\{u_1,u_2\}^{1-\frac{2\beta}{(1+\eta)\,(2+\eta)}}, \end{aligned}$$

where the last equality uses the parameter restriction $\mu = 1-\beta/(1+\eta)$. This equals the bivariate Cuadras–Augé copula C_α from Example 1.4 with dependence parameter $\alpha := 2\,\beta/((1+\eta)\,(2+\eta)) \in [0,1]$. Explaining the matrix scatterplot in Figure 3.6 further, the two numbers in each panel of the diagonal denote the respective indices j,i of the component $\max\big\{U_{j,i}^{(0)},U_i^{(j)}\big\}^2$ of the random vector (3.21) in question. Above the diagonal, the panel in row l and column k illustrates the scatterplot corresponding to the bivariate subvector with the indices given in the diagonal in row l and column k, respectively. The pairwise Cuadras–Augé coefficients corresponding to the respective pair are given below the diagonal, depending on whether the two corresponding random variables are in the same or in different groups. The numbers in row l and column k correspond to the samples in row k and column l. The upper value is the theoretical value, whereas the lower one gives the empirical value for the specific panel based on the 1 500 samples. For this estimation the maximum likelihood estimator derived in Ruiz-Rivas and Cuadras (1988) is used. The figure illustrates the different levels of dependence between and within groups.

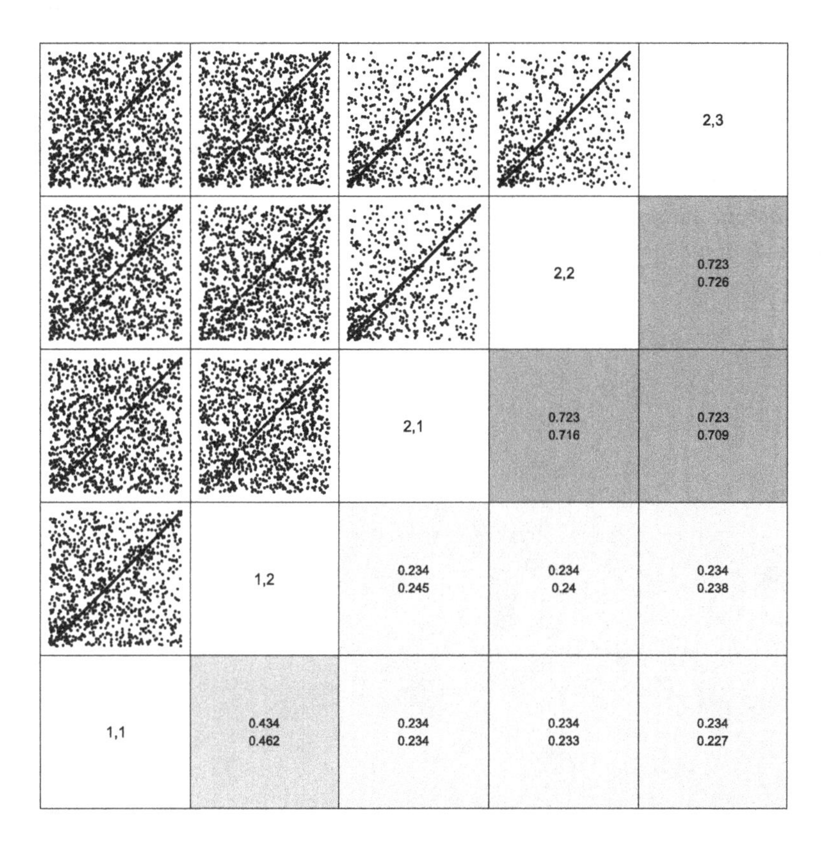

Fig. 3.6 This figure illustrates 1 500 samples (by means of pairwise scatterplots) of a five-dimensional random vector as considered in Lemma 3.10. The components are partitioned into $J = 2$ groups with dimensions $d_1 = 2, d_2 = 3$. The three involved compound Poisson subordinators are from family (3.22) with corresponding parameters $(\mu_0, \beta_0, \eta_0) = (0.2995, 1.401, 1)$, $(\mu_1, \beta_1, \eta_1) = (0.2, 2.4, 2)$, and $(\mu_2, \beta_2, \eta_2) = (0.0151, 0.994749, 0.01)$.

Remark 3.1 (Deeper Hierarchical Structures)

Of course, by iteration of Lemma 3.9 one can construct higher-order hierarchical Lévy-frailty copulas. Instead of only partitioning the dimension $d = d_1 + \ldots + d_J$, *one can do the same within the subgroups. For example,* d_1 *might be partitioned into* $d_1 = d_{1,1} + \ldots + d_{1,J_1}$, *and so on. In summary, one has to simulate one Lévy-frailty copula for each introduced subgroup. Therefore, the finer the partition, the more expensive the simulation becomes. This approach, in some sense, allows us to "interpolate" between the exchangeable (and extendible) subfamily and the most general Marshall–Olkin copula. By choosing an appropriate "interpolation", one can obtain quite flexible structures, which still can be sampled efficiently in large dimensions.*

Chapter 4

Elliptical Copulas

The class of elliptical copulas differs from most aforementioned classes of copulas in the way that only an implicit analytical expression is available. More precisely, an elliptical copula is defined via Sklar's theorem as the dependence structure of a related elliptical distribution and is obtained from the respective multivariate distribution function by standardizing the univariate marginal laws. Sampling such copulas therefore requires a sampling scheme for the respective multivariate distribution and an analytical expression of the univariate marginal distribution functions. To make this more precise, let[1] $\boldsymbol{X} = (X_1, \ldots, X_d)'$ be a sample from some elliptical distribution and let $F_1, \ldots, F_d$ be the univariate marginal laws. Then, $\big(F_1(X_1), \ldots, F_d(X_d)\big)'$ is a sample from the respective copula. The analytical form of the copula is obtained by plugging the univariate marginal quantile functions $F_1^{-1}, \ldots, F_d^{-1}$ into the multivariate distribution function F, i.e. $C(u_1, \ldots, u_d) := F\big(F_1^{-1}(u_1), \ldots, F_d^{-1}(u_d)\big)$. It is only rarely possible to simplify this expression, since the univariate quantiles are often not available in closed form and the multivariate distribution function is typically complicated. The most prominent example of the class of elliptical copulas is the Gaussian copula (see Example 1.7 of Chapter 1 and Section 4.5). Another important example is the t-copula (see Section 4.5).

Sketching a roadmap for this chapter (see Figure 4.1), we first introduce *spherical distributions.* A random vector $\boldsymbol{Y}$ with a spherical distribution has the stochastic representation

$$\boldsymbol{Y} \stackrel{d}{=} R\,\boldsymbol{S},$$

where R is a non-negative random variable (interpreted as radius) and

[1] The derivations in this chapter often involve multiplications of vectors with matrices. Unlike all other chapters in this book, in this chapter we interpret random vectors as column vectors to simplify notation.

$\boldsymbol{S}$ is a random vector (interpreted as direction) that is independent of R and uniformly distributed on the unit sphere $S_{L^2,d} := \{\boldsymbol{x} \in \mathbb{R}^d : \|\boldsymbol{x}\|^2 = \sum_{i=1}^d x_i^2 = 1\}$. In a second step, *elliptical distributions* are defined as affine transformations of spherical distributions. Hence, a random vector $\boldsymbol{X}$ of this class admits the representation

$$\boldsymbol{X} \stackrel{d}{=} \boldsymbol{\mu} + A' \boldsymbol{Y} \stackrel{d}{=} \boldsymbol{\mu} + A' R \boldsymbol{S},$$

where $\boldsymbol{\mu} \in \mathbb{R}^d$ is a linear shift and $A \in \mathbb{R}^{k \times d}$ is a linear transformation of the k-dimensional spherically distributed random vector $\boldsymbol{Y}$. Finally, elliptical copulas are obtained by standardizing the univariate marginals of such distributions. In this chapter, we use the notation and definitions of Fang et al. (1990) and McNeil et al. (2005).

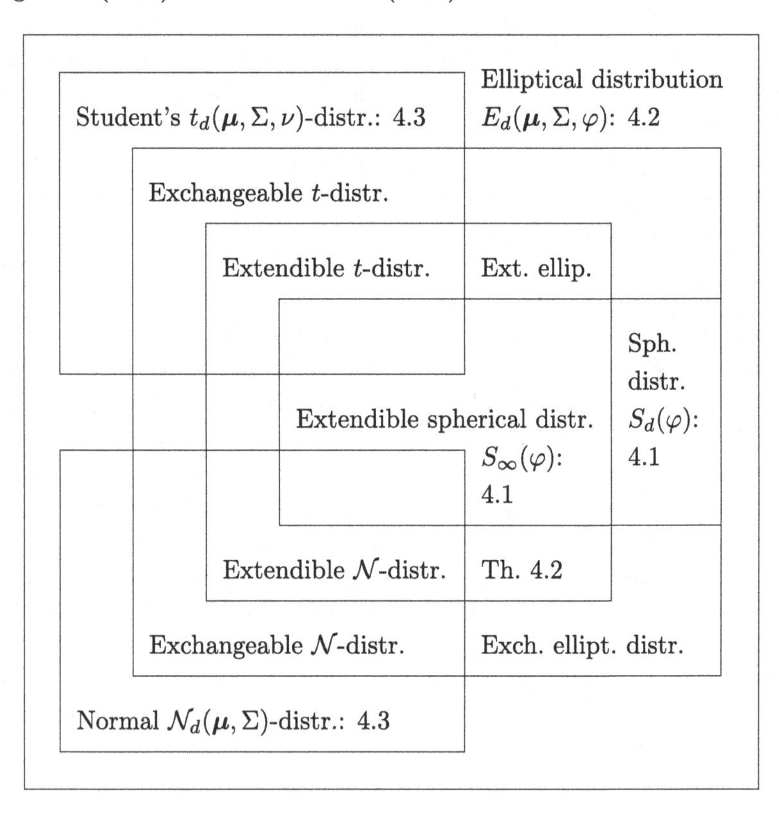

Fig. 4.1 Classification of spherical $S_d(\varphi)$ and elliptical $E_d(\boldsymbol{\mu}, \Sigma, \varphi)$ distributions and sections where these are discussed in this chapter. Associated copulas are discussed in Section 4.4.

4.1 Spherical Distributions

Definition 4.1 (Spherical Distribution)
A d-dimensional random vector $\boldsymbol{X} = (X_1, \ldots, X_d)'$ *has a* spherical distribution *if for every orthogonal*[2] *matrix* $O \in \mathbb{R}^{d\times d}$ *one has*

$$O\boldsymbol{X} \stackrel{d}{=} \boldsymbol{X}. \tag{4.1}$$

Property (4.1) means that spherical distributions are invariant under rotations and reflections. It is easy to verify that this property immediately implies exchangeability. To see this, take any permutation σ on $\{1, \ldots, d\}$ and define $O := \big(\boldsymbol{e}_{\sigma^{-1}(1)}, \ldots, \boldsymbol{e}_{\sigma^{-1}(d)}\big)$, where $\boldsymbol{e}_i$ is the ith unit vector. Then, O is orthogonal and by Equation (4.1),

$$O\boldsymbol{X} = \big(X_{\sigma(1)}, \ldots, X_{\sigma(d)}\big)' \stackrel{d}{=} \boldsymbol{X}.$$

Moreover, taking $O := -I$, where I is the identity matrix, shows that spherical distributions are radially symmetric with respect to the origin, in the sense that $\boldsymbol{X} \stackrel{d}{=} -I\boldsymbol{X} = -\boldsymbol{X}$.

The defining property (4.1) can also be used to show that characteristic functions of spherical distributions can be parameterized by means of a certain one-dimensional function φ (see Lemma 4.1(3)).

Lemma 4.1 (Spherical Distributions: Equivalent Characterization)
The following statements equivalently characterize spherical distributions:

(1) The random vector $\boldsymbol{X} = (X_1, \ldots, X_d)'$ *has a spherical distribution, i.e. (4.1) holds.*

(2) There exists a random variable $R \geq 0$ *and, independently, a random vector* $\boldsymbol{S}$ *with uniform distribution on the unit sphere* $S_{L^2,d}$ *such that*

$$\boldsymbol{X} \stackrel{d}{=} R\,\boldsymbol{S}. \tag{4.2}$$

(3) There exists some function φ *in one variable such that the characteristic function of* $\boldsymbol{X}$*, denoted* $\phi_{\boldsymbol{X}}$*, admits the representation*

$$\phi_{\boldsymbol{X}}(\boldsymbol{t}) = \mathbb{E}\big[e^{i\,\boldsymbol{t}'\boldsymbol{X}}\big] = \varphi\big(\|\boldsymbol{t}\|^2\big), \quad \boldsymbol{t} \in \mathbb{R}^d, \tag{4.3}$$

where $\|\boldsymbol{t}\| := \big(t_1^2 + \ldots + t_d^2\big)^{1/2}$ *is the usual Euclidean norm. With regard to Equation (4.3), we sometimes write* $\boldsymbol{X} \sim S_d(\varphi)$.

[2] A matrix $O \in \mathbb{R}^{d\times d}$ is called orthogonal if its columns and rows are orthonormal vectors. This is equivalent to $O'O = OO' = I$, where I is the identity matrix. Interpreted as a linear transformation, O can be seen as a rotation or reflection.

(4) For all $\boldsymbol{a} \in \mathbb{R}^d$ *one has*

$$\boldsymbol{a}'\boldsymbol{X} \stackrel{d}{=} \|\boldsymbol{a}\|\, X_1. \tag{4.4}$$

Proof. (1) $\Rightarrow$ (3): Suppose for $\boldsymbol{t}_1, \boldsymbol{t}_2 \in \mathbb{R}^d \setminus \{\boldsymbol{0}\}$ that $\|\boldsymbol{t}_1\| = \|\boldsymbol{t}_2\|$. Using the Gram–Schmidt orthogonalization procedure we find an orthonormal basis $\{\boldsymbol{t}_1/\|\boldsymbol{t}_1\|, \boldsymbol{O}_2, \ldots, \boldsymbol{O}_d\}$ of $\mathbb{R}^d$ and we define the orthogonal matrix $(\boldsymbol{t}_1/\|\boldsymbol{t}_1\|, \boldsymbol{O}_2, \ldots, \boldsymbol{O}_d)' := O_1$. Observe that this matrix is constructed to satisfy $O_1\,\boldsymbol{t}_1 = \|\boldsymbol{t}_1\|\,\boldsymbol{e}_1$, where $\boldsymbol{e}_1 = (1, 0, \ldots, 0)'$. Similarly, we construct a second orthogonal matrix O_2 with $O_2\,\boldsymbol{t}_2 = \|\boldsymbol{t}_2\|\,\boldsymbol{e}_1$. Consequently, $O_1\,\boldsymbol{t}_1 = O_2\,\boldsymbol{t}_2$, $O_1'\,O_2\,\boldsymbol{t}_2 = \boldsymbol{t}_1$, and the matrix $O_1'\,O_2$ is orthogonal. Then

$$\phi_{\boldsymbol{X}}(\boldsymbol{t}_1) = \phi_{\boldsymbol{X}}(O_1'\,O_2\,\boldsymbol{t}_2) = \mathbb{E}\big[e^{i\,\boldsymbol{t}_2'\,O_2'\,O_1\,\boldsymbol{X}}\big] \stackrel{(4.1)}{=} \mathbb{E}\big[e^{i\,\boldsymbol{t}_2'\,\boldsymbol{X}}\big] = \phi_{\boldsymbol{X}}(\boldsymbol{t}_2).$$

We conclude that $\boldsymbol{t} \mapsto \phi_{\boldsymbol{X}}(\boldsymbol{t})$ only depends on $\boldsymbol{t}$ via $\|\boldsymbol{t}\|$, which proves the claim.
(3) $\Rightarrow$ (4): Observe that

$$\begin{aligned}
\phi_{\boldsymbol{a}'\boldsymbol{X}}(t) &= \mathbb{E}\big[e^{i\,t\,\boldsymbol{a}'\boldsymbol{X}}\big] = \phi_{\boldsymbol{X}}(t\,\boldsymbol{a}) \stackrel{(3)}{=} \varphi(t^2\,\|\boldsymbol{a}\|^2) = \varphi(t^2\,\|\boldsymbol{a}\|^2\,\|\boldsymbol{e}_1\|^2) \\
&= \phi_{\boldsymbol{X}}(t\,\|\boldsymbol{a}\|\,\boldsymbol{e}_1) = \mathbb{E}\big[e^{i\,t\,\|\boldsymbol{a}\|\,\boldsymbol{e}_1'\,\boldsymbol{X}}\big] = \mathbb{E}\big[e^{i\,t\,\|\boldsymbol{a}\|\,X_1}\big] = \phi_{\|\boldsymbol{a}\|\,X_1}(t).
\end{aligned}$$

(4) $\Rightarrow$ (1): Let $O \in \mathbb{R}^{d\times d}$ be orthogonal. Then

$$\begin{aligned}
\phi_{O\,\boldsymbol{X}}(\boldsymbol{t}) &= \mathbb{E}\big[e^{i\,\boldsymbol{t}'\,O\,\boldsymbol{X}}\big] = \mathbb{E}\big[e^{i\,(O'\,\boldsymbol{t})'\,\boldsymbol{X}}\big] \stackrel{(4)}{=} \mathbb{E}\big[e^{i\,\|O'\,\boldsymbol{t}\|\,X_1}\big] \\
&= \mathbb{E}\big[e^{i\,\|\boldsymbol{t}\|\,X_1}\big] \stackrel{(4)}{=} \mathbb{E}\big[e^{i\,\boldsymbol{t}'\,\boldsymbol{X}}\big] = \phi_{\boldsymbol{X}}(\boldsymbol{t}).
\end{aligned}$$

Summing up, we have established (1) $\Leftrightarrow$ (3) $\Leftrightarrow$ (4) so far. It is left to show that (2) $\Leftrightarrow$ (1), (3), (4).

(2) $\Rightarrow$ (1), (3), (4): Clearly, $\boldsymbol{S}$ is spherically distributed. Using (3) $\Leftrightarrow$ (1) there exists some $\varphi_{\boldsymbol{S}}$ such that $\mathbb{E}\big[\exp(i\,\boldsymbol{t}'\boldsymbol{S})\big] = \varphi_{\boldsymbol{S}}\big(\|\boldsymbol{t}\|^2\big)$. More clearly,

$$\varphi_{\boldsymbol{S}}(x) = \frac{1}{\nu(S_{L^2,d})} \int_{S_{L^2,d}} e^{i\,\boldsymbol{x}'\,\boldsymbol{y}}\,\nu(d\boldsymbol{y}),$$

where $\boldsymbol{x} \in \mathbb{R}^d$ is an arbitrary vector with $\|\boldsymbol{x}\|^2 = x$ and $\nu(d\boldsymbol{y})$ denotes the surface measure on $S_{L^2,d}$. This shows that

$$\begin{aligned}
\phi_{R\,\boldsymbol{S}}(\boldsymbol{t}) &= \mathbb{E}\big[e^{i\,\boldsymbol{t}'\,R\,\boldsymbol{S}}\big] = \mathbb{E}\Big[\mathbb{E}\big[e^{i\,R\,\boldsymbol{t}'\,\boldsymbol{S}}\,|\,R\big]\Big] \\
&= \mathbb{E}\big[\varphi_{\boldsymbol{S}}(R^2\,\|\boldsymbol{t}\|^2)\big] = \int_0^\infty \varphi_{\boldsymbol{S}}(r^2\,\|\boldsymbol{t}\|^2)\,d\mathbb{P}(R \leq r)
\end{aligned}$$

is a function of $\|\boldsymbol{t}\|^2$. Hence, $R\,\boldsymbol{S}$ is spherical by (3) $\Leftrightarrow$ (1).

(1), (3), (4) $\Rightarrow$ (2): Let $\boldsymbol{X}$ be spherical with φ from part (3). For arbitrary $\boldsymbol{x} \in \mathbb{R}^d$ with unit norm $\|\boldsymbol{x}\|^2 = \boldsymbol{x}'\boldsymbol{x} = 1$ we see that

$$\phi_{\boldsymbol{X}}(\boldsymbol{t}) = \varphi(\|\boldsymbol{t}\|^2) = \varphi\big((\|\boldsymbol{t}\|\,\boldsymbol{x})'\,(\|\boldsymbol{t}\|\,\boldsymbol{x})\big) = \phi_{\boldsymbol{X}}(\|\boldsymbol{t}\|\,\boldsymbol{x}). \tag{4.5}$$

Then

$$\begin{aligned}
\phi_{\boldsymbol{X}}(\boldsymbol{t}) &= \phi_{\boldsymbol{X}}(\boldsymbol{t}) \underbrace{\frac{1}{\nu(S_{L^2,d})} \int_{S_{L^2,d}} 1\,\nu(d\boldsymbol{x})}_{=1} \\
&\overset{(4.5)}{=} \frac{1}{\nu(S_{L^2,d})} \int_{S_{L^2,d}} \phi_{\boldsymbol{X}}(\|\boldsymbol{t}\|\,\boldsymbol{x})\nu(d\boldsymbol{x}) \\
&= \frac{1}{\nu(S_{L^2,d})} \int_{S_{L^2,d}} \mathbb{E}\big[e^{i\,\|\boldsymbol{t}\|\,\boldsymbol{x}'\,\boldsymbol{X}}\big]\nu(d\boldsymbol{x}) \\
&= \mathbb{E}\Big[\frac{1}{\nu(S_{L^2,d})} \int_{S_{L^2,d}} e^{i\,\|\boldsymbol{t}\|\,\boldsymbol{X}'\,\boldsymbol{x}}\nu(d\boldsymbol{x})\Big] \\
&= \mathbb{E}\big[\varphi_{\boldsymbol{S}}(\|\boldsymbol{t}\|^2\,\|\boldsymbol{X}\|^2)\big] \\
&= \int_0^\infty \varphi_{\boldsymbol{S}}(r^2\,\|\boldsymbol{t}\|^2)\,d\mathbb{P}(\|\boldsymbol{X}\| \le r).
\end{aligned}$$

Now define a probability space supporting $R \overset{d}{=} \|\boldsymbol{X}\|$ and, independent of R, $\boldsymbol{S} \sim \text{Uniform}(S_{L^2,d})$. The claim is established by observing that $\phi_{\boldsymbol{X}}(\boldsymbol{t}) = \phi_{R\,\boldsymbol{S}}(\boldsymbol{t})$ for all $\boldsymbol{t} \in \mathbb{R}^d$. □

Lemma 4.1 establishes that spherical distributions can be characterized by means of φ; this function is consequently called the *characteristic generator* and the notation $\boldsymbol{X} \sim S_d(\varphi)$ is used to describe a d-dimensional spherical random vector with characteristic function (4.3). When used for sampling, the stochastic representation (4.2) allows for the following interpretation: First, a direction $\boldsymbol{S} \in S_{L^2,d}$ is drawn. Being uniformly distributed on the unit d-sphere, all directions are *equally likely* (compare Figure 4.2). In a second step, independent of the direction, a *radius* (distance to the origin) R is drawn and multiplied by the direction $\boldsymbol{S}$, yielding the required sample in $\mathbb{R}^d$.

The following corollary to Lemma 4.1 lists some statements on the characteristic function (c.f.) of spherical distributions, which are almost immediately obtained from Lemma 4.1.

Corollary 4.1 (The Characteristic Function Revisited)
Define the set of all characteristic generators of spherical distributions by

$$\Phi_d := \{\varphi : \boldsymbol{t} \mapsto \varphi(\|\boldsymbol{t}\|^2) \text{ is the c.f. of a } d\text{-dimensional random vector}\}.$$

Then the following assertions hold:

(1) $\Phi_2 \supset \Phi_3 \supset \Phi_4 \supset \ldots$
(2) $\varphi \in \Phi_d$ *if and only if* $\varphi(x) = \int_{[0,\infty)} \varphi_{\boldsymbol{S}}(x\, r^2)\, dF(r)$, *with* $\varphi_{\boldsymbol{S}}$ *as defined in the proof of Lemma 4.1 and* F *an arbitrary distribution function on* $[0, \infty)$.
(3) Any $\varphi \in \Phi_d$ *is real-valued.*

Proof.

(1) Let $\varphi \in \Phi_{d+1}$. This implies that there exists some spherical random vector $(X_1, \ldots, X_{d+1})'$ such that

$$\varphi(t_1^2 + \ldots + t_{d+1}^2) = \mathbb{E}\big[e^{i\, \sum_{k=1}^{d+1} t_k\, X_k}\big].$$

Clearly,

$$\mathbb{E}\big[e^{i\, \sum_{k=1}^{d} t_k\, X_k}\big] = \varphi(t_1^2 + \ldots + t_d^2)$$

is a characteristic function as well.
(2) Clear by Lemma 4.1(2).
(3) Let $\boldsymbol{X}$ be spherically distributed. Then by Lemma 4.1(4) and Euler's formula $\exp(i\, x) = \cos(x) + i\, \sin(x)$, we obtain

$$\begin{aligned}\phi_{\boldsymbol{X}}(\boldsymbol{t}) &= \mathbb{E}\big[e^{i\, \boldsymbol{t}'\boldsymbol{X}}\big] = \mathbb{E}\big[e^{i\, \|\boldsymbol{t}\|\, X_1}\big] \\ &= \mathbb{E}\big[\cos\big(\|\boldsymbol{t}\|\, X_1\big)\big] + i\, \mathbb{E}\big[\sin\big(\|\boldsymbol{t}\|\, X_1\big)\big].\end{aligned}$$

Since $-X_1 \stackrel{d}{=} X_1$, $\sin(-x) = -\sin(x)$, and $\mathbb{P}(X_1 = 0) = 0$, one has

$$\begin{aligned}&\mathbb{E}\big[\sin\big(\|\boldsymbol{t}\|\, X_1\big)\big] \\ &\quad = \mathbb{E}\big[\sin\big(\|\boldsymbol{t}\|\, X_1\big)\mathbb{1}_{\{X_1>0\}}\big] + \mathbb{E}\big[\sin\big(\|\boldsymbol{t}\|\, X_1\big)\mathbb{1}_{\{X_1<0\}}\big] + 0 \\ &\quad = \mathbb{E}\big[\sin\big(\|\boldsymbol{t}\|\, X_1\big)\mathbb{1}_{\{X_1>0\}}\big] + \mathbb{E}\big[\sin\big(-\|\boldsymbol{t}\|\, X_1\big)\mathbb{1}_{\{-X_1<0\}}\big] \\ &\quad = \mathbb{E}\big[\sin\big(\|\boldsymbol{t}\|\, X_1\big)\mathbb{1}_{\{X_1>0\}}\big] - \mathbb{E}\big[\sin\big(+\|\boldsymbol{t}\|\, X_1\big)\mathbb{1}_{\{X_1>0\}}\big] \\ &\quad = 0.\end{aligned}$$

□

Example 4.1 (Prominent Spherical Distributions)

(1) Let $\boldsymbol{Z} = (Z_1, \ldots, Z_d)'$ be multivariate normally distributed (formally defined in Definition 4.3) with zero mean vector and covariance matrix $\sigma^2 I$ for some $\sigma > 0$, i.e. $Z_1, \ldots, Z_d$ are i.i.d. $\mathcal{N}(0, \sigma^2)$-distributed. It is well known that for each $Z_k \sim \mathcal{N}(0, \sigma^2)$, one has $\mathbb{E}[\exp(i\, t_k\, Z_k)] = \exp(-\sigma^2\, t_k^2/2)$ (see Billingsley (1995, Example 26.1, p. 344)). It follows by the independence of the components that

$$\mathbb{E}[\exp(i\, \boldsymbol{t}'\, \boldsymbol{Z})] = \exp(-\sigma^2\, \|\boldsymbol{t}\|^2/2).$$

Hence, $\boldsymbol{Z}$ has a spherical distribution by Lemma 4.1(3).

(2) Let $\boldsymbol{Z}$ be defined as in the previous example with $\sigma = 1$ and consider an independent random variable $W \geq 0$. Then it follows that

$$\begin{aligned}
\phi_{\sqrt{W}\,\boldsymbol{Z}}(\boldsymbol{t}) &= \mathbb{E}\Big[e^{i\,\boldsymbol{t}'\,\sqrt{W}\,\boldsymbol{Z}}\Big] \\
&= \mathbb{E}\Big[\mathbb{E}\Big[e^{i\,\boldsymbol{t}'\,\sqrt{W}\,\boldsymbol{Z}}\,\Big|\,W\Big]\Big] \\
&= \mathbb{E}\Big[e^{-W\,\|\boldsymbol{t}\|^2/2}\Big] = \int_{[0,\infty)} e^{-w\,\|\boldsymbol{t}\|^2/2}\, d\mathbb{P}(W \leq w) \\
&= \varphi(\|\boldsymbol{t}\|^2), \quad \varphi(x) := \int_{[0,\infty)} e^{-w\,x/2}\, d\mathbb{P}(W \leq w),
\end{aligned}$$

showing with Lemma 4.1(3) that

$$\sqrt{W}\,\boldsymbol{Z}$$

also has a spherical distribution.

Distributions obtained as in Example 4.1(2) are called *mixtures of normal distributions*. The following theorem, which is due to Schoenberg (1938), shows that this construction comprises all extendible spherical distributions.

Theorem 4.1 (Extendible Spherical Distributions)
$\boldsymbol{X}$ has a spherical distribution and is extendible if and only if there exists a random variable $W \geq 0$ and an independent random vector $\boldsymbol{Z}$ with independent and standard normally distributed components such that $\boldsymbol{X} \stackrel{d}{=} \sqrt{W}\,\boldsymbol{Z}$.

Proof.

$\Rightarrow$: If $\boldsymbol{X}$ is extendible with associated characteristic generator φ, then $\varphi \in \bigcap_{d \in \mathbb{N}} \Phi_d$. Up to the substitution $r^2 \mapsto w/2$, it is shown in Fang et al.

(1990, Theorem 2.21, p. 48) that

$$\varphi \in \bigcap_{d \in \mathbb{N}} \Phi_d \Leftrightarrow \varphi(x) = \int_0^\infty e^{-x\,w/2}\, d\mathbb{P}(W \leq w), \text{ for some r.v. } W \geq 0.$$

Comparing this to Example 4.1(2) establishes the claim.

$\Leftarrow$: The random vector $\sqrt{W}\,\boldsymbol{Z}$ is spherical by Example 4.1(2). It is clearly extendible; one simply has to extend $\boldsymbol{Z}$ by adding independent standard normal components.

□

At first glance, it is somewhat surprising that the normal distribution appears in the characterization of the extendible subclass in Theorem 4.1. In contrast, the canonical stochastic representation of spherical distributions contains a uniformly distributed random vector $\boldsymbol{S}$ on $S_{L^2,d}$. The connection between these two stochastic representations is clarified by the following lemma.

Lemma 4.2 (Spherical Decomposition of i.i.d. Normals)
Let $\boldsymbol{S}$ be uniformly distributed on $S_{L^2,d}$ and let $R \geq 0$ be an independent random variable such that $R^2 \sim \chi^2(d)$. Then $R\,\boldsymbol{S} \stackrel{d}{=} \boldsymbol{Z}$, where $\boldsymbol{Z} = (Z_1, \ldots, Z_d)'$ is normally distributed with zero mean vector, zero correlations, and unit variances. We write $\boldsymbol{Z} \sim \mathcal{N}_d(\boldsymbol{0}, I)$.

Proof. Let $\boldsymbol{Z} \sim \mathcal{N}_d(\boldsymbol{0}, I)$. By Example 4.1(1), $\boldsymbol{Z}$ is spherical. By Lemma 4.1(2), one can represent $\boldsymbol{Z}$ via $\boldsymbol{Z} \stackrel{d}{=} R\,\boldsymbol{S}$, where $R \stackrel{d}{=} \|\boldsymbol{Z}\|$ and R is independent of $\boldsymbol{S}$, which is uniformly distributed on $S_{L^2,d}$. Since

$$\|\boldsymbol{Z}\|^2 = Z_1^2 + \ldots + Z_d^2 \sim \chi^2(d),$$

the claim follows. □

4.2 Elliptical Distributions

The family of elliptical distributions is a generalization of the class of spherical distributions. Each elliptical distribution is obtained as a linear transformation of spherical distributions. Formally, it is defined as follows.

Definition 4.2 (Elliptical Distribution)
The random vector $\boldsymbol{X} = (X_1, \ldots, X_d)'$ has an elliptical distribution *if*

$$\boldsymbol{X} \stackrel{d}{=} \boldsymbol{\mu} + A'\,\boldsymbol{Y} \stackrel{d}{=} \boldsymbol{\mu} + A'\,R\,\boldsymbol{S}, \tag{4.6}$$

where $\boldsymbol{Y}$ *has a spherical distribution on* $\mathbb{R}^k$, $A \in \mathbb{R}^{k\times d}$ *with* $\Sigma := A'A \in \mathbb{R}^{d\times d}$, *and* $rk(\Sigma) = k \le d$, $\boldsymbol{\mu} \in \mathbb{R}^d$. *The second equality recalls the canonical representation* $\boldsymbol{Y} \stackrel{d}{=} R\,\boldsymbol{S}$ *of spherical distributions in Lemma 4.1(2).*

Provided the existence of a density, an elliptical distribution has elliptically contoured density level surfaces, which explains the name. Besides the transformation via A, the random variable R introduces additional dependence to the components and influences in particular the (joint) tail behavior of the distribution (see Schmidt (2002)).

Remark 4.1 (Notation and Parameterization)
For elliptical distributions, we use the notation $\boldsymbol{X} \sim E_d(\boldsymbol{\mu}, \Sigma, \varphi)$, *where* $\varphi \in \Phi_k$ *is the function characterizing the spherical distribution of* $\boldsymbol{Y}$. *The distribution of* $\boldsymbol{X}$ *depends on* A *only via* Σ, *i.e. it makes no difference (in distribution) if we replace* A *with some possibly different matrix* $B \in \mathbb{R}^{k\times d}$ *with* $B'B = \Sigma = A'A$. *Moreover,* $\boldsymbol{\mu}$ *is unique, but* φ *and* Σ *are not uniquely determined by the distribution (see McNeil et al. (2005, Remark 3.27, p. 93)).*

The most important properties of elliptical distributions are listed in the sequel. Notice that the spherical distributions arise as a special case with $\Sigma = I$ and $\boldsymbol{\mu} = \boldsymbol{0}$.

Lemma 4.3 (Properties of Elliptical Distributions)
Let $\boldsymbol{X} \sim E_d(\boldsymbol{\mu}, \Sigma, \varphi)$ *be defined as in Equation (4.6). Then:*

(1) The characteristic function *of* $\boldsymbol{X}$ *is*

$$\phi_{\boldsymbol{X}}(\boldsymbol{t}) = \mathbb{E}\big[e^{i\,\boldsymbol{t}'\boldsymbol{X}}\big] = e^{i\,\boldsymbol{t}'\boldsymbol{\mu}}\varphi(\boldsymbol{t}'\,\Sigma\,\boldsymbol{t}), \quad \boldsymbol{t} \in \mathbb{R}^d. \tag{4.7}$$

(2) Elliptical distributions are stable under affine transformations *in the following sense. Let* $B \in \mathbb{R}^{d\times m}$ *and* $\boldsymbol{\nu} \in \mathbb{R}^m$. *Then*

$$\boldsymbol{\nu} + B'\,\boldsymbol{X} \sim E_m(\boldsymbol{\nu} + B'\,\boldsymbol{\mu}, B'\,\Sigma\,B, \varphi).$$

(3) Marginal laws*: Take any* l*-dimensional subvector* $(X_{i_1}, \ldots, X_{i_l})'$ *from* $\boldsymbol{X}$, *for* $1 \le i_1 < \ldots < i_l \le d$. *This subvector is then distributed as* $E_l(\boldsymbol{\mu}_l, \Sigma_l, \varphi)$, *where* $\boldsymbol{\mu}_l := (\mu_{i_1}, \ldots, \mu_{i_l})'$ *is the respective subvector of* $\boldsymbol{\mu}$ *and* $\Sigma_l \in \mathbb{R}^{l\times l}$ *is the block-matrix containing the respective elements* $\{\Sigma_{mj}\}_{m,j=i_1,\ldots,i_l}$ *of* Σ.

(4) The covariance matrix *is influenced by both* Σ *and* φ. *Given that* $\mathbb{E}[\|\boldsymbol{Y}\|^2] < \infty$ *and* $\mathbb{P}(\|\boldsymbol{Y}\| > 0) = 1$, *it holds that*

$$\mathbb{E}[\boldsymbol{X}] = \boldsymbol{\mu}, \quad Cov(\boldsymbol{X}) = \frac{1}{k}\,\mathbb{E}[\|\boldsymbol{Y}\|^2]\,\Sigma. \tag{4.8}$$

Note that $\mathbb{E}[\|\boldsymbol{Y}\|^2]$ *depends on* φ.

Proof.

(1) Follows from Lemma 4.1(3) and basic linear algebra.

(2) Follows from part (1) by computing the characteristic function of the linear transform.

(3) Consider some subset $\emptyset \neq I \subset \{1, \ldots, d\}$ with $|I| = l$. Define $B \in \mathbb{R}^{d \times |I|}$ to be the matrix whose columns are the unit vectors $\boldsymbol{e}_i$ for $i \in I$ and apply it along with $\boldsymbol{\nu} = \boldsymbol{0}$ in (2).

(4) Let $\boldsymbol{Z} = (Z_1, \ldots, Z_k)' \sim \mathcal{N}_k(\boldsymbol{0}, I)$. By Lemma 4.2 there exists some random variable R_Z such that $R_Z^2 \sim \chi^2(k)$ and some random vector $\boldsymbol{S}$ independent of R_Z and uniformly distributed on the k-dimensional unit sphere, such that $\boldsymbol{Z} \stackrel{d}{=} R_Z\, \boldsymbol{S}$. From $\mathbb{E}[R_Z] > 0$, we conclude for $i = 1, \ldots, k$ that

$$0 = \mathbb{E}[Z_i] = \mathbb{E}[R_Z]\, \mathbb{E}[S_i] \quad \Rightarrow \quad \mathbb{E}[\boldsymbol{S}] = \boldsymbol{0}.$$

Moreover, from $\mathbb{E}[R_Z^2] = k$ one has for all $i = 1, \ldots, k$

$$1 = \mathbb{E}[Z_i^2] = \mathbb{E}[R_Z^2]\, \mathbb{E}[S_i^2] \quad \Rightarrow \quad \mathrm{Var}(S_i) = \frac{1}{k}.$$

Finally, for $i \neq j$

$$0 = \mathbb{E}[Z_i\, Z_j] = \mathbb{E}[R_Z^2]\, \mathbb{E}[S_i\, S_j] \quad \Rightarrow \quad \mathrm{Cov}(\boldsymbol{S}) = \frac{1}{k}\, I.$$

Now consider the spherically distributed random vector $\boldsymbol{Y}$ with decomposition $\boldsymbol{Y} \stackrel{d}{=} R\, \boldsymbol{S}$, $\mathbb{E}[R^2] < \infty$. Then

$$\begin{aligned} \mathbb{E}[\boldsymbol{X}] &= \mathbb{E}[\boldsymbol{\mu} + A'\, \boldsymbol{Y}] = \mathbb{E}[\boldsymbol{\mu} + A'\, R\, \boldsymbol{S}] \\ &= \boldsymbol{\mu} + \mathbb{E}[R]\, A'\, \mathbb{E}[\boldsymbol{S}] = \boldsymbol{\mu} + \boldsymbol{0}. \end{aligned}$$

Moreover, denoting by $\{a_{ij}\}$ the entries of A, one finds using $\mathbb{E}[S_{i_1}\, S_{i_2}] = \mathbb{1}_{\{i_1 = i_2\}}/k$ that

$$\begin{aligned} \mathbb{E}[X_j^2] &= \mathbb{E}\big[(\mu_j + R \sum_{i=1}^{k} a_{ij}\, S_i)^2\big] \\ &= \mu_j^2 + 2\, \mathbb{E}[R]\, 0 + \mathbb{E}[R^2]\, \mathbb{E}\big[(\sum_{i=1}^{k} a_{ij}\, S_i)^2\big] \\ &= \mu_j^2 + \mathbb{E}[R^2] \sum_{i_1=1}^{k} \sum_{i_2=1}^{k} a_{i_1 j}\, a_{i_2 j} \mathbb{E}[S_{i_1}\, S_{i_2}] \\ &= \mu_j^2 + \mathbb{E}[R^2] \sum_{i=1}^{k} a_{ij}^2 \frac{1}{k}, \end{aligned}$$

establishing $\text{Var}(X_j) = \mathbb{E}[R^2]/k \sum_{i=1}^k a_{ij}^2$. Considering the mixed moments, we find for $l \neq j$ that

$$\begin{aligned}
\mathbb{E}[X_l X_j] &= \mathbb{E}\big[(\mu_l + R \sum_{i=1}^k a_{il} S_i)\,(\mu_j + R \sum_{i=1}^k a_{ij} S_i)\big] \\
&= \mu_l\,\mu_j + 0 + 0 + \mathbb{E}[R^2] \sum_{i_1=1}^k \sum_{i_2=1}^k a_{i_1 l}\, a_{i_2 j} \mathbb{E}[S_{i_1} S_{i_2}] \\
&= \mu_l\,\mu_j + \mathbb{E}[R^2]\,\frac{1}{k} \sum_{i=1}^k a_{il}\, a_{ij},
\end{aligned}$$

implying $\text{Cov}(X_l, X_j) = \mathbb{E}[R^2]/k \sum_{i=1}^k a_{il}\, a_{ij}$ and finally

$$\text{Cov}(\boldsymbol{X}) = \frac{\mathbb{E}[R^2]}{k}\,\Sigma.$$

□

We are now in the position to lift Theorem 4.1 from the spherical to the elliptical case, providing a characterization of the subclass of extendible elliptical distributions.

Theorem 4.2 (Extendible Elliptical Distributions)
Let $\boldsymbol{X}$ be defined as in Equation (4.6) with $\mathbb{E}[\|\boldsymbol{Y}\|^2] < \infty$ and $\mathbb{P}(\|\boldsymbol{Y}\| > 0) = 1$. Then $\boldsymbol{X}$ is extendible if and only if $\mu_1 = \ldots = \mu_d$ and $\boldsymbol{X} \stackrel{d}{=} \boldsymbol{\mu} + R\,\boldsymbol{Z}_*$ for $R > 0$, with $\mathbb{E}[R^2] < \infty$ and $\boldsymbol{Z}_*$ multivariate normal with zero mean vector and covariance matrix Σ, independent of R. The matrix Σ is given by

$$\Sigma = \sigma^2 \begin{pmatrix} 1 & \ldots & \rho \\ \vdots & \ddots & \vdots \\ \rho & \ldots & 1 \end{pmatrix}, \quad \sigma^2 > 0,\, \rho \geq 0.$$

Proof.

$\Leftarrow$: A conditionally i.i.d. construction of the distribution in question is the following. Take independent standard normal random variables $M, \epsilon_1, \ldots, \epsilon_d \sim \mathcal{N}(0,1)$. Then define

$$W_i := \sigma\left(\sqrt{\rho}\,M + \sqrt{1-\rho}\,\epsilon_i\right) \sim \mathcal{N}(0, \sigma^2), \quad i = 1, \ldots, d.$$

One easily verifies that $\boldsymbol{W} = (W_1 \ldots, W_d)' \stackrel{d}{=} \boldsymbol{Z}_* \sim \mathcal{N}_d(\boldsymbol{0}, \Sigma)$. This shows that $\boldsymbol{W}$ (resp. $\boldsymbol{Z}_*$) is extendible, since $W_1, \ldots, W_d$ are i.i.d. given the sigma algebra generated by M. Moreover, $R\,\boldsymbol{Z}_*$ is also extendible, since $R\,W_1,\ldots,R\,W_d$ are i.i.d. given the sigma algebra generated by M and R. Shifting $R\,\boldsymbol{Z}_*$ by $\boldsymbol{\mu}$ with $\mu_1=\ldots=\mu_d$ does not affect extendibility. It is left to show that $\boldsymbol{X}=\boldsymbol{\mu}+R\,\boldsymbol{Z}_*\sim E_d(\boldsymbol{\mu},\Sigma,\varphi)$. With $A'\,A=\Sigma$ and $\boldsymbol{Z}\stackrel{d}{=}R_Z\,\boldsymbol{S}\sim\mathcal{N}_d(\boldsymbol{0},I)$ from Lemma 4.2, one has

$$\begin{aligned}\boldsymbol{X}&=\boldsymbol{\mu}+R\,\boldsymbol{Z}_*\stackrel{d}{=}\boldsymbol{\mu}+R\,A'\,\boldsymbol{Z}\\&\stackrel{d}{=}\boldsymbol{\mu}+R\,A'\,(R_Z\,\boldsymbol{S})=\boldsymbol{\mu}+(R\,R_Z)A'\,\boldsymbol{S}.\end{aligned}$$

We obtain the required representation with $\boldsymbol{\mu}$, $\Sigma=A'\,A$ as claimed, and φ stemming from the radius $R\,R_Z$.

$\Rightarrow$: Necessary for extendibility is exchangeability. Hence, $\mathbb{E}[\boldsymbol{X}]=\boldsymbol{\mu}$ implies $\mu_1=\ldots=\mu_d$, as otherwise the univariate marginal laws could not be the same. Similarly, we obtain a homogeneous variance. Since (by exchangeability) all bivariate marginals must also be the same, we obtain the claimed homogeneous structure of Σ. It is left, however, to show that the off-diagonal entries ρ of Σ are non-negative. This follows from the fact that extendible random vectors must have non-negative pairwise correlations (resp. covariances) (see Lemma 1.9) and in the present case we have $\mathrm{Cov}(\boldsymbol{X})=\mathbb{E}[R^2]\,\Sigma/d$ (see Lemma 4.3(4)).
Moreover, $\boldsymbol{X}\sim E_d(\boldsymbol{\mu},\Sigma,\varphi)$ implies $\boldsymbol{X}\stackrel{d}{=}\boldsymbol{\mu}+A'\,\boldsymbol{Y}$ for $\boldsymbol{Y}$ spherically distributed on $\mathbb{R}^k$ with $k=\mathrm{rk}(\Sigma)\leq d$. We already have $\boldsymbol{\mu}$ and $A'\,A=\Sigma$. It is left to verify that $\boldsymbol{Y}$ is related to i.i.d. normals. This, however, was solved in Theorem 4.1, providing $\boldsymbol{Y}\stackrel{d}{=}R\,\boldsymbol{Z}$ with $R>0$ and $\boldsymbol{Z}\sim\mathcal{N}_d(\boldsymbol{0},I)$ (independent). Hence, we have $\boldsymbol{X}\stackrel{d}{=}\boldsymbol{\mu}+R\,A'\,\boldsymbol{Z}$, establishing the claim since $A'\,\boldsymbol{Z}\stackrel{d}{=}\boldsymbol{Z}_*$.

□

4.3 Parametric Families of Elliptical Distributions

Definition 4.3 (Multivariate Normal Distribution)
The random vector $\boldsymbol{X}=(X_1,\ldots,X_d)'$ follows a multivariate normal distribution, *abbreviated $\boldsymbol{X}\sim\mathcal{N}_d(\boldsymbol{\mu},\Sigma)$, if*

$$\boldsymbol{X}\stackrel{d}{=}\boldsymbol{\mu}+A'\,\boldsymbol{Z}\stackrel{d}{=}\boldsymbol{\mu}+A'\,R\,\boldsymbol{S},\tag{4.9}$$

with mean vector $\boldsymbol{\mu}\in\mathbb{R}^d$, (positive-semidefinite) covariance matrix $\Sigma=A'\,A\in\mathbb{R}^{d\times d}$, where $A\in\mathbb{R}^{k\times d}$ with rk$(A)=k\leq d$, and k-dimensional vector $\boldsymbol{Z}\sim\mathcal{N}_k(\boldsymbol{0},I)$, whose independent components have univariate standard normal law, i.e. $Z_i\sim\mathcal{N}(0,1)$ for $i=1,\ldots,k$. The distribution of

R^2 equals that of $Z_1^2 + \ldots + Z_k^2$, so it is a χ^2-distribution with k degrees of freedom.

Lemma 4.4 (Properties of the Multivariate Normal Distribution) *Let $X \sim \mathcal{N}_d(\boldsymbol{\mu}, \Sigma)$. Then:*

(1) The characteristic function *of $\boldsymbol{X}$ is given by*

$$\phi_{\boldsymbol{X}}(\boldsymbol{t}) = \mathbb{E}\big[e^{i\,\boldsymbol{t}'\boldsymbol{X}}\big] = e^{i\,\boldsymbol{t}'\,\boldsymbol{\mu} - \frac{1}{2}\,\boldsymbol{t}'\,\Sigma\,\boldsymbol{t}}, \quad \boldsymbol{t} \in \mathbb{R}^d. \tag{4.10}$$

(2) For positive-definite Σ, the density *of $\boldsymbol{X}$ is given by*

$$f(\boldsymbol{x}) = \frac{e^{-\frac{1}{2}(\boldsymbol{x}-\boldsymbol{\mu})'\Sigma^{-1}(\boldsymbol{x}-\boldsymbol{\mu})}}{(2\,\pi)^{d/2}\det(\Sigma)^{1/2}}, \quad \boldsymbol{x} \in \mathbb{R}^d. \tag{4.11}$$

Note that the level curves of f are ellipsoids with

$$(\boldsymbol{x} - \boldsymbol{\mu})'\,\Sigma^{-1}\,(\boldsymbol{x} - \boldsymbol{\mu}) \equiv const.$$

(3) Stability under convolution*: Given independent $\boldsymbol{X} \sim \mathcal{N}_d(\boldsymbol{\mu}, \Sigma)$, $\tilde{\boldsymbol{X}} \sim \mathcal{N}_d(\tilde{\boldsymbol{\mu}}, \tilde{\Sigma})$, the sum of both random vectors is again multivariate normal, i.e.*

$$\boldsymbol{X} + \tilde{\boldsymbol{X}} \sim \mathcal{N}_d\big(\boldsymbol{\mu} + \tilde{\boldsymbol{\mu}}, \Sigma + \tilde{\Sigma}\big).$$

(4) Uncorrelated corresponds to independent*: A normally distributed random vector with uncorrelated components has independent components.*

Proof.

(1) Represent $\boldsymbol{X}$ by $\boldsymbol{X} \stackrel{d}{=} \boldsymbol{\mu} + A'\boldsymbol{Z}$, where $\boldsymbol{Z} \sim \mathcal{N}_k(\boldsymbol{0}, I)$ and $\Sigma = A'\,A$. Then apply Example 4.1(1) and Equation (4.7).
(2) Most textbooks, e.g. DeGroot (2004, Section 5.4, p. 51), define the multivariate normal distribution via its density (4.11) and derive the characteristic function (4.10).
(3) Follows from Lemma 4.3(2) with $(\boldsymbol{X}', \tilde{\boldsymbol{X}}')' \in \mathbb{R}^{2\,d}$, $B' := (I, I) \in \mathbb{R}^{d \times 2\,d}$, and $\boldsymbol{\nu} := \boldsymbol{0} \in \mathbb{R}^d$.
(4) Zero correlation implying independence can be deduced either from the characteristic function, the density (both factorize), or the stochastic representation (4.9).

□

Lemma 4.4(3) and Lemma 4.3(2) show that the multivariate normal distribution with mean vector $\boldsymbol{\mu} = \boldsymbol{0}$ and d-dimensional identity matrix I as covariance, i.e. $\boldsymbol{X} \sim \mathcal{N}_d(\boldsymbol{0}, I)$, satisfies the defining property (4.1) of spherical distributions. For any orthogonal matrix O the distribution of $O\,\boldsymbol{X}$ is

$\mathcal{N}_d(O\,\mathbf{0}, O'\,I\,O) = \mathcal{N}_d(\mathbf{0}, I)$. Alternatively, this might also be shown using Lemma 4.1. The characteristic function of $\boldsymbol{X} \sim \mathcal{N}_d(\mathbf{0}, I)$, obtained from (4.10) with $\Sigma = I$ and $\boldsymbol{\mu} = \mathbf{0}$, is easily seen to be a function of $\|\boldsymbol{t}\|$ alone.

Besides the multivariate normal distribution, another popular elliptical distribution is the multivariate Student's t-distribution.

Definition 4.4 (Multivariate Student's t-Distribution)
Let $\boldsymbol{X} = (X_1, \ldots, X_d)'$ be constructed as

$$\boldsymbol{X} \stackrel{d}{=} \boldsymbol{\mu} + \Sigma^{1/2}\sqrt{W}\,\boldsymbol{Z} \stackrel{d}{=} \boldsymbol{\mu} + \Sigma^{1/2}\,\frac{\boldsymbol{Z}\,\sqrt{\nu}}{\sqrt{\chi^2_\nu}}, \tag{4.12}$$

where $\boldsymbol{Z} \sim \mathcal{N}_d(\mathbf{0}, I)$ and $\Sigma^{1/2}\,\Sigma^{1/2} = \Sigma$ is positive definite.[3] *All random variables and vectors, i.e. W, $\boldsymbol{Z}$, and χ^2_ν, used in Equation (4.12) are independent, and the mixing variable W is inverse Gamma distributed with parameters $W \sim Inv\Gamma(\nu/2, \nu/2)$ (see Remark 4.2). Then $\boldsymbol{X}$ is said to have the multivariate Student's t-distribution with ν degrees of freedom and location parameter $\boldsymbol{\mu}$, abbreviated $\boldsymbol{X} \sim t_d(\boldsymbol{\mu}, \Sigma, \nu)$. Note that the $t_d(\mathbf{0}, I, \nu)$-distribution can be seen as a mixture of normal distributions (see Example 4.1(2)).*

Alternatively, the multivariate Student's t-distribution can be constructed as an elliptical distribution via the representation

$$\boldsymbol{X} \stackrel{d}{=} \boldsymbol{\mu} + \Sigma^{1/2}\,R\,\boldsymbol{S}, \tag{4.13}$$

where R^2/d is distributed as an $F(d, \nu)$-distribution[4] *and $\Sigma^{1/2}\,\Sigma^{1/2} = \Sigma$. Again, R and $\boldsymbol{S}$ are independent.*

Remark 4.2 (Inverse Gamma Distribution)
The inverse Gamma distribution $Inv\Gamma(\beta, \eta)$ is the distribution of $Y := 1/X$, where $X \sim \Gamma(\beta, \eta)$. Its density f is obtained from the density of the Gamma distribution via the density transformation theorem. One finds

$$f(x) = \mathbb{1}_{\{x>0\}}\,\frac{\eta^\beta\,x^{-(\beta+1)}\,e^{-\eta/x}}{\Gamma(\beta)}.$$

The characteristic function of $Y \sim Inv\Gamma(\beta, \eta)$ is given by

$$\phi(x) = \frac{2(-i\,\eta\,x)^{\beta/2}K_\beta\big(\sqrt{-4\,i\,\eta\,x}\big)}{\Gamma(\beta)}$$

[3] The matrix $\Sigma^{1/2}$ can be found via $\Sigma^{1/2} = O\,\Lambda^{1/2}\,O'$, where the orthogonal matrix O contains the standardized eigenvectors of Σ and $\Lambda^{1/2}$ is a diagonal matrix containing the square roots of the respective eigenvalues. Defined in this way, it is the unique symmetric root of Σ. Note that $O\,\Lambda^{1/2}$ is an alternative (but non-symmetric) root of Σ.

[4] The $F(d, \nu)$-distribution is the distribution of $(X_1/d)/(X_2/\nu)$, where $X_1 \sim \chi^2(d)$ and $X_2 \sim \chi^2(\nu)$ are independent (see DeGroot (2004, p. 42)).

(see Witkovský (2001)), where K_β is the modified Bessel function of the second kind. The choice $W \sim Inv\Gamma(\nu/2, \nu/2)$ in Definition 4.4 implies that $W \stackrel{d}{=} \nu/\chi^2_\nu$, showing the equivalence of the first two representations in Equation (4.12).

Lemma 4.5 (Properties of the Multivariate t-Distribution)
Let $\boldsymbol{X} \sim t_d(\boldsymbol{\mu}, \Sigma, \nu)$. Then:

(1) The characteristic function *of $\boldsymbol{X}$ is given by $e^{i\,\boldsymbol{t}'\boldsymbol{\mu}}\varphi(\boldsymbol{t}'\Sigma\,\boldsymbol{t})$, where*

$$\varphi(x) = \frac{2\,\Gamma\big((\nu+1)/2\big)}{\sqrt{\pi}\,\Gamma\big(\nu/2\big)} \int_0^\infty \cos(\,s\,\sqrt{\nu\,x})\,(1+s^2)^{-(\nu+1)/2}\,ds.$$

(2) The density *of $\boldsymbol{X}$ is given by*

$$f(\boldsymbol{x}) = \frac{\Gamma\big(\frac{1}{2}(\nu+d)\big)\big(1+\frac{1}{\nu}(\boldsymbol{x}-\boldsymbol{\mu})'\Sigma^{-1}(\boldsymbol{x}-\boldsymbol{\mu})\big)^{-(\nu+d)/2}}{\Gamma\big(\frac{\nu}{2}\big)(\pi\,\nu)^{d/2}\det(\Sigma)^{1/2}}, \tag{4.14}$$

where $\boldsymbol{x} \in \mathbb{R}^d$. Note that the level curves of f are ellipsoids with

$$(\boldsymbol{x}-\boldsymbol{\mu})'\,\Sigma^{-1}\,(\boldsymbol{x}-\boldsymbol{\mu}) \equiv const.$$

Further, we observe that for $\Sigma = I$, the density does not factorize in univariate marginal densities, showing that $\Sigma = I$ does not imply independence, in contrast to the multivariate normal distribution.

(3) The covariance matrix *of $\boldsymbol{X}$ is given by $\Sigma\,\nu/(\nu-2)$, provided that $\nu > 2$. Clearly, Σ is the limiting covariance matrix as $\nu \to \infty$.*

(4) The limit distribution of $t_d(\boldsymbol{\mu}, \Sigma, \nu)$ for $\nu \to \infty$ is the normal distribution $\mathcal{N}_d(\boldsymbol{\mu}, \Sigma)$.

Proof.

(1) See Fang et al. (1990, Theorem 3.9, p. 85).
(2) See Fang et al. (1990, p. 82) or DeGroot (2004, p. 60).
(3) Note that $R^2/d \sim F(d, \nu)$ (see, e.g., DeGroot (2004, p. 43)), so $\mathbb{E}[R^2/\mathrm{rk}(\Sigma)] = \mathbb{E}[R^2/d] = \nu/(\nu-2)$ for $\nu > 2$. The result then follows from the general covariance formula for elliptical distributions, i.e. Equation (4.8).
(4) This follows from taking the limits in the density or the characteristic function, respectively.

□

4.4 Elliptical Copulas

Recall that a d-dimensional random vector $\boldsymbol{X} = (X_1, \ldots, X_d)' \sim E_d(\boldsymbol{\mu}, \Sigma, \varphi)$ has an elliptical distribution if $\boldsymbol{X} \stackrel{d}{=} \boldsymbol{\mu} + A'\boldsymbol{Y} \stackrel{d}{=} \boldsymbol{\mu} + A'R\boldsymbol{S}$, where $\boldsymbol{Y} \sim S_k(\varphi)$, $A \in \mathbb{R}^{k \times d}$, $\Sigma := A'A \in \mathbb{R}^{d \times d}$, $rk(\Sigma) = k \leq d$, $\boldsymbol{\mu} \in \mathbb{R}^d$, and R, $\boldsymbol{S}$ are given as in Equation (4.2).

Definition 4.5 (Elliptical Copula)
An elliptical copula *is defined as the copula of the related elliptical distribution F. Its analytical form is obtained via Sklar's theorem (see Theorem 1.2) from the distribution function F, i.e.*

$$C(u_1, \ldots, u_d) := F\big(F_1^{-1}(u_1), \ldots, F_d^{-1}(u_d)\big), \quad (u_1, \ldots, u_d) \in [0,1]^d,$$

where F_k^{-1} are the univariate quantile functions, $k = 1, \ldots, d$.

Remark 4.3 (Parameterization of Elliptical Copulas)
A problem with the notation $E_d(\boldsymbol{\mu}, \Sigma, \varphi)$ is that this characterization of an elliptical distribution is not unique. For any $c > 0$, we can take $\tilde{\Sigma} := c\,\Sigma$, $\tilde{\varphi}(t) := \varphi(t/c)$ and obtain the same distribution. Moreover, applying a strictly increasing transformation to all univariate marginals of $\boldsymbol{X}$ does not change its copula (see Lemma 1.5). Hence, for the copula of an elliptical distribution, the shift by $\boldsymbol{\mu}$ is not relevant and we might always assume $\boldsymbol{\mu} = \boldsymbol{0}$. Moreover, we can further standardize the matrix Σ by using the following normalization: introduce $P \in \mathbb{R}^{d \times d}$ with $p_{ij} := \Sigma_{ij}/(\sqrt{\Sigma_{ii}}\sqrt{\Sigma_{jj}})$. Then, P is a correlation matrix and the copula of $\boldsymbol{X} \sim E_d(\boldsymbol{\mu}, \Sigma, \varphi)$ is the same as the one of $\boldsymbol{Y} \sim E_d(\boldsymbol{0}, P, \varphi)$, since the marginals just undergo a strictly increasing affine transformation from one distribution to the other, which does not affect the copula. Obviously, the parameterization with P has fewer parameters and is therefore more convenient, so we denote by $C_{P,\varphi}$ the copula of $E_d(\boldsymbol{0}, P, \varphi)$.

Lemma 4.6 (Properties of Elliptical Copulas)
Let the elliptical copula $C_{P,\varphi}$ be given. Then the following properties hold:

(1) Elliptical copulas are radially symmetric, *i.e. $C_{P,\varphi} = \hat{C}_{P,\varphi}$.*
(2) The upper- and lower-tail dependence *coefficients, being the same by radial symmetry, depend on the choice of R. Loosely speaking, heavy tails of R translate into positive tail dependence.*
(3) For a bivariate elliptical copula $C_{P,\varphi}$ with P having off-diagonal entries $p := p_{12} = p_{21}$, stemming from a non-degenerate elliptical distribution

$\boldsymbol{X} = (X_1, X_2)'$ *with* $\mathbb{P}(X_1 = \mu_1) = \mathbb{P}(X_2 = \mu_2) = 0$, Kendall's tau *is given by*

$$\tau_{C_{P,\varphi}} = \frac{2}{\pi}\, arcsin(p) \in [-1, 1].$$

It is worth mentioning that this measure of dependence does not depend on the choice of R*, respectively* φ*.*

Proof.

(1) Let $C_{P,\varphi}$ be the copula of the random vector $\boldsymbol{\mu} + A'\boldsymbol{Y}$, using the notation of Equation (4.6) (we assume w.l.o.g. that $\boldsymbol{\mu} = \boldsymbol{0}$ and $A'A = P$; a correlation matrix). Multiplying this random vector componentwise by (-1) gives the random vector $A'(-1)\boldsymbol{Y}$, whose copula is $\hat{C}_{P,\varphi}$ according to Corollary 1.1. Since $\boldsymbol{Y} \stackrel{d}{=} -I\boldsymbol{Y} = (-1)\boldsymbol{Y}$ for spherically distributed $\boldsymbol{Y}$, we finally have $C_{P,\varphi} = \hat{C}_{P,\varphi}$.

(2) A deep investigation of this subject, including quantitative results for regularly varying R (at infinity), is provided in Schmidt (2002). The specific cases of the Gaussian and the t-copula are investigated in Lemmas 4.7 and 4.8.

(3) A proof and an extension to spherical distributions with $\mathbb{P}(X_i = \mu_i) \in (0, 1)$, $i = 1, 2$, is given in Lindskog et al. (2002).

□

4.5 Parametric Families of Elliptical Copulas

Definition 4.6 (Gaussian Copula)
The normal *or* Gaussian copula C_P^{Gauss} *is the copula of* $\boldsymbol{X} \sim \mathcal{N}_d(0, P)$*, where* P *is a correlation matrix. The functional form is obtained by*

$$C_P^{Gauss}(u_1, \ldots, u_d) := F_P\big(\Phi^{-1}(u_1), \ldots, \Phi^{-1}(u_d)\big), \tag{4.15}$$

where $(u_1, \ldots, u_d) \in [0, 1]^d$*,* F_P *is the joint distribution function of* $\boldsymbol{X}$*, and* Φ^{-1} *is the quantile function of the univariate standard normal law.*

Instead of defining C_P^{Gauss} using the correlation matrix P, one could equivalently use the covariance matrix Σ. In this case, the quantile functions must be adjusted accordingly. Both choices imply the very same copula; the choice P, however, is the more intuitive parameterization and does not contain redundant parameters.

Lemma 4.7 (Gaussian Copula: Tail Independence)
Both tail dependence coefficients of the bivariate Gaussian copula with correlation parameter $p \in (-1, 1)$ are 0.

Proof. We consider w.l.o.g. the lower-tail dependence, which agrees with the upper-tail dependence by radial symmetry. The proof requires a reformulation of the tail dependence formula. Following Embrechts et al. (2003), we find for some exchangeable copula C

$$\begin{aligned} LTD_C &= \lim_{u \downarrow 0} \frac{C(u,u)}{u} \stackrel{(*)}{=} \lim_{u \downarrow 0} \frac{d}{du} C(u,u) \\ &= \lim_{u \downarrow 0} \Big(\frac{\partial}{\partial u_1} C(u_1,u_2)\big|_{u_1=u_2=u} + \frac{\partial}{\partial u_2} C(u_1,u_2)\big|_{u_1=u_2=u} \Big) \\ &= \lim_{u \downarrow 0} \big(\mathbb{P}(U_2 \leq u | U_1 = u) + \mathbb{P}(U_1 \leq u | U_2 = u) \big) \\ &\stackrel{(**)}{=} 2 \lim_{u \downarrow 0} \mathbb{P}(U_2 \leq u | U_1 = u) = 2 \lim_{u \downarrow 0} \mathbb{P}(U_1 \leq u | U_2 = u), \end{aligned}$$

where $(*)$ follows from de l'Hospital's rule and $(**)$ requires exchangeability. For the Gaussian copula, a simple transformation with Φ^{-1} shows that the above limit agrees with $2 \lim_{x \downarrow -\infty} \mathbb{P}(X_2 \leq x | X_1 = x)$ for $(X_1, X_2) := \big(\Phi^{-1}(U_1), \Phi^{-1}(U_2)\big) \sim \mathcal{N}_2(\mathbf{0}, p)$, i.e. $(X_1, X_2)'$ is bivariate standard normally distributed with correlation p. For the next step, we need the conditional distribution of X_2 given $X_1 = x$. This can be found by computing the conditional density, which involves dividing the joint density by the density of X_1 at x (see, e.g., Fahrmeir and Hamerle (1984, Satz 3.6, p. 27)). We find $X_2 | X_1 = x \sim \mathcal{N}(p\,x, 1-p^2)$. Hence

$$\begin{aligned} LTD_{C_p^{Gauss}} &= 2 \lim_{x \downarrow -\infty} \mathbb{P}(X_2 \leq x | X_1 = x) \\ &= 2 \lim_{x \downarrow -\infty} \Phi\Big(x \frac{1-p}{\sqrt{1-p^2}} \Big) = 0. \end{aligned}$$

□

Example 4.2 (H-Extendible Gaussian Copula)
Consider the Gaussian copula

$$C_P^{Gauss}(u_1, \ldots, u_d) = F_P\big(\Phi^{-1}(u_1), \ldots, \Phi^{-1}(u_d)\big),$$

where P is a correlation matrix and F_P is the distribution function of the $\mathcal{N}_d(\mathbf{0}, P)$ law. The extendible Gaussian copula corresponding to P has a homogeneous correlation ρ with $\rho \in [0,1]$ (see Example 1.17). A conditional i.i.d. construction for the extendible Gaussian copula is (similar to the proof of Theorem 4.2) given by

$$U_i := \Phi\big(\sqrt{\rho}\, M^{(0)} + \sqrt{1-\rho}\, \epsilon_i\big), \quad i = 1, \ldots, d,$$

where $M^{(0)}, \epsilon_1, \ldots, \epsilon_d \overset{i.i.d.}{\sim} \mathcal{N}(0,1)$. *It is clear that with* $Z_i := \sqrt{\rho}\, M^{(0)} + \sqrt{1-\rho}\, \epsilon_i \sim \mathcal{N}(0,1)$, *the vector* $\boldsymbol{Z} = (Z_1, \ldots, Z_d)'$ *is multivariate normal, and the correlation of* Z_k *with* Z_l *is* ρ, $k \neq l$. Φ *transforms the univariate marginal laws to* $U[0,1]$. *All components* $U_1, \ldots, U_d$ *are i.i.d. given* $\mathcal{G}_1 := \sigma(M^{(0)})$, *the* σ*-algebra generated by* $M^{(0)}$. *On the second level, an h-extendible Gaussian copula with two levels of hierarchy and* J *groups is constructed as follows: the correlation matrix* P *is now a block matrix; between different groups with homogeneous* $\rho_0 \in [0,1]$, *inside group* j *with homogeneous* $\rho_j \in [\rho_0, 1]$. *A stochastic model generating this copula is*

$$U_k := \Phi\big(\sqrt{\rho_0}\, M^{(0)} + \sqrt{\rho_j - \rho_0}\, M^{(j)} + \sqrt{1-\rho_j}\, \epsilon_k\big), \tag{4.16}$$

for $k \in \{1 + \sum_{l=1}^{j-1} d_l, \ldots, \sum_{l=1}^{j} d_l\}$, $j = 1, \ldots J$, *where* $M^{(0)}$, $M^{(1)}, \ldots, M^{(J)}$, $\epsilon_1, \ldots, \epsilon_d \overset{i.i.d.}{\sim} \mathcal{N}(0,1)$ *and the partition into groups is as in Section 1.2.4. Extensions to more levels of hierarchy are possible by introducing additional factors.*

An important advantage of h-extendible Gaussian copulas is that sampling, along the lines of construction (4.16), is extremely fast and conveniently simple to implement. Note that for two levels of hierarchy (dimension d and J groups), only $d + J + 1$ random variables have to be drawn and a simple transformation must be applied to all components. Since $J \leq d$, we have an overall effort of linear order $\mathcal{O}(d)$ in the dimension, which is smaller than the effort in the general case (see Algorithm 4.3).

Another important elliptical copula is the t-copula, which we now discuss.

Definition 4.7 (**t-Copula**)
The t-copula $C^t_{P,\nu}$ *is the copula of* $\boldsymbol{X} \sim t_d(\boldsymbol{0}, P, \nu)$, *where* P *is a correlation matrix. The analytical form is obtained by*

$$C^t_{P,\nu}(u_1, \ldots, u_d) := t_{\nu,P}\big(t_\nu^{-1}(u_1), \ldots, t_\nu^{-1}(u_d)\big), \tag{4.17}$$

where $(u_1, \ldots, u_d) \in [0,1]^d$, $t_{\nu,P}$ *is the joint distribution function of* $\boldsymbol{X}$ *and* t_ν^{-1} *is the quantile function of the univariate standard* t*-distribution with* ν *degrees of freedom.*

Lemma 4.8 (**t-Copula: Tail Dependence**)
The upper- (and lower-, by radial symmetry) tail dependence coefficient of a bivariate t*-copula* $C^t_{P,\nu}$ *with* ν *degrees of freedom and* P *having off-*

diagonal entries $p \in (-1,1)$ is given by

$$UTD_{C^t_{\nu,p}} = LTD_{C^t_{\nu,p}} = 2\,t_{\nu+1}\Big(-\sqrt{\frac{(\nu+1)(1-p)}{1+p}}\Big),$$

where $t_{\nu+1}$ is the distribution function of the univariate t-distribution with $\nu+1$ degrees of freedom, zero mean, and unit variance.

Proof. The proof is mostly similar to the derivation of the tail dependence coefficients in the Gaussian case. The difference, however, is that for X_1, X_2 being standard t-distributed with ν degrees of freedom and copula $C^t_{P,\nu}$, the conditional distribution of $X_2|X_1 = x$ is again a t-distribution with $\nu+1$ degrees of freedom, expectation $p\,x$, and variance $(\nu + x^2)\,(1-p^2)/(\nu+1)$. This can be found from the conditional density of $X_2|X_1 = x$ (see DeGroot (2004, p. 62)). Then

$$\begin{aligned}
LTD_{C^t_{\nu,\rho}} &= 2 \lim_{x\downarrow -\infty} \mathbb{P}(X_2 \leq x | X_1 = x)\\
&= 2 \lim_{x\downarrow -\infty} t_{\nu+1}\Big(\frac{(x-p\,x)\sqrt{\nu+1}}{\sqrt{(\nu+x^2)\,(1-p^2)}}\Big)\\
&= 2 \lim_{x\downarrow -\infty} t_{\nu+1}\Big(\text{sign}(x)\,\frac{\sqrt{\nu+1}\,\sqrt{1-p}}{\sqrt{\nu/x^2+1}\,\sqrt{1+p}}\Big)\\
&= 2\,t_{\nu+1}\Big(-\sqrt{\frac{(\nu+1)(1-p)}{1+p}}\Big).
\end{aligned}$$

□

Example 4.3 (H-Extendible Grouped t-Copula)

The notion of a grouped t-copula is originally due to Daul et al. (2003); a modified version that fulfills the h-extendibility criterion is presented in Mai and Scherer (2011b). One defines $(Z_1,\ldots,Z_d)' \sim \mathcal{N}_d(\mathbf{0},\Sigma)$ using the construction $Z_i := \sqrt{\rho}\,M^{(0)} + \sqrt{1-\rho}\,\epsilon_i$ as in Theorem 4.2, i.e. the Z_i's are standard normal with pairwise correlation $\rho \in [0,1]$. Independent of all previously defined random variables, take a list of independent random variables $R^{(1)},\ldots,R^{(J)}$ with distribution $R^{(j)} \stackrel{d}{=} \sqrt{\nu_j/\chi^2(\nu_j)}$. Consider the random vector

$$\Big(\underbrace{t_{\nu_1}(R^{(1)}Z_1),\ldots}_{group\ 1},\underbrace{t_{\nu_2}(R^{(2)}Z_{d_1+1}),\ldots}_{group\ 2},\ldots,\underbrace{t_{\nu_J}(R^{(J)}Z_{d_1+\ldots+d_{J-1}+1}),\ldots}_{group\ J}\Big),$$

where t_{ν_j} denotes the distribution function of the components $R^{(j)}Z_l$ in group j, which is a $t(\nu_j)$-distribution. This transforms $R^{(j)}Z_l$ to $U[0,1]$-distributed marginals. The resulting random vector is h-extendible: conditioned on $\mathcal{G}_1 = \sigma(M^{(0)})$, the groups are independent. Conditioned on

$\mathcal{G}_2 = \sigma(M^{(0)}, R^{(1)}, \dots, R^{(J)})$, all components are independent. Note that the dependence within group j is a t-copula with homogeneous correlation matrix and ν_j degrees of freedom. Between the groups, the dependence is of the Gaussian kind with correlation matrix Σ. Hence, we have positive tail dependence within each group and zero tail dependence between the groups.

Similar to the h-extendible Gaussian copula, the h-extendible grouped t-copula is very efficient to simulate. Using the above stochastic model, only $d + J + 1$ random variables must be drawn and suitably combined in order to obtain a sample of a d-dimensional grouped t-copula with J groups. Clearly, this is of linear order in d.

4.6 Sampling Algorithms

In this section, we state a generic sampling scheme for elliptical distributions based on their stochastic representation. This requires a sampling scheme for the respective radius variable R and a sampling scheme for a uniform distribution on the unit sphere. Besides this generic scheme, we present specific algorithms for the normal distribution and the t-distribution. In each case, a sample of the associated copula is obtained by standardizing the marginal laws. This requires the univariate marginal distribution functions.

4.6.1 *A Generic Sampling Scheme*

Algorithm 4.1 (Sampling Elliptical Distributions and Copulas)
To simulate a random vector from (the copula of) an elliptical distribution $\boldsymbol{X} \sim E_d(\boldsymbol{\mu}, \Sigma, \varphi)$, corresponding to the stochastic representation $\boldsymbol{X} \stackrel{d}{=} \boldsymbol{\mu} + A' R \boldsymbol{S}$ with $A' A = \Sigma$, perform the following steps:

(1) Simulate the radius R.
(2) Simulate $\boldsymbol{S}$, uniformly distributed on the k-dimensional unit sphere (see Algorithm 4.2).
(3) Compute (and return, if $\boldsymbol{X}$ is required) $\boldsymbol{X} := \boldsymbol{\mu} + A' R \boldsymbol{S}$.
(4) If a sample from the copula is to be returned, perform steps (1) to (3) with $\boldsymbol{\mu} = \boldsymbol{0}$ and standardize the univariate marginals via $U_i := F_i(X_i)$, $i = 1, \dots, d$, where F_i is the univariate marginal distribution function of X_i. Then return $\boldsymbol{U} := (U_1, \dots, U_d)'$.

Algorithm 4.2 (Sampling Uniformly on the d-Sphere)
A simple approach to sample uniformly on the d-dimensional unit sphere is the following rejection scheme. Repeat simulating $\boldsymbol{V} := (V_1, \ldots, V_d)'$, *where the* V_i*'s are i.i.d. with* $U[-1,1]$*-distribution, until* $\|\boldsymbol{V}\|^2 = V_1^2 + \ldots + V_d^2 \leq 1$. *Then take* $\boldsymbol{S} := \boldsymbol{V}/\|\mathbf{V}\|$. *This becomes slow in high dimensions, since the probability of accepting a sample is precisely the size of the d-dimensional unit ball divided by the size of the d-dimensional cube with side-length 2, which tends to 0 in d. A faster scheme for high-dimensional cases is given in Muller (1959), which consists of generating* $\boldsymbol{Z} = (Z_1, \ldots, Z_d)'$ *from a* $\mathcal{N}_d(\mathbf{0}, I)$*-distribution and taking* $\boldsymbol{S} := \boldsymbol{Z}/\|\mathbf{Z}\|$. *A faster scheme for low-dimensional cases is given in Marsaglia (1972). An example for* $d \in \{2,3\}$ *is provided in Figure 4.2.*

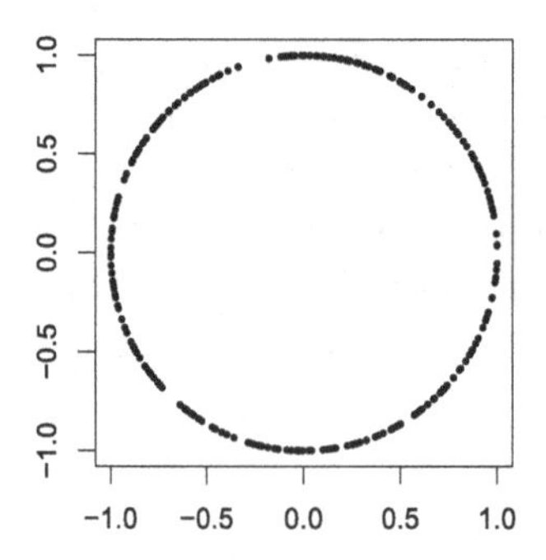

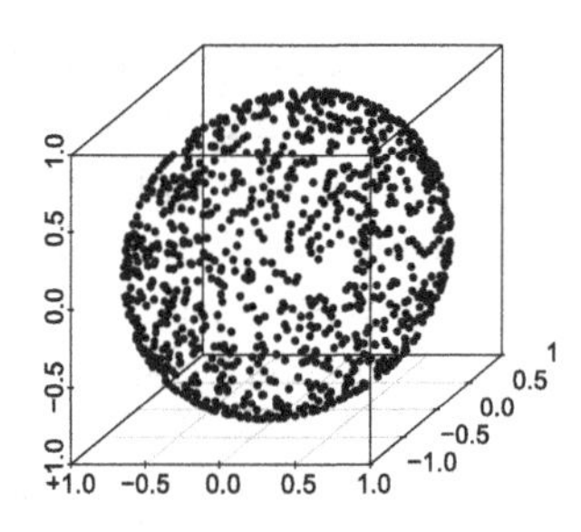

Fig. 4.2 Scatterplot of 250 samples from the uniform distribution on the two-sphere (left) and 1 000 samples from the uniform distribution on the unit three-sphere (right).

Example 4.4 ($R \equiv 1$ and $R \sim Geo(\vartheta)$)
Let us first visualize as a starting point the case $R \equiv 1$ *in dimension two. To simulate from the respective copula, we need the marginal law of* S_i, $i = 1, 2$. *This is found to be*[5]

$$F(x) := \mathbb{P}(S_1 \leq x) = \int_0^{2\pi} \mathbb{1}_{\{\cos(y) \leq x\}} \frac{dy}{2\pi} = \int_0^{\pi} \mathbb{1}_{\{y \geq arccos(x)\}} \frac{dy}{\pi}$$
$$= 1 - arccos(x)/\pi, \quad x \in (-1, 1).$$

[5]Note that $(S_1, S_2) \stackrel{d}{=} \big(\cos(U), \sin(U)\big)$ for $U \sim U[0, 2\pi]$. Moreover, we have identical marginal distributions $F_1 = F_2$ by exchangeability.

Given this, we simply have to simulate $\boldsymbol{S}$ *uniformly on the two-dimensional unit sphere and then transform the univariate marginals using* F*. The result is visualized in Figure 4.3 (left). To make the example more interesting, we now choose* $R \sim Geo(\vartheta)$, $\vartheta \in (0,1)$*. We compute the marginal law* G *of* $X_i = R\,S_i$, $i = 1, 2$, *by conditioning on* R *and obtain*

$$G(x) = \sum_{i=1}^{\infty} \mathbb{P}(R = i)\, F(x/i).$$

Then, we sample $R\,\boldsymbol{S}$ *and standardize the marginals using* G*. The result is displayed in Figure 4.3 (middle) and (right).*

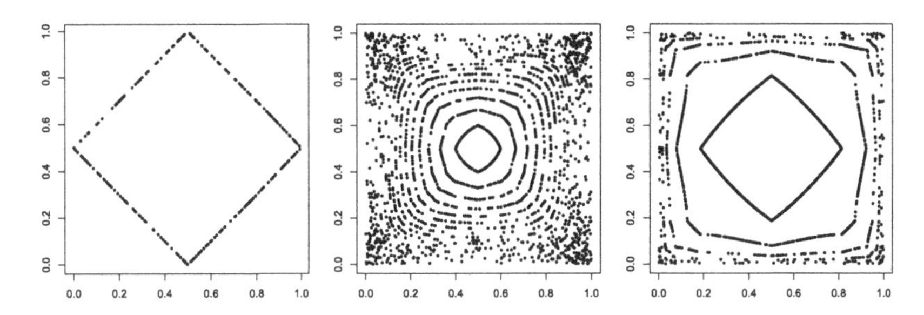

Fig. 4.3 Scatterplot of 250 samples from the bivariate copula of $\boldsymbol{S}$, the uniform distribution on the two-sphere (left), and 2 500 samples from the bivariate copula of $R\,\boldsymbol{S}$, where $R \sim \text{Geo}(0.1)$ (middle) and $R \sim \text{Geo}(0.5)$ (right).

4.6.2 *Sampling Important Parametric Families*

For the multivariate normal distribution (and the associated Gaussian copula) we have two stochastic representations at hand (compare Equation (4.9)). More convenient to implement is the representations based on a list of i.i.d. standard normal variates $(Z_1, \ldots, Z_d)' =: \boldsymbol{Z} \sim \mathcal{N}_d(\boldsymbol{0}, I)$, since it only requires a sampling scheme for independent univariate standard normal random variables. Hence, Algorithm 4.3 is formulated in this way, and we do not suggest using the generic Algorithm 4.1 for the Gaussian copula.

Algorithm 4.3 (Sampling the Gaussian Copula)
To simulate a random vector from C_P^{Gauss}, where $P \in \mathbb{R}^{d\times d}$ is a positive-definite correlation matrix, perform the following steps:

(1) Compute the Cholesky decomposition[6] of P, i.e. compute the matrix $L \in \mathbb{R}^{d\times d}$ with $L\,L' = P$, where $L \in \mathbb{R}^{d\times d}$ is a lower triangular matrix. Alternatively, one can compute $P^{1/2}$, where $P^{1/2}\,P^{1/2} = P$, and use it later on instead of L.
(2) Simulate a vector of independent standard normal random variables $\boldsymbol{Z} \sim \mathcal{N}_d(\boldsymbol{0}, I)$.
(3) Compute $\boldsymbol{X} := L\,\boldsymbol{Z} \sim \mathcal{N}_d(L\,\boldsymbol{0}, L\,I\,L') = \mathcal{N}_d(\boldsymbol{0}, P)$.
(4) Return the vector $\big(\Phi(X_1), \ldots, \Phi(X_d)\big)'$, where Φ is the distribution function of a univariate standard normal distribution.[7]

For an efficient implementation it is evident that if several random vectors have to be drawn, the required Cholesky decomposition needs to be computed only once. Scatterplots based on different parameters are visualized in Figure 4.4.

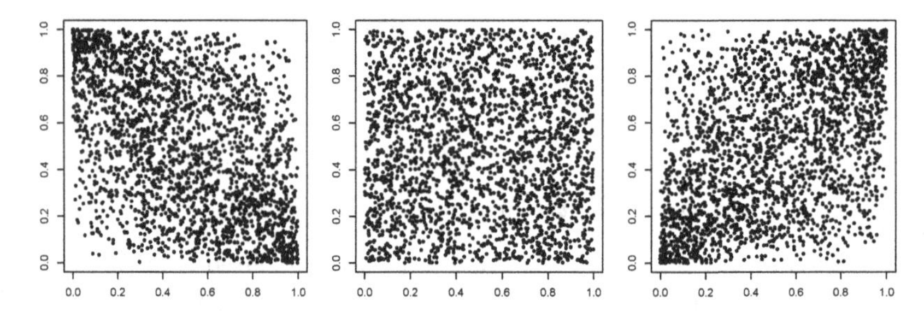

Fig. 4.4 Scatterplot of 2 500 samples from the bivariate Gaussian copula with $p = -0.5$ (left), 0 (middle), and 0.5 (right). Note that the dependence is increasing with p, and the case 0 corresponds to independence. Also note that the scatterplots have two axes of symmetry: (1) symmetry around the first diagonal (corresponding to exchangeability) and (2) symmetry around the second diagonal (corresponding to the fact that the Gaussian copula is radially symmetric).

---

[6]A derivation of this decomposition and explicit algorithms are given, e.g., in Golub and van Loan (1989, p. 97ff). Finding this decomposition requires effort in the order of $\mathcal{O}(d^3)$. Efficient schemes for sparse matrices and parallel computing might even improve the performance.

[7]Note that since P is a correlation matrix, the univariate marginal laws are $X_i \sim \mathcal{N}(0,1)$, $i = 1, \ldots, d$.

To sample from the t-copula $C^t_{P,\nu}$, we again start with independent $\mathcal{N}(0,1)$-distributed random variables. These are made dependent via (1) a linear transformation and (2) a mixing variable (see Equation (4.12)). In a final step, the marginals are transformed to $U[0,1]$-distributions.

Algorithm 4.4 (Sampling the t-Copula)
To simulate a random vector from $C^t_{P,\nu}$, where $P \in \mathbb{R}^{d\times d}$ is a positive-definite correlation matrix and ν the degrees of freedom, perform the following steps:

(1) Compute the Cholesky decomposition of P, i.e. compute the matrix $L \in \mathbb{R}^{d\times d}$ with $L\,L' = P$, where $L \in \mathbb{R}^{d\times d}$ is a lower triangular matrix. Alternatively, one can compute $P^{1/2}$, where $P^{1/2}\,P^{1/2} = P$, and use it later on instead of L.
(2) Simulate a vector $\boldsymbol{Z} = (Z_1,\dots,Z_d)'$ of independent $\mathcal{N}(0,1)$-distributed random variables.
(3) Simulate W from an $Inv\Gamma(\nu/2,\nu/2)$-distribution. Alternatively, one can draw $V \sim \chi^2(\nu)$.
(4) Compute $\boldsymbol{X} := \sqrt{W}\,L\,\boldsymbol{Z}$. Alternatively, one might use $\boldsymbol{X} := \sqrt{\nu/V}\,L\,\boldsymbol{Z}$. Both representations are equal in distribution.
(5) Return the vector $\big(t_\nu(X_1),\dots,t_\nu(X_d)\big)'$, where t_ν is the distribution function of a univariate t-distribution with ν degrees of freedom and mean 0.

Similar to the Gaussian case, when several random variables have to be drawn, the required Cholesky decomposition (alternatively the root of P) must only be computed once. Scatterplots of the t-copula are provided in Figure 4.5.

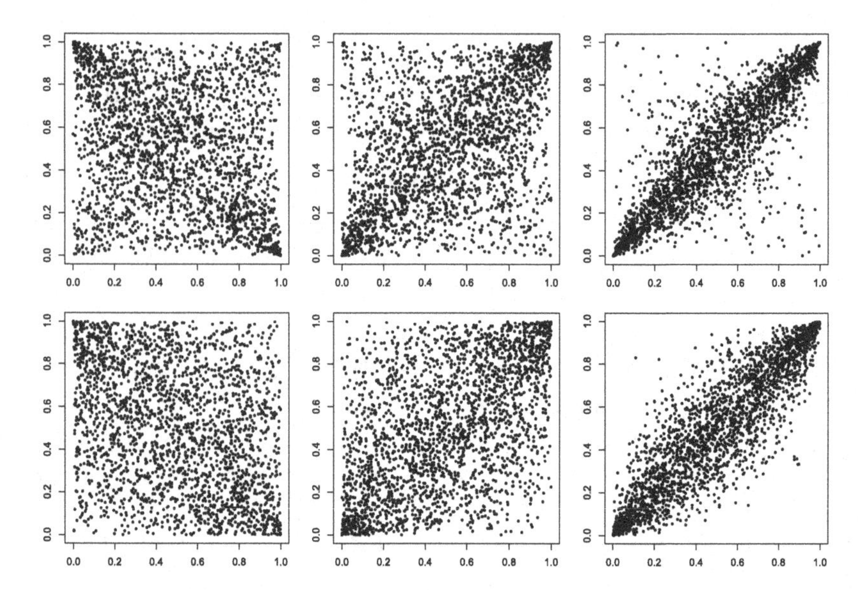

Fig. 4.5 Scatterplot of 2 500 samples from the bivariate t-copula with $p = -0.3$ (left), 0.5 (middle), and 0.9 (right). The degrees of freedom are set to $\nu = 2$ in the upper panel, and to $\nu = 10$ in the lower panel. Note that the dependence is increasing with p. Also note that the scatterplot has two axes of symmetry: (1) symmetry around the first diagonal (corresponding to exchangeability) and (2) symmetry around the second diagonal (corresponding to the fact that the Student's t-copula is radially symmetric).

Chapter 5

Pair Copula Constructions

This chapter was contributed by **Claudia Czado** and **Jakob Stöber**,[1] for which we would like to thank them very much.[2]

In the introductory section to this book it was noted that compared to the scarceness of work on multivariate copulas, there is an extensive literature on bivariate copulas and their properties. Pair copula constructions (PCCs) build high-dimensional copulas out of bivariate ones, thus exploiting the richness of the class of bivariate copulas and providing a flexible and convenient way to extend the bivariate theory to arbitrary dimensions. This construction principle was introduced in the pioneering work of Bedford and Cooke (2001a) and Bedford and Cooke (2002), generalizing an approach by Joe (1996). Aas et al. (2009) were the first to use PCCs in an inferential context and presented a more practical approach with algorithms to calculate likelihoods and to do simulations.

The earliest research on pair copulas arose from questions related to the *construction* of a multivariate distribution and sought to derive a general principle for finding distributions fulfilling certain given specifications. This view on the subject is an appropriate starting point for obtaining existence theorems and motivating the graphical description using vines. From a statistical point of view, however, more important issues include whether a multivariate distribution can be *decomposed* into bivariate (pair) copulas and the distributions of the margins and which assumptions need to be made in order to do so. Starting with the decomposition of a given distribution has the benefit that it is not only very intuitive but also shows the restrictions of regular vine constructions and why not all multivariate copulas can be built using a PCC.

[1]Lehrstuhl für Mathematische Statistik, Technische Universität München, Parkring 13, 85748 Garching-Hochbrück, Germany, `cczado@ma.tum.de` and `stoeber@ma.tum.de`.

[2]The numerical computations were performed on a Linux cluster supported by DFG grant INST 95/919-1 FUGG.

This chapter is organized as follows. First, Section 5.1 introduces PCCs. Section 5.2 formalizes the description of PCCs by regular vines and introduces a matrix notation which can conveniently be used to describe algorithms for the generation of random samples and facilitate their implementation in computer code. After that, some further remarks on notation and on bivariate copulas are made in Section 5.3, before outlining an algorithm to simulate from PCCs. Section 5.4 discusses the dependence properties realizable by PCCs and Section 5.5 concludes with an application to Value at Risk (VaR) forecasting.

5.1 Introduction to Pair Copula Constructions

Consider d random variables $\mathbf{X} = (X_1, \ldots, X_d)$ with joint density function $f_{1:d}(x_1, \ldots, x_d)$. This density can be recursively factorized using conditional densities:

$$f_{1:d}(x_1, \ldots x_d) = f_1(x_1) \cdot f_{2|1}(x_2|x_1) \cdot f_{3|2,1}(x_3|x_1, x_2) \cdot \ldots \\ \cdot f_{d|1:(d-1)}(x_d|x_1, \ldots, x_{d-1}). \tag{5.1}$$

This expression is unique up to relabeling. We will now apply Sklar's theorem (Theorem 1.2) to the conditional densities and thereby decompose a multivariate density into bivariate copula densities and densities of one-dimensional margins. By differentiating formula (1.6) corresponding to a distribution with joint density $f(x_1, \ldots, x_d)$, marginal densities f_j and marginal cumulative distribution functions (cdfs) F_j, $j = 1, \ldots, d$, we obtain

$$f_{1:d}(x_1, \ldots x_d) = \\ c_{1:d}(F_1(x_1), F_2(x_2), \ldots, F_d(x_d)) \cdot f_1(x_1) \cdot f_2(x_2) \cdot \cdots \cdot f_d(x_d)$$

for some d-variate copula density $c_{1:d}(.)$. In the bivariate case, this simplifies to

$$f_{1,2}(x_1, x_2) = c_{1,2}(F_1(x_1), F_2(x_2)) \cdot f_1(x_1) \cdot f_2(x_2)$$

and yields

$$f_{1|2}(x_1|x_2) = c_{1,2}(F_1(x_1), F_2(x_2)) \cdot f_1(x_1) \tag{5.2}$$

for the conditional density of X_1, given $X_2 = x_2$. Equation (5.2) can be stepwisely applied to each factor on the right-hand side of (5.1) in order to decompose the multivariate density into bivariate copula densities and densities of one-dimensional margins as desired. For example, the factor $f(x_3|x_1, x_2)$ can be decomposed into

$$f_{3|2,1}(x_3|x_1, x_2) = c_{3,2|1}(F_{3|1}(x_3|x_1), F_{2|1}(x_2|x_1)) \cdot f_{3|1}(x_3|x_1),$$

and generally we have for $X \notin \mathbf{X} = (X_1, \ldots, X_d)$

$$f_{X|\mathbf{X}}(x|\mathbf{x}) = c_{X,X_j|\mathbf{X}_{-j}}(F_{X|\mathbf{X}_{-j}}(x|\mathbf{x}_{-j}), F_{X_j|\mathbf{X}_{-j}}(x_j|\mathbf{x}_{-j})) \cdot f_{X|\mathbf{X}_{-j}}(x|\mathbf{x}_{-j}),$$

where $\mathbf{x}_{-j}$ denotes the vector $\mathbf{x}$ without the jth component and $F_{X|\mathbf{X}_{-j}}(x|\mathbf{x}_{-j})$ the conditional distribution function of X, given $\mathbf{X}_{-j} = \mathbf{x}_{-j}$ evaluated at x. For this decomposition we assume, as already suggested by the notation, that the copula $c_{X,X_j|\mathbf{X}_{-j}}$ of the conditional distribution depends on the realization $\boldsymbol{x}_{-j}$ of $\boldsymbol{X}_{-j}$ only through its arguments. In general, $c_{X,X_j|\mathbf{X}_{-j}}$ will of course also depend on $\boldsymbol{x}_{-j}$ but we have to make this assumption in order to keep inference and in particular model selection possible, fast, and tractable. For the existence theorem of Section 5.2 this assumption is not necessary. Practitioners, however, will almost always be interested in estimation problems, too, and for this reason we keep the assumption throughout this chapter and use it to simplify our notation where possible. Because of this assumption, not all multivariate distributions (with densities) can be modeled using a (simplified) pair copula construction as we understand it. For examples on which distributions can be expressed using a PCC and which not, we refer the reader to Haff et al. (2010).

Let us now illustrate a complete decomposition of (5.1) in five dimensions given as

$$\begin{aligned} f(x_1, \ldots x_5) = f(x_1) \cdot f(x_2|x_1) \cdot f(x_3|x_1, x_2) \cdot \\ \cdot f(x_4|x_1, x_2, x_3) \cdot f(x_5|x_1, x_2, x_3, x_4). \end{aligned} \tag{5.3}$$

To shorten the notation, we define

$$\begin{aligned} F_{i|i_1,\ldots,i_r} &:= F_{i|i_1,\ldots,i_r}(x_i|x_{i_1}, \ldots, x_{i_r}), \\ f_{i|i_1,\ldots,i_r} &:= f_{i|i_1,\ldots,i_r}(x_i|x_{i_1}, \ldots, x_{i_r}), \\ c_{i,j|i_1,\ldots,i_r} &:= c_{i,j|i_1,\ldots,i_r}\left(F_{i|i_1,\ldots,i_r}, F_{j|i_1,\ldots,i_r}\right), \end{aligned}$$

for $i \neq j$ and additional distinct indices $i_1, \ldots, i_r$. If the arguments of a density are denoted and it is clear from the context which density we refer to, the notation of indices is omitted.

Example 5.1 (A Five-Dimensional Example)
Applying Equation (5.2), Expression (5.3) is assumed to be further decomposable as follows:

$$
\begin{aligned}
f(x_2|x_1) &= c_{2,1}(F_2, F_1) \cdot f(x_2), && (5.4a)\\
f(x_3|x_1, x_2) &= c_{3,2|1}(F_{3|1}, F_{2|1}) \cdot f(x_3|x_1) \\
&= c_{3,2|1}(F_{3|1}, F_{2|1}) \cdot c_{3,1}(F_3, F_1) \cdot f(x_3), && (5.4b)\\
f(x_4|x_1, x_2, x_3) &= c_{4,2|1,3} \cdot f(x_4|x_1, x_3) = c_{4,2|1,3} \cdot c_{4,1|3} \cdot f(x_4|x_3) \\
&= c_{4,2|1,3} \cdot c_{4,1|3} \cdot c_{4,3} \cdot f(x_4), && (5.4c)\\
f(x_5|x_1, x_2, x_3, x_4) &= c_{5,2|1,3,4} \cdot f(x_5|x_1, x_3, x_4) \\
&= c_{5,2|1,3,4} \cdot c_{5,4|1,3} \cdot f(x_5|x_1, x_3) \\
&= c_{5,2|1,3,4} \cdot c_{5,4|1,3} \cdot c_{5,3|1} \cdot f(x_5|x_1) \\
&= c_{5,2|1,3,4} \cdot c_{5,4|1,3} \cdot c_{5,3|1} \cdot c_{5,1} \cdot f(x_5). && (5.4d)
\end{aligned}
$$

This gives the expression for the joint density

$$
\begin{aligned}
f(x_1, \dots, x_5) = \; & c_{5,2|4,3,1} \cdot c_{5,4|3,1} \cdot c_{4,2|3,1} \cdot c_{5,3|1} \cdot c_{4,1|3} \\
& \cdot \; c_{3,2|1} \cdot c_{5,1} \cdot c_{4,3} \cdot c_{3,1} \cdot c_{2,1} \qquad (5.5)\\
& \cdot \; f(x_5) \cdot f(x_4) \cdot f(x_3) \cdot f(x_2) \cdot f(x_1).
\end{aligned}
$$

Hereby the conditional distribution functions which form the copula densities' arguments can themselves be obtained from copula cdfs. For example, the second argument of $c_{4,2|3,1}$ is $F_{2|3,1}$ and can be expressed using only lower-level copulas (i.e. copulas in which we condition on fewer variables) in the following way:

$$
F(x_2|x_1, x_3) = \frac{\partial C_{2,3|1}(F(x_2|x_1), F(x_3|x_1))}{\partial F(x_3|x_1)}, \qquad (5.6)
$$

where $F(x_3|x_1)$ and $F(x_2|x_1)$ can be expressed similarly using $c_{3,1}$ and $c_{2,1}$. To see this, we determine, using the chain rule for differentiation, the rule for differentiation under the integral sign, and Equation (5.2), that

$$\begin{aligned}
F(x_2|x_1,x_3) &= \int_{-\infty}^{x_2} f(y_2|x_1,x_3)dy_2 \\
&= \int_{-\infty}^{x_2} c_{2,3|1}(F(y_2|x_1),F(x_3|x_1))f(y_2|x_1)dy_2 \\
&= \int_{-\infty}^{x_2} \frac{\partial}{\partial F(y_2|x_1)} \frac{\partial}{\partial F(x_3|x_1)} C_{2,3|1}(F(y_2|x_1),F(x_3|x_1))f(y_2|x_1)dy_2 \\
&= \frac{\partial}{\partial F(x_3|x_1)} \int_{-\infty}^{x_2} \frac{\partial}{\partial y_2} C_{2,3|1}(F(y_2|x_1),F(x_3|x_1))dy_2 \\
&= \frac{\partial}{\partial F(x_3|x_1)} C_{2,3|1}(F(x_2|x_1),F(x_3|x_1)).
\end{aligned}$$

Obviously, this also holds when conditioning on more than one variable and

$$F(x|\mathbf{x}) = \frac{\partial C_{X,X_j|\mathbf{X}_{-j}}\left(F(x|\mathbf{x}_{-j}),F(x_j|\mathbf{x}_{-j})\right)}{\partial F(x_j|\mathbf{x}_{-j})} \tag{5.7}$$

does hold in general for $X \notin \mathbf{X} = (X_1,\ldots,X_d)$. If we want to obtain expressions for the arguments of the bivariate densities which only contain bivariate copulas already used in the decomposition and which do not require integration, we will have to make a "clever choice" in each step. How this recursive calculation works in Example 5.1 is illustrated in Figure 5.1. For the first arguments of the conditional copula densities in (5.4), marked with circles in Figure 5.1, it is clear that the copulas (marked with diamonds) from which they can be calculated using Expression (5.7) are available. They are obtained from decomposing the remaining conditional density in Sklar's theorem (5.2), e.g. $f(x_5|x_1,x_3,x_4)$ after applying Sklar's theorem on $f(x_5|x_1,\ldots,x_4)$ in (5.4d).

For the second arguments, which are left unmarked in Figure 5.1, however, the necessary copulas are in general not available. For example, $F_{2|1,3}$, the second argument of $c_{4,2|1,3}$ in (5.4c), is obtained from the copula with density $c_{3,2|1}$, chosen in (5.4b).

If we had chosen the copula density $c_{4,1|2,3}$ instead of $c_{4,2|1,3}$ in (5.4c), the density of neither $C_{3,1|2}$ nor $C_{2,1|3}$, from which the argument $F_{1|2,3}$ can be calculated, would have been included in the decomposition (Figure 5.2). But how to systematically make a "clever choice" in each step? The arguments of copulas conditioned on d variables will always have to be expressed by copulas conditioned on $d-1,\ldots,1$ variables. It is thus natural to use a bottom-up approach, since possible choices on the "higher" levels depend on the choices made on the "lower" levels. To illustrate how this

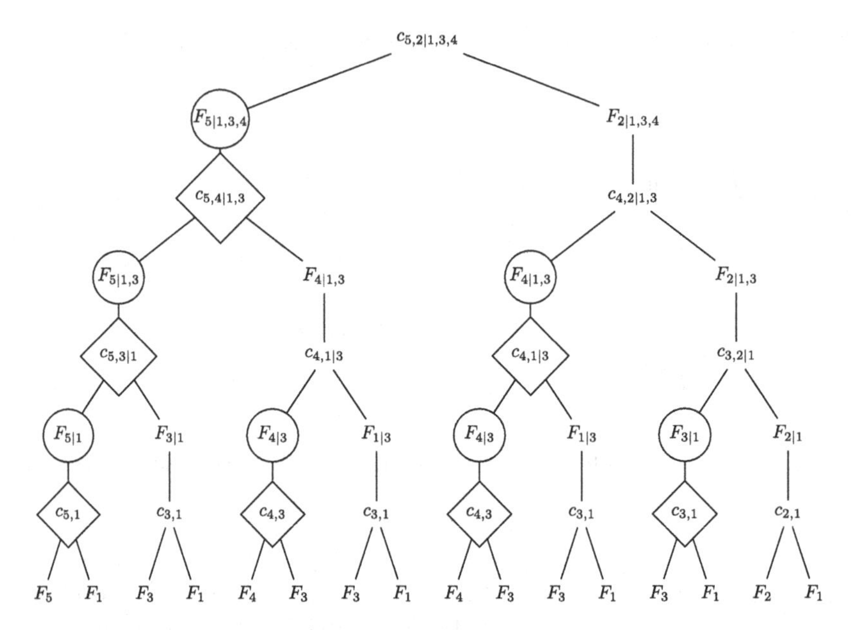

Fig. 5.1 Following this chart, the arguments of all copulas in decomposition (5.5) for Example 5.1 can be obtained recursively.

works, the structure of the decomposition in Example 5.1 is displayed by a sequence of trees in Figure 5.3.

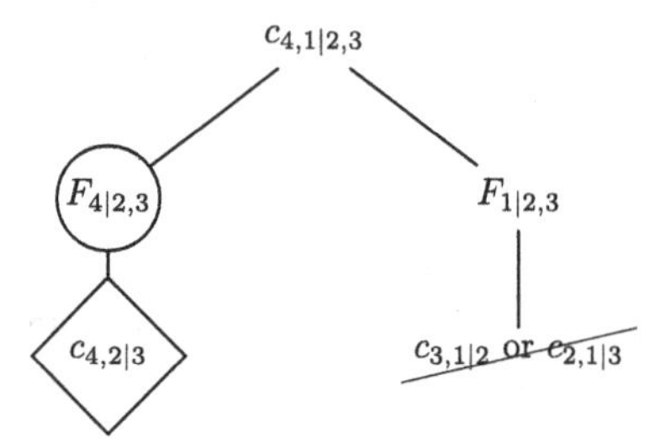

Fig. 5.2 If we choose $c_{4,1|3,2}$ instead of $c_{4,2|3,1}$ in (5.4c), the copulas required for obtaining the second argument $F_{1|3,2}$ are not included in (5.4a,b,d).

In the first tree, i.e. T_1, we have all the bivariate copulas of decomposition (5.5), each $C_{i,j}$ pictured as an edge linking the univariate margins i and j (pictured as nodes). For T_2 we have the bivariate copula indices which were the edges in tree T_1 as nodes. The edges are the indices of the conditional copulas conditioned on one variable. They link the nodes from which their arguments are obtained using an expression as in (5.7). Continuing in a similar fashion, all bivariate copulas required can be identified by four trees. Consider, for example, the density

$$c_{4,2|1,3}(F(x_4|x_1,x_3),F(x_2|x_1,x_3)),$$

where the arguments are obtained from

$$F(x_4|x_1,x_3)=\frac{\partial C_{4,1|3}(F(x_4|x_3),F(x_1|x_3))}{\partial F(x_1|x_3)}$$

and

$$F(x_2|x_1,x_3)=\frac{\partial C_{2,3|1}(F(x_2|x_1),F(x_3|x_1))}{\partial F(x_3|x_1)}.$$

It is displayed as edge $2,4|1,3$ between nodes $2,3|1$ and $1,4|3$. Note that we will usually denote edges in ascending order, e.g. $2,4|1,3$. For copulas, the ordering in the notation is always chosen such that $C_{i,j|i_1,\dots,i_r}$ has first argument $F_{i|i_1,\dots,i_r}$ and second argument $F_{j|i_1,\dots,i_r}$, for $i\neq j$ and additional distinct indices $i_1,\dots,i_r$.

In the following chapter regular vine tree sequences which are sequences of trees as in Figure 5.3 and which correspond to valid decompositions having the desired properties are defined.

5.2 Copula Construction by Regular Vine Trees

We begin with the graph theoretical definition of our main object, the regular vine (R-vine) tree sequence, in Section 5.2.1, where we also state the general existence result for R-vine distributions. Section 5.2.2 continues by introducing a matrix notation for storing R-vine tree sequences. For the general theory on vines our presentation follows Bedford and Cooke (2001a) and Bedford and Cooke (2002). The introduction to the matrix notation is taken from Dißmann (2010).

5.2.1 *Regular Vines*

To define R-vines, we recall some basic definitions from graph theory. Let N be a set and E be a set of possible combinations from N, i.e.

$$E\subset\{\{n_1,n_2\}|n_1,n_2\in N\}.$$

Then we call the tuple $G=(N,E)$ an *undirected graph*, the elements of N *nodes*, and the elements of E *edges*. A *path* in a graph can be defined

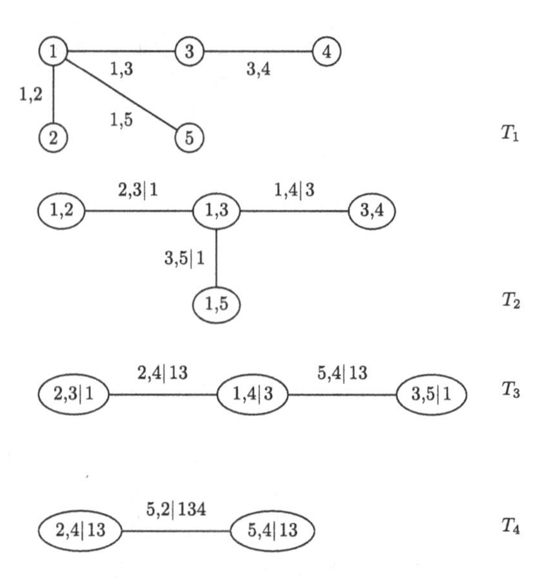

Fig. 5.3 Graphical illustration of decomposition (5.5) for Example 5.1, from the lowest level in tree T_1 to the highest level in tree T_4. A line in the graph corresponds to the indices of a copula linking two distributions on a lower level through conditioning.

as a sequence of nodes $(n_1, \ldots, n_k) \in N^k$, $k \geq 2$, where N^k denotes the k-fold Cartesian product, such that for each node n_i there exists an edge connecting it to the next. That means

$$\{n_i, n_{i+1}\} \in E \quad \text{for} \quad i = 1, \ldots, k-1,$$

and we call the sequence a *cycle* if $n_1 = n_k$. A *tree* $T = (N, E)$ is an acyclic undirected graph (i.e. we cannot find any sequence from N that is a cycle), which is *connected*, i.e. for all nodes $a, b \in T_j$, $j = 1, \ldots, d-1$, there exists a path $(n_1, \ldots, n_k) \in N_j^k$ with $a = n_1$, $b = n_k$.

Based on the above graph theoretic notation, R-vine tree sequences can be defined as follows.

Definition 5.1 (Regular Vine Tree Sequence)
$\mathcal{V} = (T_1, \ldots, T_{d-1})$ *is a regular vine tree sequence on d elements if:*

(1) T_1 is a tree with nodes $N_1 = \{1, \ldots, d\}$ and a set of edges E_1.
(2) For $j \geq 2$, T_j is a tree with nodes $N_j = E_{j-1}$ and edges E_j.
(3) For $j = 2, \ldots, d-1$ and $\{a, b\} \in E_j$ it must hold that $|a \cap b| = 1$.

The last property is called the *proximity condition.* It ensures that if there is an edge e connecting a and b in T_j, $j \geq 2$, then a and b must share a common node in T_{j-1}.

Figure 5.4 shows the corresponding R-vine tree sequence of Example 5.1, this time using the set formalism from Definition 5.1 instead of the indices in the pair copula decomposition. Understanding how the different forms of labeling correspond to each other is the key to understanding how an R-vine corresponds to a copula decomposition.

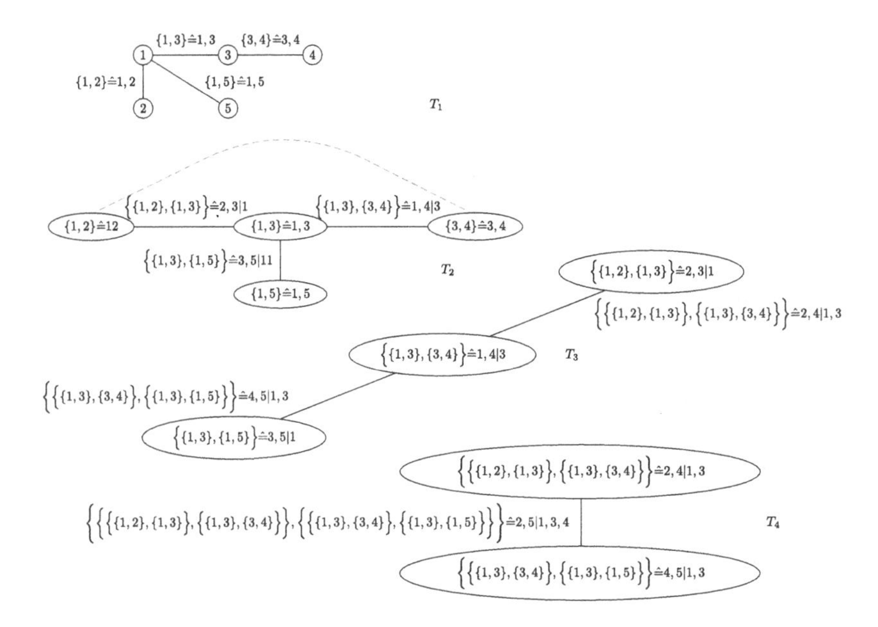

Fig. 5.4 The R-vine from Example 5.1, displayed using the set formalism. The dashed line shows an edge which is not allowed due to the proximity condition, since edges $\{1, 2\}$ and $\{3, 4\}$ do not share a common node in T_1.

We will see that if we specify a bivariate copula for every edge in an R-vine we find a multivariate copula where the copulas of this R-vine specification occur as conditional copulas. To assign copulas to edges, some further notation is needed. For $e \in E_i$ define

$$A_e := \left\{ j \in N_1 \middle| \exists\ e_1 \in E_1, \ldots, e_{i-1} \in E_{i-1} : j \in e_1 \in \ldots \in e_{i-1} \in e \right\}.$$

A_e is called the *complete union* of e. The *conditioning set* of an edge $e = \{a, b\}$ is

$$D_e := A_a \cap A_b$$

and the *conditioned sets* are given by

$$\mathcal{C}_{e,a} := A_a \setminus D_e, \quad \mathcal{C}_{e,b} := A_b \setminus D_e, \quad \mathcal{C}_e := \mathcal{C}_{e,a} \cup \mathcal{C}_{e,b}.$$

Let us illustrate these sets using edge $e = \{\{1,2\},\{1,3\}\}$ of Example 5.1. The complete union consists of all nodes on the level of T_1 which are contained in e, i.e. $A_e = \{1,2,3\}$. We further have $D_e = \{1\}$, since $A_a = \{1,2\}$ and $A_b = \{1,3\}$ and subsequently $\mathcal{C}_e = \{2,3\}$.

Until this point the R-vine tree sequence defined earlier is solely a graph theoretic object and does not include any stochastic component. We establish this by defining what we mean when we say that a vector of random variables has an R-vine distribution.

Definition 5.2 (Regular Vine Distribution)
A joint distribution F on d random variables $X_1,\dots,X_d$ is called a regular vine distribution, if we can find a tuple $(\mathcal{F},\mathcal{V},B)$ such that:

(1) ***Marginal distributions:*** *$\mathcal{F} = (F_1,\dots,F_d)$ is a vector of continuous invertible marginal distribution functions, representing the marginal distribution functions of X_i, $i = 1,\dots,d$.*
(2) ***Regular vine tree sequence:*** *$\mathcal{V}$ is an R-vine tree sequence on d elements.*
(3) ***Bivariate conditional distributions:***

$$B = \{B_e | i = 1,\dots,d-1; e \in E_i\}$$

where B_e is a symmetric bivariate copula with density and E_i are the edge sets of the R-vine.
(4) ***Connection between tree sequence and bivariate (conditional) distributions:*** *For each $e \in E_i$, $i = 1,\dots,d-1$, $e = \{a,b\}$, B_e is the corresponding copula to the conditional distribution of $X_{\mathcal{C}_{e,a}}$ and $X_{\mathcal{C}_{e,b}}$ given $\boldsymbol{X}_{D_e} = \boldsymbol{x}_{D_e}$. $B_e(.,.)$ does not depend on $\boldsymbol{x}_{D_e}$.*

The copula B_e for edge $e = \{\{1,2\},\{1,3\}\}$ of Example 5.1 means that the conditional bivariate distribution function of (X_2,X_3), given $X_1 = x_1$, is given by

$$F(x_2,x_3|x_1) = B_e\left(F(x_2|x_1), F(x_3|x_1)\right).$$

From now on, we will denote the copula B_e corresponding to edge e by $C_{\mathcal{C}_{e,a}\mathcal{C}_{e,b}|D_e}$ (and the corresponding density by $c_{\mathcal{C}_{e,a}\mathcal{C}_{e,b}|D_e}$), as we already did for the density $c_{i,j|i_1,\dots,i_r}$ of the copula corresponding to the conditional distribution of X_i and X_j given $X_{i_1} = x_{i_1},\dots,X_{i_r} = x_{i_r}$ in the introduction.

Remark 5.1 (Non-Symmetric Copulas)
Having defined the regular vine tree sequence as a collection of undirected graphs in Definition 5.1, we have to restrict ourselves to symmetric copulas in Definition 5.2. This is, however, only a formal restriction as the theory presented in this chapter remains true when introducing directed edges, but it is a common assumption in the literature on vines in order to be able to use the convenient set notation. In simulations and applications we will be able to use non-symmetric copulas, and we will provide advice on how to treat this case throughout the text.

If we start with a tuple $(\mathcal{F}, \mathcal{V}, B)$ which has properties (1)–(3) of Definition 5.2, we can always go in the inverse direction and find a distribution F to which they correspond.

Theorem 5.1 (Existence of Regular Vine Distributions)
Let $(\mathcal{F}, \mathcal{V}, B)$ have properties (1)–(3) of Definition 5.2. Then there is a unique distribution with density

$$f_{1,\dots,d} = f_1 \cdot \dots \cdot f_d \cdot \prod_{i=1}^{d-1} \prod_{e \in E_i} c_{C_{e,a}C_{e,b}|D_e}(F_{C_{e,a}|D_e}, F_{C_{e,b}|D_e}) \tag{5.8}$$

such that for each $e \in E_i$, $i = 1, \dots, d-1$, with $e = \{a, b\}$, we have for the cdf of $X_{C_{e,a}}$ and $X_{C_{e,b}}$, given $\boldsymbol{X}_{D_e}$,

$$F\left(x_{C_{e,a}}, x_{C_{e,b}} | \boldsymbol{x}_{D_e}\right) = B_e\left(F(x_{C_{e,a}} | \boldsymbol{x}_{D_e}), F(x_{C_{e,b}} | \boldsymbol{x}_{D_e})\right),$$

and that the one-dimensional margins are given by $F(x_i) = F_i(x_i)$, $i = 1, \dots, d$.

Proof. The proof can be found in Bedford and Cooke (2001a). The mere existence can be shown more elegantly by generalizing regular vines to Cantor trees (which allow for more general forms of dependence) as demonstrated in Bedford and Cooke (2002). □

This existence and uniqueness result justifies the definition of regular vine distributions in terms of the tuple $(\mathcal{F}, \mathcal{V}, B)$ from which they can always be regained. A *regular vine copula* is a regular vine distribution where all margins are uniformly distributed on $[0, 1]$. As the pair copula constructions and decompositions from the introduction are realized as regular vine copulas, these terms will from now on be used equivalently.

Remark 5.2 (Directed Graphs)
The existence result stated in Theorem 5.1 still holds when we introduce ordered pairs and switch to directed graphs. This influences Equation (5.8)

only by specifying which element of an edge is "a" and which is "b". Without explicitly saying so, Bedford and Cooke (2001a) already showed this. Theorem 3 of their paper is applicable to the new situation without changes.

Counting all possibilities of choosing edges, Morales-Nápoles (2010) shows that there are $d!/2 \cdot 2^{\binom{d-2}{2}}$ R-vines in d dimensions. In many applications, however, only two subclasses of R-vines are used.

Definition 5.3 (C-Vine Tree Sequence, D-Vine Tree Sequence) *A regular vine tree sequence $\mathcal{V} = (T_1, \ldots, T_{d-1})$ is called:*

(1) A D-vine *tree sequence if for each node $n \in N_i$ we have*

$$|\{e \in E_i | n \in e\}| \leq 2.$$

(2) A C-vine *tree sequence if in each tree T_i there is one node $n \in N_i$ such that $|\{e \in E_i | n \in e\}| = d - i$.*

In particular Aas et al. (2009) have developed inference methods for D- and C-vine distributions. Applying (5.8) to these special cases, the density of a C-vine distribution can be expressed as

$$\prod_{k=1}^{d} f(x_k) \prod_{j=1}^{d-1} \prod_{i=1}^{d-j} c_{j,j+i|1,\ldots,j-1} \left(F(x_j|x_1,\ldots,x_{j-1}), F(x_{i+j}|x_1,\ldots,x_{j-1})\right)$$

and for the D-vine distribution as

$$\prod_{k=1}^{d} f(x_k) \prod_{j=1}^{d-1} \prod_{i=1}^{d-j} \times$$
$$\times\, c_{i,j+i|i+1,\ldots,i+j-1} \left(F(x_i|x_{i+1},\ldots,x_{i+j-1}), F(x_{i+j}|x_{i+1},\ldots,x_{i+j-1})\right).$$

These expressions are unique up to relabeling, and there are $d!/2$ distinct D-vines and $d!/2$ distinct C-vines in d dimensions. A graphical illustration of these special tree structures can be seen in Figure 5.5. For D-vine tree sequences, it is sufficient to require $|\{e \in E_i | n \in e\}| \leq 2$ for $n \in N_1$. The condition for $i \geq 1$ follows then by the proximity condition.

5.2.2 *Regular Vine Matrices*

While the notation from graph theory is convenient for giving an intuitive picture of a regular vine, it is less useful for describing algorithms and their implementation into computer code. We will now develop a shorter matrix notation describing the tree structure of an R-vine to be applied in

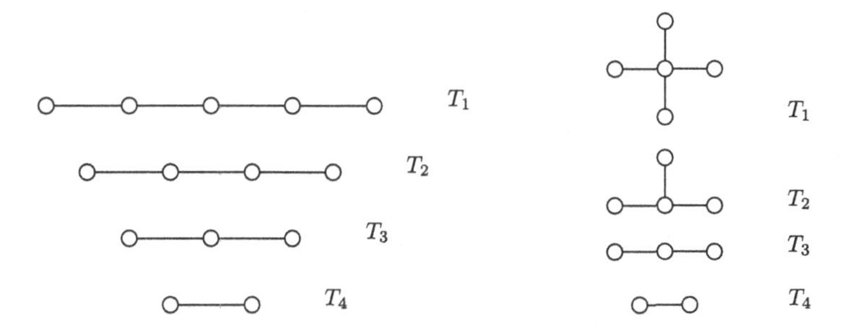

Fig. 5.5 The tree structures of a D-vine (on the left) and of a C-vine (on the right) in four dimensions. Since the structures are unique up to permutations on the basic level, labeling is omitted.

programming. Such a matrix language was first introduced in Kurowicka (2009). Our discussion follows Dißmann (2010).

The idea is to exploit the tree structure in order to store the sets $\{\{\mathcal{C}_{e,a}, \mathcal{C}_{e,b}\}, D_e\}$ in an upper triangular matrix. These sets contain all information about the R-vine tree sequence. For this let M be an upper triangular matrix with entries $m_{i,j}, i \leq j$. Each entry $m_{i,j}$ is allowed to take integer values from 1 to d.

Definition 5.4 (Regular Vine Matrix)
M is called a regular vine matrix *if it satisfies the following conditions:*

(1) $\{m_{1,i}, \ldots, m_{i,i}\} \subset \{m_{1,j}, \ldots, m_{j,j}\}$ *for* $1 \leq i < j \leq d$ *(the entries of a selected column are also contained in all columns to the right of that column).*
(2) $m_{i,i} \notin \{m_{1,i-1}, \ldots, m_{i-1,i-1}\}$ *(the diagonal entry of a column is not contained in any column further to the left).*
(3) For all $i = 3, \ldots, d$, $k = 1, \ldots, i-1$ *there exist* (j, l) *with* $j < i$ *and* $l < j$ *such that*

$$\left\{m_{k,i}, \{m_{1,i}, \ldots, m_{k-1,i}\}\right\} = \left\{m_{j,j}, \{m_{1,j}, \ldots, m_{l,j}\}\right\} \quad or$$
$$\left\{m_{k,i}, \{m_{1,i}, \ldots, m_{k-1,i}\}\right\} = \left\{m_{l,j}, \{m_{1,j}, \ldots, m_{l-1,j}, m_{j,j}\}\right\}.$$

The last property in this definition is the counterpart of the proximity condition for regular vine trees. Using this formal definition of an R-vine matrix one can prove that there is a bijection between regular vine trees and regular vine matrices, given by the algorithms we will outline next.

For the practitioner, Definition 5.4 is mainly relevant for constructing a valid R-vine matrix directly or when wanting to perform a tree structure selection on the matrix level. While we believe it to be very important for a mathematically sound introduction to R-vine copula modeling to introduce all needed definitions, we will, within the application-oriented scope of this book, once again omit proofs in the following section and focus on an informal description of the algorithms.

Algorithm 5.1 (Finding a Regular Vine Matrix for a Regular Vine)

The input for the algorithm is a regular vine $\mathcal{V} = (T_1, \ldots, T_{d-1})$ and the output will be a regular vine matrix M.

$\mathcal{X} := \{\}$ (1)

FOR $i = d, \ldots, 3$ (2)

  Choose $x, \tilde{x}, D$ *with* $\tilde{x} \notin \mathcal{X}$ *and* $|D| = i - 2$ *s.t.* $\exists$ *an edge* e *with* $C_e = \{x, \tilde{x}\}, D_e = D$ (3a)

  $m_{i,i} := x, m_{i-1,i} := \tilde{x}$ (3b)

  FOR $k = i - 2, \ldots, 1$ (4)

    Choose $\hat{x}$ *s.t.* $\exists$ *an edge* e *with* $C_e = \{x, \hat{x}\}$ *and* $|D_e| = k - 1$ (5a)

    $m_{k,i} := \hat{x}$ (5b)

  END FOR

  $\mathcal{X} := \mathcal{X} \cup \{x\}$ (6)

END FOR

Choose $x, \tilde{x} \in \{1, \ldots, d\} \setminus \mathcal{X}$ (7a)

$m_{2,2} := x, m_{1,2} := \tilde{x}, m_{1,1} := \tilde{x}$ (7b)

RETURN $M := (m_{k,i} | k = 1, \ldots, d,\ k \leq i)$ (8)

Intuitively this means the following. In the first step we select one of the elements of the conditioned set of the single edge in tree T_{d-1}, i.e. 5 or 2 from edge $5, 2|1, 3, 4$ in Example 5.1, and put it in the lower right corner of a d-dimensional matrix. Selecting, for example, the element 5 we write down all numbers which are in the conditioned sets of an edge together with 5 (bolded in Figure 5.6) ordered by the levels of the tree above this entry. In particular this identifies the edges $5, 2|1, 3, 4$; $5, 4|1, 3$; $3, 5|1$; and $1, 5$.

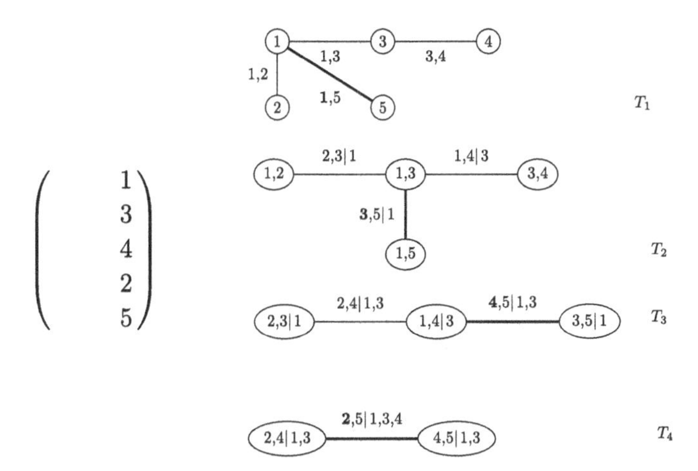

Fig. 5.6 Construction of the right column of the corresponding R-vine matrix for Example 5.1. We select 5 in the highest edge, and all entries which are in a conditioned set together with 5.

Thereby all information about how X_5 depends on $X_1, \ldots, X_4$ is stored in the right column of the matrix and we remove all nodes and edges of the vine containing 5. These are exactly the ones which we have just identified for 5 in the conditioned set, and we end up with a reduced vine tree sequence as given in Figure 5.7.

With this second vine we repeat the described procedure, selecting, e.g., 2 in the highest tree and putting it on the diagonal of the matrix. Adding the entries which are in the conditioned sets together with 2, the matrix becomes the following:

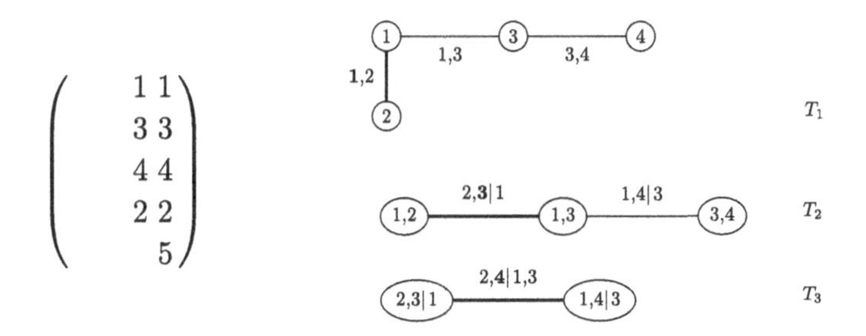

Fig. 5.7 The reduced vine after the first step.

After this, the selected nodes are removed and the resulting reduced vine tree sequence is displayed in Figure 5.8. The steps outlined are repeated

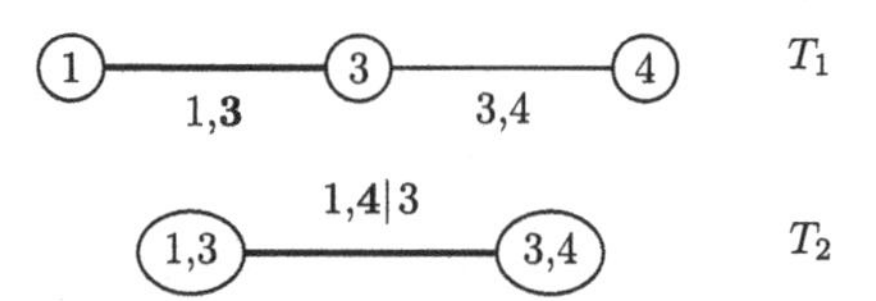

Fig. 5.8 Construction of the next column in the R-vine matrix: Here we select 1 in the highest edge and then all entries which are in a conditioned set together with 1.

until all nodes of the original vine have been removed and we obtain the final matrix

$$\begin{pmatrix} 4 & 4 & 3 & 1 & 1 \\ & 3 & 4 & 3 & 3 \\ & & 1 & 4 & 4 \\ & & & 2 & 2 \\ & & & & 5 \end{pmatrix}. \tag{5.9}$$

Following this procedure we can always compute an R-vine matrix when a regular vine tree sequence is given. Hereby the conditioned and conditioning sets of the edges are stored in the following way. The diagonal entry $m_{k,k}$ of each row k is the first entry of all conditioning sets of the entries which have been deleted from the vine filling up the row. The entries above the diagonal are added corresponding to an edge e_i with conditioned set $C_{e_i} = \{m_{k,k}, m_{i,k}\}$, $i < k$. The proximity condition implies that the conditioning set of e_i is $D_{e_i} = \{m_{i-1,k}, \ldots, m_{1,k}\}$.

Keeping that in mind, all information about the R-vine can be read from the regular vine matrix and the vine tree sequence can easily be drawn. The algorithm for the inverse direction inverts the steps of Algorithm 5.1 by adding nodes on the basic level together with corresponding edges one after another.

Algorithm 5.2 (Tree Sequence from an R-Vine Matrix M)
The input for the algorithm is a d-dimensional regular vine matrix M and the output will be a regular vine $\mathcal{V} = (T_1, \ldots, T_{d-1})$.

$$N_1 := \{1, \ldots, d\} \tag{1a}$$

$$E_2 := \{\}, \ldots, E_{d-1} := \{\} \tag{1b}$$

$$E_1 := \{m_{2,2}, m_{1,2}\} \tag{1c}$$

$$
\begin{aligned}
&FOR\ i = 3, \ldots, d && (2)\\
&\quad e_1^i := \{m_{i,i}, m_{1,i}\} && (3a)\\
&\quad E_1 := E_1 \cup \{e_1^i\} && (3b)\\
&\quad FOR\ k = 1, \ldots, i-2 && (4)\\
&\quad\quad Select\ a_k \in E_k\ with\ A_{a_k} = \{m_{1,i}, \ldots, m_{1+k,i}\} && (5a)\\
&\quad\quad e_{k+1}^i := \{e_k^i, a_k\} && (5b)\\
&\quad\quad E_{k+1} := E_{k+1} \cup \{e_{k+1}^i\} && (5c)\\
&\quad END\ FOR &&\\
&END\ FOR &&\\
&RETURN &&\\
&\quad \mathcal{V} := \Big(T_1 := (N_1, E_1), T_2 := (E_1, E_2), \ldots, T_{d-1} := (E_{d-2}, E_{d-1})\Big) && (6)
\end{aligned}
$$

The specific order in which edges are added in Algorithm 5.2 is chosen such that it coincides with the set notation.

The algorithm starts in row 2, adding an edge between the two entries in this row, in our example 3 and 4 (see Figure 5.9(a)). Further, node 3, 4 is added to tree T_2. It then moves one row to the right, adding edge 1, 3 in tree T_1, as well as node 1, 3 to tree T_2 (see Figure 5.9(b)). Then the edge $1,4|3$ between 1, 3 and 3, 4 in T_2 and the node $1,4|3$ in tree T_3 are added. These steps are repeated until the whole R-vine tree sequence is rebuilt after row d.

Remark 5.3 (Directed Graphs)
All algorithms presented in this section work exactly the same way if we replace unordered sets with ordered pairs, i.e. if we switch to directed graphs. Introducing ordering, we get a one-to-one correspondence between R-vine matrices and directed regular vines (since x and $\tilde{x}$ in Algorithm 5.1 become distinguishable). Therefore we can also assign non-symmetric copulas to an R-vine as soon as it is described by an R-vine matrix by using the convention that the first argument of the copula always corresponds to the diagonal entry of the R-vine matrix.

With the algorithms outlined, all necessary tools to describe the structure of a pair copula construction have now been developed. To conclude this section, Example 5.2 shows the corresponding matrices for the special vine tree sequences from Definition 5.3.

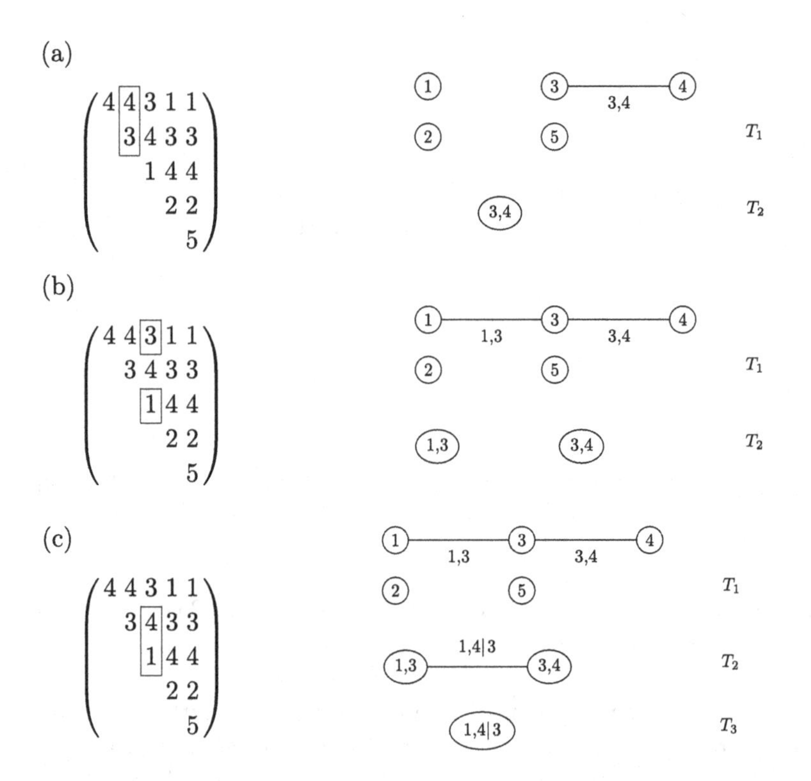

Fig. 5.9 Graphical illustration of the steps in Algorithm 5.2 to rebuild the R-vine tree sequence of Example 5.1.

Example 5.2 (C-Vine and D-Vine Matrix)

After permutation of $(1, \ldots, d)$ *the R-vine matrix of a d-dimensional C-vine tree sequence can always be expressed as*

$$\begin{pmatrix} 1 \cdots & 1 & 1 & 1 \\ \ddots & \vdots & \vdots & \vdots \\ & d-2 & d-2 & d-2 \\ & & d-1 & d-1 \\ & & & d \end{pmatrix}.$$

The matrix for a D-vine can be written as

$$\begin{pmatrix} 1 & \cdots & d-4 & d-3 & d-2 & d-1 \\ & \ddots & \vdots & \vdots & \vdots & \vdots \\ & & d-3 & 1 & 2 & 3 \\ & & & d-2 & 1 & 2 \\ & & & & d-1 & 1 \\ & & & & & d \end{pmatrix}.$$

5.3 Simulation from Regular Vine Distributions

As we have seen in the decomposition of Section 5.1, evaluating the density of an R-vine copula involves evaluating expressions of the form (5.7) given by

$$F(x|\mathbf{x}) = \frac{\partial C_{x,x_j|\mathbf{x}_{-j}}\left(F(x|\mathbf{x}_{-j}), F(x_j|\mathbf{x}_{-j})\right)}{\partial F(x_j|\mathbf{x}_{-j})}.$$

Following Aas et al. (2009), we will use the notation $h(u; v, \Theta)$ to represent this conditional distribution function in the bivariate case when $X_1 = U_1$ and $X_2 = U_2$ are uniform random variables, i.e. $F_1(u_1) = u_1$ and $F_2(u_2) = u_2$ for $u_1, u_2 \in [0, 1]$. That is,

$$h(u_1; u_2, \Theta) = F(u_1|u_2) = \frac{\partial C_{u_1,u_2}(u_1, u_2; \Theta)}{\partial u_2},$$

where the second parameter u_2 corresponds to the conditioning variable and Θ denotes the set of parameters for the bivariate copula C_{u_1,u_2}. Since sampling from a PCC will be performed by an algorithm based on inverse transformation sampling, we also need the inverse of the conditional distribution function, i.e. $h^{-1}(u_1; u_2, \Theta)$ which denotes the inverse of the h-function with respect to the first variable u_1. For readers unfamiliar with inverse transformation sampling, Fishman (1996) and Rizzo (2007) offer a detailed introduction of this concept.

The remainder of this section is structured as follows. The h- and h^{-1}-functions will be derived for several bivariate copulas in Section 5.3.1 where we will also show how they have to be modified for rotated copulas. Based on that, the main subject of this chapter, namely the simulation algorithms, are treated in Section 5.3.2.

We will only show how to sample from an R-vine copula, since a realization from an R-vine distribution in the sense of Definition 5.2 with marginal cdfs $F_1, \ldots, F_d$ is then accomplished by setting

$$X_1 := F_1^{-1}(U_1),\ X_2 := F_2^{-1}(U_2),\ \ldots, X_d := F_d^{-1}(U_d),$$

where $(U_1, \ldots, U_d)$ is a realization from the corresponding R-vine copula.

5.3.1 *h*-*Functions for Bivariate Copulas and Their Rotated Versions*

The examples we consider for calculating the h-function and h^{-1} are the Gaussian copula and general Archimedean copulas. To illustrate the general case we give also precise expressions for the Frank and the Gumbel copulas. Further calculations for the Clayton/MTCJ and Joe copulas (families B4 and B5 in Joe (1997, p. 141)), the BB1 and BB7 copulas (Joe (1997, p. 151) and Joe (1997, p. 153)) and the Student's t-copula can be found in the appendix of Aas et al. (2009) and in Schepsmeier (2010). In some cases there is no closed-form expression for the inverse h-function and it has to be evaluated numerically.

Example 5.3 (Gaussian Copula)
The cdf of the Gaussian copula is given by

$$\begin{aligned} C^{Gauss}_{P(\rho_{12})}(u_1, u_2; \rho_{12}) &= F_{P(\rho_{12})}(\Phi^{-1}(u_1), \Phi^{-1}(u_2); \rho_{12}) \\ &= \int_{-\infty}^{\Phi^{-1}(u_1)} \int_{-\infty}^{\Phi^{-1}(u_2)} \phi_2(x_1, x_2; \rho_{12}) dx_1 dx_2, \end{aligned}$$

where

$$P(\rho) := \begin{pmatrix} 1 & \rho \\ \rho & 1 \end{pmatrix}, \quad \phi_2(x_1, x_2; \rho_{12}) = \frac{1}{2\pi(1-\rho_{12}^2)^{\frac{1}{2}}} \cdot e^{-\frac{x_1^2 - 2\rho_{12}x_1x_2 + x_2^2}{2(1-\rho_{12}^2)}}$$

is the bivariate Gaussian density with zero means, unit variances, and correlation parameter ρ_{12}, and $F_{P(\rho_{12})}$ denotes the corresponding cdf. Further set $q_1 := \Phi^{-1}(u_1)$ and $q_2 := \Phi^{-1}(u_2)$. Then, following Aas et al. (2009), the h-function can be calculated as

$$
\begin{aligned}
h(u_1; u_2, \rho_{12}) =& \frac{\partial}{\partial u_2} C(u_1, u_2; \rho_{12}) \\
=& \frac{\partial}{\partial u_2} \int_{-\infty}^{q_2} \int_{-\infty}^{q_1} \phi_2(x_1, x_2; \rho_{12}) dx_1 dx_2 \\
=& \frac{\partial q_2}{\partial u_2} \frac{\partial}{\partial q_2} \int_{-\infty}^{q_2} \int_{-\infty}^{q_1} \phi_2(x_1, x_2; \rho_{12}) dx_1 dx_2 \\
=& \frac{1}{\phi(q_2)} \int_{-\infty}^{q_1} \phi_2(x_1, q_2; \rho_{12}) dx_1 \\
=& \frac{1}{\phi(q_2)} \int_{-\infty}^{q_1} \frac{1}{2\pi(1-\rho_{12}^2)^{\frac{1}{2}}} \cdot e^{-\frac{x_1^2 - 2\rho_{12} x_1 q_2 + q_2^2}{2(1-\rho_{12}^2)}} dx_1 \\
=& \frac{1}{\phi(q_2)} \int_{-\infty}^{q_1} \frac{1}{2\pi(1-\rho_{12}^2)^{\frac{1}{2}}} \cdot e^{-\frac{(x_1 - \rho_{12} q_2)^2 + (q_2^2 - \rho_{12}^2 q_2^2)}{2(1-\rho_{12}^2)}} dx_1 \\
=& \frac{1}{\phi(q_2)} \cdot \frac{1}{(2\pi)^{\frac{1}{2}}} \cdot e^{\frac{-q_2^2}{2}} \int_{-\infty}^{q_1} \frac{1}{(2\pi)^{\frac{1}{2}}(1-\rho_{12}^2)^{\frac{1}{2}}} e^{-\frac{(x_1 - \rho_{12} q_2)^2}{2(1-\rho_{12}^2)}} dx_1 \\
=& \frac{1}{\phi(q_2)} \cdot \phi(q_2) \cdot \Phi\left(\frac{q_1 - \rho_{12} q_2}{(1-\rho_{12}^2)^{\frac{1}{2}}}\right) = \Phi\left(\frac{q_1 - \rho_{12} q_2}{(1-\rho_{12}^2)^{\frac{1}{2}}}\right).
\end{aligned}
$$

Furthermore, this function can be easily inverted using the quantile function of the standard normal distribution, yielding the corresponding inverse h-function

$$
h^{-1}(u_1; u_2, \rho_{12}) = \Phi\left(\Phi^{-1}(u_1) \cdot (1-\rho_{12}^2)^{\frac{1}{2}} + \rho_{12} q_2\right).
$$

Example 5.4 (Bivariate Archimedean Copula)
Bivariate Archimedean copulas are given in the form

$$
C(u_1, u_2; \varphi) = \varphi(\varphi^{-1}(u_1) + \varphi^{-1}(u_2)) \tag{5.10}
$$

(see Equation (2.2)). From (5.10) the general form of the h-function for Archimedean copulas can be derived as

$$
\begin{aligned}
h(u_1; u_2, \varphi) &= \frac{\partial\Big(\varphi(\varphi^{-1}(u_1) + \varphi^{-1}(u_2))\Big)}{\partial u_2} \\
&= \varphi'(\varphi^{-1}(u_1) + \varphi^{-1}(u_2)) \cdot \frac{1}{\varphi'(\varphi^{-1}(u_2))}
\end{aligned} \tag{5.11}
$$

by the chain rule, and simple inversion of (5.11) as in Algorithm 2.6 gives

$$
h^{-1}(u_1; u_2, \varphi) = \varphi\Big((\varphi')^{-1}\Big(u_1 \varphi'\left(\varphi^{-1}(u_2)\right)\Big) - \varphi^{-1}(u_2)\Big). \tag{5.12}
$$

Example 5.5 (Gumbel Copula)

An important bivariate Archimedean copula is the Gumbel copula *(cf. Equation (2.12)) with generator function*

$$\varphi(x) = e^{-x^{1/\vartheta}}, \quad \vartheta \in (1, \infty).$$

For the inverse of φ and required derivatives we have

$$\varphi^{-1}(u) = (-\log(u))^{\vartheta}, \quad \varphi'(x) = e^{-x^{1/\vartheta}} \left(-\frac{1}{\vartheta} x^{(1/\vartheta)-1} \right),$$

$$(\varphi')^{-1}(y) = \left((\vartheta - 1) \cdot W \left(\frac{(-\vartheta y)^{\frac{1}{(1-\vartheta)}}}{(\vartheta - 1)} \right) \right)^{\vartheta}. \tag{5.13}$$

Hereby, W in (5.13) denotes the Lambert W function (see Corless et al. (1996) for a definition and properties). To derive (5.13), let

$$\varphi'(x) = e^{-x^{1/\vartheta}} \left(-\frac{1}{\vartheta} x^{(1/\vartheta)-1} \right) \stackrel{!}{=} y.$$

Defining $z := x^{1/\vartheta}$, we have

$$\begin{aligned} e^{-z} z^{1-\vartheta} \left(-\frac{1}{\vartheta} \right) = y \quad &\Leftrightarrow \quad e^{\frac{z}{(\vartheta-1)}} \frac{z}{(\vartheta - 1)} = \frac{(-\vartheta y)^{\frac{1}{(1-\vartheta)}}}{(\vartheta - 1)} \\ &\Leftrightarrow \quad \frac{z}{(\vartheta - 1)} = W \left(\frac{(-\vartheta y)^{\frac{1}{(1-\vartheta)}}}{(\vartheta - 1)} \right) \\ &\Leftrightarrow \quad z = (\vartheta - 1) W \left(\frac{(-\vartheta y)^{\frac{1}{(1-\vartheta)}}}{(\vartheta - 1)} \right), \end{aligned}$$

since, by definition of the Lambert W function, $W(z)e^{W(z)} = z$. From this, we get (5.13) by resubstitution of x for y.

Example 5.6 (Frank Copula)

The bivariate Frank copula *family, cf. Equation (2.9), has the generator function*

$$\varphi(x) = -\frac{1}{\vartheta} \cdot \log \left(e^{-x} (e^{-\vartheta} - 1) + 1 \right), \quad \vartheta \in (-\infty, \infty) \setminus \{0\}.$$

Inverting φ and calculating the derivative yields

$$\begin{aligned} \varphi^{-1}(u) &= -\log \left(\frac{e^{-\vartheta u} - 1}{e^{-\vartheta} - 1} \right), \\ \varphi'(x) &= \frac{1}{\vartheta} \cdot \frac{1}{e^{-x} (e^{-\vartheta} - 1) + 1} \left(e^{-\vartheta} - 1 \right) e^{-x} \\ &= \frac{1}{\vartheta} \cdot \left(1 - \frac{1}{e^{-x} (e^{-\vartheta} - 1) + 1} \right), \end{aligned}$$

and thus determining the inverse $(\varphi')^{-1}(y)$ *is straightforward:*

$$(\varphi')^{-1}(y) = -\log\left(\frac{\vartheta y}{(1-\vartheta y)(e^{-\vartheta}-1)}\right).$$

Given the cdf, h-function, or inverse h-function for a copula, these functions can also be calculated for rotated versions of the bivariate copula by $90\,^\circ$, $180\,^\circ$, and $270\,^\circ$. These *rotated copulas* are interesting because they allow for an easy modification of common parametric families to describe more general forms of dependence. For example, using bivariate Gumbel, Clayton, and Student's t-copulas, only tail dependence in the "lower left" and the "upper right" corners of the unit cube can be covered. Rotating the bivariate copulas does make it possible to model tail dependence in the "upper left" and "lower right" corners as well.

Rotating a bivariate, absolutely continuous copula means rotating the corresponding density c:

$$\begin{aligned}
90\,^\circ &: c_{90}(u_1,u_2) := c(1-u_1,u_2)\\
180\,^\circ &: c_{180}(u_1,u_2) := c(1-u_1,1-u_2)\\
270\,^\circ &: c_{270}(u_1,u_2) := c(u_1,1-u_2).
\end{aligned}$$

The $180\,^\circ$-rotated copula is exactly the survival copula of C since $c_{180}(u_1,u_2) = c(1-u_1,1-u_2)$ means $(1-U_1,1-U_2) \sim C$ and thus $(U_1,U_2) \sim \hat{C}$ in the notation of Section 1.1.2. As shown in Equation (1.10), we have

$$C_{180}(u_1,u_2) = C(1-u_1,1-u_2) + u_1 + u_2 - 1$$

for the cdf. For $90\,^\circ$ we calculate

$$\begin{aligned}
C_{90}(u_1,u_2) &= \int_0^{u_2}\int_0^{u_1} c(1-v_1,v_2)dv_1dv_2\\
&= \int_0^{u_2}\left[-\frac{\partial C(1-v_1,v_2)}{\partial v_2}\right]_0^{u_1} dv_2\\
&= \int_0^{u_2}\left(1-\frac{\partial C(1-u_1,v_2)}{\partial v_2}\right)dv_2 = u_2 - C(1-u_1,u_2),
\end{aligned}$$

and similarly $C_{270}(u_1,u_2) = u_1 - C(u_1,1-u_2)$. From these, the corresponding h-functions

$$\begin{aligned}
h_{90}(u_1;u_2) &= 1 - h(1-u_1;u_2)\\
h_{180}(u_1;u_2) &= 1 - h(1-u_1;1-u_2)\\
h_{270}(u_1;u_2) &= h(u_1;1-u_2)
\end{aligned}$$

and inverse h-functions

$$h_{90}^{-1}(u_1; u_2) = 1 - h^{-1}(1 - u_1; u_2)$$
$$h_{180}^{-1}(u_1; u_2) = 1 - h^{-1}(1 - u_1; 1 - u_2)$$
$$h_{270}^{-1}(u_1; u_2) = h^{-1}(u_1; 1 - u_2)$$

are easily derived by differentiation and by applying the inverse h-function of the original copula. The calculations are left to the reader.

5.3.2 *The Sampling Algorithms*

For the remainder of this chapter let us assume that all one-dimensional margins are uniform, i.e. we sample from a d-dimensional copula. Other distributions for the margins can then be included by transforming the uniform data to the desired distribution using inverse cdfs.

Simulations from vines were first discussed in Bedford and Cooke (2001a) and Bedford and Cooke (2001b) but without explicitly stating programmable algorithms. Aas et al. (2009) and Kurowicka and Cooke (2007) showed sampling algorithms for C-vines and D-vines. While Kurowicka and Cooke (2007) also gave hints on how to treat the general R-vine case, it was Dißmann (2010) who demonstrated how to write a sampling algorithm for the general R-vine using the matrix notation from Section 5.2.2. The algorithms we are going to present here are improved versions of these, where some redundant calculations have been omitted.

The general algorithm to draw a sample from a d-dimensional vine copula will be the following stepwise inverse transformation procedure:

$$\textbf{First: } \text{Sample } W_i \overset{\text{i.i.d.}}{\sim} U[0,1], \quad i = 1, \ldots, d.$$
$$\textbf{Then: } U_1 := W_1$$
$$U_2 := F_{2|1}^{-1}(W_2 | U_1)$$
$$\vdots$$
$$U_d := F_{d|d-1,\ldots,1}^{-1}(W_d | U_{d-1}, \ldots, U_1).$$

To determine the conditional cdfs $F_{i|i-1,\ldots,1}$, $i = 1, \ldots, d$, we will use Relation (5.7) together with the definition of the h-function to calculate it iteratively using h-functions. We can then invert it by applying inverse h-functions. For this, we use the expression $h(u; v, \theta)$ for a general h-function, with parameters θ of a specified bivariate copula. Before deriving algorithms for arbitrary dimensions, let us first consider a three-dimensional example.

Example 5.7 (Three-Dimensional R-Vine)
In three dimensions, there is (again, up to permutations in the one-dimensional margins) only one possible R-vine structure, displayed in Figure 5.10. Having drawn W_1, W_2, W_3 we can directly set $U_1 := W_1$. For U_2,

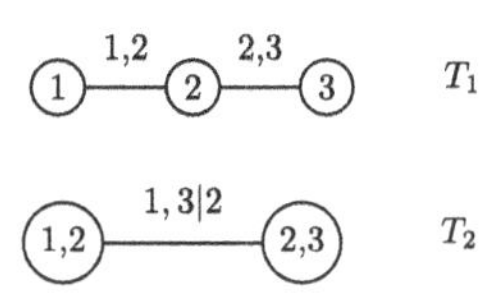

Fig. 5.10 The only possible three-dimensional R-vine (up to permutations).

we have

$$F(u_2|u_1) = \frac{\partial C_{2,1}(u_2, u_1)}{\partial u_1} = h(u_2; u_1, \theta_{1,2}).$$

Thus, $U_2 := h^{-1}(W_2; U_1, \theta_{1,2})$. The calculation for U_3 is similar:

$$\begin{aligned} F(u_3|u_1, u_2) &= \frac{\partial C_{3,1|2}\left(F(u_3|u_2), F(u_1|u_2)\right)}{\partial F(u_1|u_2)} \\ &= h\left(F(u_3|u_2); F(u_1|u_2), \theta_{1,3|2}\right) \\ &= h\left(h(u_3; u_2, \theta_{3,2}); h(u_1; u_2, \theta_{1,2}), \theta_{1,3|2}\right). \end{aligned}$$

This implies

$$F^{-1}_{3|2,1}(w_3|u_1, u_2) = h^{-1}\left(h^{-1}\left(w_3; h(u_2; u_1, \theta_{1,2}), \theta_{1,3|2}\right); u_1, \theta_{1,3}\right).$$

To generalize this procedure we need to find an algorithmic way of selecting the right arguments for the inverse h-functions in each step.

C-Vines and D-Vines

This is particularly easy for the C-vine, which always has (up to relabeling) the structure shown in Figure 5.11 for five dimensions.

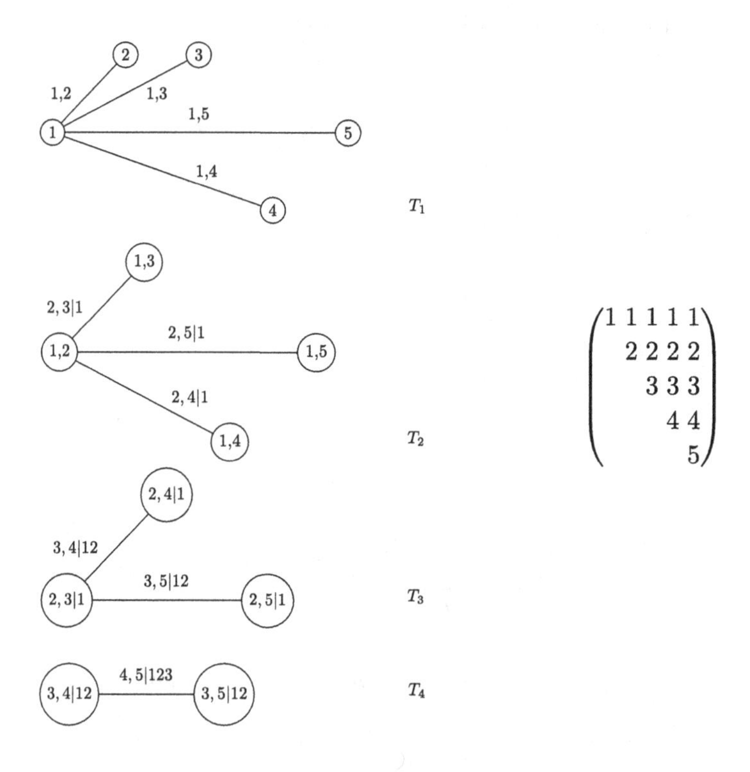

Fig. 5.11 The five-dimensional C-vine tree sequence with its corresponding C-vine matrix.

In this special case, we can express $F(u_i|u_1,\ldots,u_{i-k})$ as

$$F(u_i|u_1,\ldots,u_{i-k}) = \tag{5.14}$$
$$\frac{\partial C_{i,i-k|1,\ldots,i-k-1}\left(F(u_i|u_1,\ldots,u_{i-k-1}),F(u_{i-k}|u_1,\ldots,u_{i-k-1})\right)}{\partial F(u_{i-k}|u_1,\ldots,u_{i-k-1})}$$

for $i = 1,\ldots,d$ and $k = 1,\ldots,i-1$. The sampling algorithm uses the copula parameters stored in matrix

$$\Theta = \begin{pmatrix} \theta_{1,2} & \theta_{1,3} & \theta_{1,4} & \cdots \\ & \theta_{2,3|1} & \theta_{2,4|1} & \cdots \\ & & \theta_{3,4|1,2} & \cdots \\ & & & \cdots \end{pmatrix}$$

to calculate the entries of the following $d \times d$ matrix:

$$V = \begin{pmatrix} u_1 & u_2 & u_3 & u_4 & \dots \\ & F(u_2|u_1) & F(u_3|u_1) & F(u_4|u_1) & \dots \\ & & F(u_3|u_1, u_2) & F(u_4|u_1, u_2) & \dots \\ & & & F(u_4|u_1, u_2, u_3) & \dots \\ & & & & \dots \end{pmatrix}. \tag{5.15}$$

For this, we use that $F(u_j|u_1, \dots, u_{j-1}) = w_j$ for $j = 1, \dots, d$, and that rewriting (5.14) in terms of h-functions as

$$F(u_i|u_1, \dots, u_{i-k}) = h((F(u_i|u_1, \dots, u_{i-k-1}), F(u_{i-k}|u_1, \dots, u_{i-k-1}))$$

and inverting it for $i = 1, \dots, d$ and $k = 1, \dots, i-1$ yields $F(u_i|u_1, \dots, u_{i-k-1}) =$

$$h^{-1}\left(F(u_i|, u_1, \dots u_{i-k}); F(u_{i-k}|u_1, \dots, u_{i-k-1}), \theta_{i,i-k|1,\dots,i-k-1}\right).$$

Algorithm 5.3 applies this relationship recursively to calculate the entries of matrix (5.15) and generate a sample $(U_1, \dots, U_d)$, which we can find in the first row of matrix (5.15) from the C-vine copula.

Algorithm 5.3 (Sampling from a C-Vine Copula)

The input for the algorithm is a matrix Θ of copula parameters for the d-dimensional C-vine; the output will be a sample from the C-vine copula.

$$\begin{aligned}
&\textit{Sample } W_i \overset{i.i.d.}{\sim} U[0,1], \quad i = 1, \dots, d && (1)\\
&V_{1,1} := W_1 && (2)\\
&\textit{FOR } i = 2, \dots, d && (3)\\
&\quad V_{i,i} := W_i && (4)\\
&\quad \textit{FOR } k = i-1, \dots, 1 && (5)\\
&\qquad V_{k,i} := h^{-1}(V_{k+1,i}; V_{k,k}, \Theta_{k,i}) && (6)\\
&\quad \textit{END FOR} && \\
&\textit{END FOR} && \\
&\textit{RETURN } U_i := V_{1,i}, \quad i = 1, \dots, d && (7)
\end{aligned}$$

Remark 5.4 (Storing the Matrix)

We do not need to store the whole matrix in this algorithm, since the only entries we are going to use more than once are $V_{i,i} = W_i$. Thus, we can always delete or overwrite the other entries after they have been used as *input for the next recursion. The matrix structure is chosen to illustrate the iterative procedure, and because it is needed to understand the general regular vine case presented later.*

For sampling from a D-vine copula (Figure 5.12) the relationship which we use instead of (5.14) is

$$F(u_i|u_k, u_{k+1}, \dots, u_{i-1}) = \frac{\partial C_{i,k|k+1,\dots,i-1}\left(F(u_i|u_{k+1}, \dots u_{i-1}), F(u_k|u_{k+1}, \dots, u_{i-1})\right)}{\partial F(u_k|u_{k+1}, \dots, u_{i-1})}, \tag{5.16}$$

where $i = 1, \dots, d$ and $k = 1, \dots, i-1$. In contrast to the C-vine, we do not automatically obtain the second argument in (5.16), which we need to further use the equality, during the recursion. This means that an extra step for its computation has to be added and that we have to calculate two matrices

$$V = \begin{pmatrix} u_1 & u_2 & u_3 & u_4 & \dots \\ & F(u_2|u_1) & F(u_3|u_2) & F(u_4|u_3) & \dots \\ & & F(u_3|u_2, u_1) & F(u_4|u_3, u_2) & \dots \\ & & & F(u_4|u_3, u_2, u_1) & \dots \\ & & & & \dots \end{pmatrix}, \tag{5.17}$$

$$V^2 = \begin{pmatrix} u_1 & u_2 & u_3 & u_4 & \dots \\ & F(u_1|u_2) & F(u_2|u_3) & F(u_3|u_4) & \dots \\ & & F(u_1|u_2, u_3) & F(u_2|u_3, u_4) & \dots \\ & & & F(u_1|u_4, u_3, u_2) & \dots \\ & & & & \dots \end{pmatrix}, \tag{5.18}$$

using the matrix of copula parameters

$$\Theta = \begin{pmatrix} & \theta_{1,2} & \theta_{2,3} & \theta_{3,4} & \dots \\ & & \theta_{3,1|2} & \theta_{4,2|3} & \dots \\ & & & \theta_{4,1|3,2} & \dots \\ & & & & \dots \end{pmatrix}.$$

This is done recursively in Algorithm 5.4.

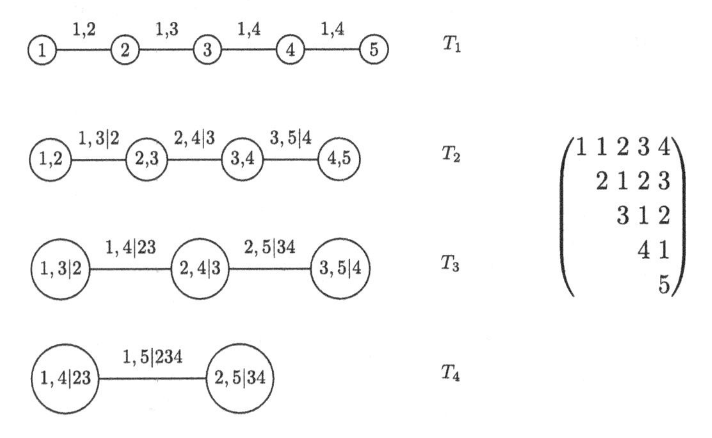

Fig. 5.12 The five-dimensional D-vine and the five-dimensional D-vine matrix.

Algorithm 5.4 (Sampling from a D-Vine Copula)
The input for the algorithm is a matrix Θ of copula parameters for the d-dimensional D-vine; the output will be a sample from the D-vine distribution.

$$\begin{array}{ll}
\textit{Sample } W_i \overset{i.i.d.}{\sim} U[0,1], \quad i = 1, \dots, d & (1) \\
V_{1,1} := W_1;\ V_{1,1}^2 := W_1 & (2) \\
\textit{FOR } i = 2, \dots, d & (3) \\
\quad V_{i,i} := W_i & (4) \\
\quad \textit{FOR } k = i-1, \dots, 1 & (5) \\
\quad\quad V_{k,i} := h^{-1}(V_{k+1,i}; V_{k,i-1}^2, \Theta_{k,i}) & (6) \\
\quad\quad \textit{IF } i < d & (7) \\
\quad\quad\quad V_{k+1,i}^2 := h(V_{k,i-1}^2; V_{k,i}, \Theta_{k+1,i}) & (8) \\
\quad\quad \textit{END IF} & \\
\quad \textit{END FOR} & \\
\quad V_{1,i}^2 := V_{1,i} & (9) \\
\textit{END FOR} & \\
\textit{RETURN } U_i := V_{1,i}, \quad i = 1, \dots, d & (10)
\end{array}$$

Just as we noted for the C-vine copula in Remark 5.4, we do not need to store all matrix entries calculated during the recursion. The entries in row

i of the first matrix can always be deleted after they have been used to calculate u_i and the entries of the second matrix.

R-Vines

Having considered simulations for these two special structures, we can develop the sampling procedure for the general R-vine copula. The C- and D-vine tree sequences can be seen as extreme cases: In the C-vine case there is one node in each tree which shares edges with *all* other nodes, while in the D-vine case each node has joint edges at most with *two* other nodes. In simulations, we always have to select an entry from the second matrix (5.18) as the second argument for the inverse h-function in the D-vine case. For the C-vine case we do not even need the second matrix. Similar to the number of joint edges for each node, the sampling procedure for the R-vine copula will be a mixture of the two extreme cases.

Let a d-dimensional R-vine matrix $M = (m_{i,j})_{i,j=1,\ldots,d}$ as defined in Section 5.2.2 be given. Without loss of generality, we assume that the entries on the diagonal are ordered as $1, 2, \ldots, d$. This can always be realized by permuting the labeling of the dimensions. Furthermore, let us assume that the parameters for the pair copulas are given in a matrix Θ, where $\Theta_{i,j}$ is the parameter of $c_{m_{j,j},m_{i,j}|m_{1,j},\ldots,m_{i-1,j}}$, i.e.

$$\Theta = \begin{pmatrix} \theta_{m_{1,2},2} & \theta_{m_{1,3},3} & \theta_{m_{1,4},4} & \cdots \\ & \theta_{m_{2,3},3|m_{1,3}} & \theta_{m_{2,4},4|m_{1,4}} & \cdots \\ & & \theta_{m_{3,4},4|m_{2,4},m_{1,4}} & \cdots \\ & & & \cdots \end{pmatrix}. \tag{5.19}$$

For the general R-vine copula, where, other than for the C-vine and D-vine copulas, different tree structures are possible, the arguments of all functions have to be expressed in terms of the R-vine matrix M. In particular, we determine the conditional cdf corresponding to (5.14) and (5.16) as

$$F(u_{m_{i,i}}|u_{m_{k,i}}, u_{m_{k-1,i}}, \ldots, u_{m_{1,i}}) = \frac{\partial C_{m_{i,i},m_{k,i}|m_{k-1,i},\ldots,m_{1,i}}(y)}{\partial F(u_{m_{k,i}}|u_{m_{k-1,i}}, \ldots, u_{m_{1,i}})}, \tag{5.20}$$

where $y := F(u_{m_{i,i}}|u_{m_{k-1,i}}, \ldots, u_{m_{1,i}}), F(u_{m_{k,i}}|u_{m_{k-1,i}}, \ldots, u_{m_{1,i}})$ for $i = 1, \ldots, d$ and $k = 1, \ldots, i-1$.

Again, we will calculate two matrices

$$V = \begin{pmatrix} u_1 & u_2 & u_3 & u_4 & \cdots \\ & F(u_2|u_{m_{1,2}}) & F(u_3|u_{m_{1,3}}) & F(u_4|u_{m_{1,4}}) & \cdots \\ & & F(u_3|u_{m_{1,3}}, u_{m_{2,3}}) & F(u_4|u_{m_{1,4}}, u_{m_{2,4}}) & \cdots \\ & & & F(u_4|u_{m_{1,4}}, u_{m_{2,4}}, u_{m_{3,4}}) & \cdots \\ & & & & \cdots \end{pmatrix}, \quad (5.21)$$

$$V^2 = \begin{pmatrix} & & & & \cdots \\ & F(u_{m_{1,2}}|u_2) & F(u_{m_{1,3}}|u_3) & F(u_{m_{1,4}}|u_4) & \cdots \\ & & F(u_{m_{2,3}}|u_{m_{1,3}}, u_3) & F(u_{m_{2,4}}|u_{m_{1,4}}, u_4) & \cdots \\ & & & F(u_{m_{3,4}}|u_{m_{1,4}}, u_{m_{2,4}}, u_4) & \cdots \\ & & & & \cdots \end{pmatrix}, \quad (5.22)$$

just as for the D-vine copula.

There are two questions which have to be addressed before a sampling algorithm can be constructed: From which column do we have to select the second argument of the inverse h-function (and the first argument of the h-function)? Do we have to select it from the first matrix (5.21) or from the second matrix (5.22)?

To answer these, let us recall the R-vine matrices M corresponding to the C-vine case

$$\begin{pmatrix} 1 \cdots & 1 & 1 & 1 \\ \ddots & \vdots & \vdots & \vdots \\ & d-2 & d-2 & d-2 \\ & & d-1 & d-1 \\ & & & d \end{pmatrix}$$

and to the D-vine case

$$\begin{pmatrix} 1 & \cdots & d-4 & d-3 & d-2 & d-1 \\ & \ddots & \vdots & \vdots & \vdots & \vdots \\ & & d-3 & 1 & 2 & 3 \\ & & & d-2 & 1 & 2 \\ & & & & d-1 & 1 \\ & & & & & d \end{pmatrix}.$$

Using these we can understand how the choices in Algorithms 5.3 and 5.4 can be expressed in the form of R-vine matrices, which will give us an idea of how to choose the right arguments in the general framework. Furthermore, let us introduce the matrix $\mathcal{M} = (\tilde{m}_{k,i})$, $k \leq i$, which is defined as $\tilde{m}_{k,i} := \max\{m_{k,i}, m_{k-1,i}, \ldots, m_{1,i}\}$. Note that $\mathcal{M} = M$ holds for the C-vine matrix, i.e. we have $\tilde{m}_{k,i} = k$ for all entries of $\mathcal{M}$. For the D-vine, $\mathcal{M} \neq M$, in particular $\tilde{m}_{k,i} = i - 1$ for all off-diagonal entries of $\mathcal{M}$.

For the C-vine copula, we always select an entry from the kth column within Step (6) of Algorithm 5.3. This corresponds to column $m_{k,i} = \tilde{m}_{k,i}$. For the D-vine copula we stay in column $i-1$ ($= \tilde{m}_{k,i}$) in Step (6) of Algorithm 5.4. Similarly, the sampling algorithm for an R-vine copula will always choose $\tilde{m}_{k,i}$. The entry, which is needed as the second argument for the inverse h-function *has* to be in this column, since the second argument in (5.20) is $F(u_{m_{k,i}}|u_{m_{k-1,i}}, \ldots, u_{m_{1,i}})$, and $\tilde{m}_{k,i}$ is the largest index in this expression. The first row where an entry containing index $\tilde{m}_{k,i}$ can be located is row $\tilde{m}_{k,i}$, since the diagonals of M are arranged in increasing order and since $m_{l,h} \leq h$ for $l = 1, \ldots, h$ by property (3) of Definition 5.4. Furthermore, each element in column h of (5.21) and (5.22) contains the index h, which means that the entry we are looking for cannot be found in a column to the right of column $\tilde{m}_{k,i}$.

In matrix V (5.21), all entries in column $\tilde{m}_{k,i}$ are conditional cdfs of $U_{\tilde{m}_{k,i}}$ given other variables, and in matrix V^2 (5.22) $U_{\tilde{m}_{k,i}}$ is part of the conditioned variables. Thus, we only need to check whether $\tilde{m}_{k,i} = m_{k,i}$ to choose from the appropriate matrix.

Algorithm 5.5 summarizes the results of the preceding paragraph. An inductive proof of the fact that at each step of the algorithm the appropriate entry is selected is straightforward by showing that the calculated matrices take the form of (5.21) and (5.22). A more formal proof can be found in Dißmann (2010, Chapter 5).

Algorithm 5.5 (Sampling from an R-Vine Copula)
The input for the algorithm is a matrix Θ, given in (5.19), of copula parameters for the d-dimensional R-vine copula. The output will be a sample from the R-vine copula.

$$
\begin{array}{ll}
\textit{Sample } W_i \overset{i.i.d.}{\sim} U[0,1], \quad i = 1, \ldots, d & (1) \\
V_{1,1} := W_1 & (2) \\
\textit{FOR } i = 2, \ldots, d & (3) \\
\quad V_{i,i} := W_i & (4) \\
\quad \textit{FOR } k = i-1, \ldots, 1 & (5) \\
\quad\quad \textit{IF } (m_{k,i} = \tilde{m}_{k,i}) & (6) \\
\quad\quad\quad V_{k,i} := h^{-1}(V_{k+1,i}; V_{k,\tilde{m}_{k,i}}, \Theta_{k,i}) & (7a) \\
\quad\quad \textit{ELSE} &
\end{array}
$$

$$V_{k,i} := h^{-1}(V_{k+1,i}; V^2_{k,\tilde{m}_{k,i}}, \Theta_{k,i}) \qquad (7b)$$

END IF ELSE

IF $(i < d)$ (8)

IF $(m_{k,i} = \tilde{m}_{k,i})$ (9)

$$V^2_{k+1,i} := h(V_{k,\tilde{m}_{k,i}}; V_{k,i}, \Theta_{k+1,i}) \qquad (10a)$$

ELSE

$$V^2_{k+1,i} := h(V^2_{k,\tilde{m}_{k,i}}; V_{k,i}, \Theta_{k+1,i}) \qquad (10b)$$

END IF ELSE

END IF

END FOR

END FOR

RETURN $U_i := V_{1,i}, \quad i = 1, \ldots, d$ (11)

To illustrate the steps of Algorithm 5.5, let us once again consider Example 5.1, where the R-vine matrix (see (5.9)) after reordering of the diagonal is given by

$$M = \begin{pmatrix} 1 & 1 & 2 & 3 & 3 \\ & 2 & 1 & 2 & 2 \\ & & 3 & 1 & 1 \\ & & & 4 & 4 \\ & & & & 5 \end{pmatrix}.$$

For columns 1–4 corresponding to $U_1, \ldots, U_4$ this matrix is the same as for a D-vine copula, which means that except for row 1 we have $m_{k,i} \neq \tilde{m}_{k,i}$ and that we select the second entry of the inverse h-function from the second matrix (5.22). In row 1, $m_{1,i} = \tilde{m}_{1,i}$ for $i = 1, \ldots, 5$ such that in the last step of the iteration for $U_1, \ldots, U_4$ we select from (5.21).

For obtaining U_5 conditional on the other variables, we calculate first $F(u_5|u_1, u_2, u_3)$ from $F(u_5|u_1, u_2, u_3, u_4) = w_5$ as

$$F(u_5|u_1, u_2, u_3) = h^{-1}\left(w_5; F(u_4|u_1, u_2, u_3), \theta_{4,5|1,2,3}\right).$$

Here, $F(u_4|u_1, u_2, u_3)$ is given in matrix (5.21) and as $m_{45} = 4 = \tilde{m}_{45}$ it gets correctly selected.

In the next two steps of the recursion, we need $F(u_1|u_2, u_3)$ and $F(u_2|u_3)$, which are given in the third column of the second matrix (5.22). Correspondingly we have $m_{3,5} = 1 \neq \tilde{m}_{3,5} = 3$ and $m_{2,5} = 2 \neq \tilde{m}_{2,5} = 3$.

In the last step, we do select u_3 from the third column of matrix (5.21), $\tilde{m}_{1,5} = 3$.

As it was noted in Remark 5.3, an R-vine matrix corresponds to a directed vine. We can thus introduce one further dimension in the parameter of the h-function giving the angle of rotation, and still apply Algorithm 5.5. The simplified algorithms for the C-vine and D-vine copulas correspond to the special structure in the corresponding R-vine matrices from Example 5.2.

Using the results from Section 5.2 a regular vine distribution can be specified, and with Algorithm 5.5 we have also collected all necessary tools to sample from it.

5.4 Dependence Properties

Due to the high number of admissible R-vine structures in d dimensions and the freedom to select arbitrary bivariate copulas as conditional copulas for each edge, pair copula constructions are highly flexible. For practical applications it is important to understand the dependence properties of PCCs, particularly compared to other frameworks for constructing multivariate copulas such as nested Archimedean copulas (NACs) and elliptical copulas. In the following we show that important elliptical copulas are R-vine copulas and discuss the main results of Joe et al. (2010), who consider tail dependence properties for PCCs.

Multivariate Gaussian and Student's t-Copulas

The multivariate copula families which are most often used in applications are still the multivariate normal and the multivariate Student's t-copula. These copulas can also be constructed using an R-vine and are thus nested within the class of vine copulas.

Table 5.1 summarizes the properties of the bivariate conditional distributions of a multivariate normal or Student's t-distribution, which are needed to identify them as an R-vine distribution. Considering the decomposition of a multivariate normal distribution using an R-vine, we obtain the following result:

Theorem 5.2.
The multivariate Gaussian copula can be represented as a regular vine distribution where all bivariate copulas are Gaussian.

Table 5.1 Bivariate conditional distributions of the multivariate normal and the multivariate Student's t-distribution.

$$\boldsymbol{\mu} = (\mu_i)_{i=1,\dots,d} = \begin{pmatrix} \boldsymbol{\mu}_U \\ \boldsymbol{\mu}_C \end{pmatrix}, \text{ where}$$
$$\boldsymbol{\mu}_U = (\mu_i)_{i=1,2},\ \boldsymbol{\mu}_C = (\mu_i)_{i=3,\dots,d}$$
$$(\Sigma_{ij})_{i,j=1,\dots,d} = \begin{pmatrix} \Sigma_U & \Sigma_{UC} \\ \Sigma_{CU} & \Sigma_C \end{pmatrix}, \text{ where}$$
$$\Sigma_U = (\Sigma_{ij})_{i,j=1,2},\ \Sigma_{CU} = (\Sigma_{ij})_{i=3,\dots,d;j=1,2} = \Sigma_{UC}^T,\ \Sigma_C = (\Sigma_{ij})_{i,j=3,\dots,d}$$

Multivariate Gaussian distribution in d dimensions	
conditional distribution $X_1, X_2 \vert \mathbf{X}_{3,\dots,d} = \mathbf{x}_C$	multivariate Gaussian with $\boldsymbol{\mu}_{U\vert C} = \boldsymbol{\mu}_U + \Sigma_{UC}\Sigma_C^{-1}(\mathbf{x}_C - \boldsymbol{\mu}_C)$, $\Sigma_{U\vert C} = \Sigma_U - \Sigma_{UC}\Sigma_C^{-1}\Sigma_{CU}$
copula corresponding to $X_1, X_2 \vert \mathbf{X}_{3,\dots,d} = \mathbf{x}_C$	bivariate Gaussian copula $C^{Gauss}_{\mathcal{R}_{U\vert C}}$, with correlation $\mathcal{R}_{U\vert C} = diag(\Sigma_{U\vert C})^{-\frac{1}{2}}\ \Sigma_{U\vert C}\ diag(\Sigma_{U\vert C})^{-\frac{1}{2}}$
Multivariate Student's t-distribution in d dimensions	
conditional distribution $X_1, X_2 \vert \mathbf{X}_{3,\dots,d} = \mathbf{x}_C$	Student's t with $\boldsymbol{\mu}_{U\vert C} = \boldsymbol{\mu}_U + \Sigma_{UC}\Sigma_C^{-1}(\mathbf{x}_C - \boldsymbol{\mu}_C)$, $\Sigma^t_{U\vert C} = \frac{(\Sigma_U - \Sigma_{UC}\Sigma_C^{-1}\Sigma_{CU})\cdot(\nu+d-2)}{\nu + (\mathbf{x}_C - \boldsymbol{\mu}_C)^T\Sigma_C^{-1}(\mathbf{x}_C - \boldsymbol{\mu}_C)}$, $\nu_{U\vert C} = \nu + d - 2$
copula corresponding to $X_1, X_2 \vert \mathbf{X}_{3,\dots,d} = \mathbf{x}_C$	bivariate Student's t-copula $C^t_{\mathcal{R}^t_{U\vert C}}$, with correlation $\mathcal{R}^t_{U\vert C} = diag(\Sigma^t_{U\vert C})^{-\frac{1}{2}}\ \Sigma^t_{U\vert C}\ diag(\Sigma^t_{U\vert C})^{-\frac{1}{2}}$

Proof. The result follows directly from the fact that the conditional bivariate copulas of a multivariate normal distribution are Gaussian copulas which do not depend on the values of the conditioning variables and that the lower-dimensional marginal distributions of a multivariate normal distribution are again normal, cf. Table 5.1. □

Remark 5.5 (Inverse Direction)

The inverse direction, i.e. that every regular vine specification where all bivariate copulas are Gaussian and we have Gaussian margins leads to a multivariate normal distribution, also holds and can be shown, e.g., by induction. Another way to prove it would be to directly write down a joint normal distribution which will decompose to a regular vine distribution with the specified copulas. To do that, we need to understand how conditional/partial correlations, which coincide for the multivariate normal distribution, specified in a vine correspond to the correlation matrix.

While we refer the reader to Bedford and Cooke (2002) for the general case, we will now define partial correlations in general and use them to demonstrate how a three-dimensional Gaussian distribution can be constructed using an R-vine.

Definition 5.5 (Partial Correlation)
Let $X_1, X_2, X_3, \ldots, X_d$ be random variables and $X^\star_{1|3,\ldots,d}$, $X^\star_{2|3,\ldots,d}$ be the linear functions of $X_3, \ldots, X_d$, minimizing $\mathbb{E}[(X_1 - X^\star_{1|3,\ldots,d})^2]$ and $\mathbb{E}[(X_2 - X^\star_{2|3,\ldots,d})^2]$, respectively. Then the partial correlation coefficient *of X_1, X_2, denoted by $\rho_{1,2;3,\ldots,d}$, is defined as the ordinary correlation coefficient between $Y_1 := X_1 - X^\star_{1|3,\ldots,d}$ and $Y_2 := X_2 - X^\star_{2|3,\ldots,d}$, i.e.*

$$\rho_{1,2;3,\ldots,d} = \frac{\mathbb{E}\Big[(Y_1 - \mathbb{E}[Y_1])(Y_2 - \mathbb{E}[Y_2])\Big]}{\sqrt{Var(Y_1)Var(Y_2)}}.$$

It can be calculated recursively by

$$\rho_{1,2;3,\ldots,d} = \frac{\rho_{1,2;3,\ldots,d-1} - \rho_{1,d;3,\ldots,d-1} \cdot \rho_{2,d;3,\ldots,d-1}}{\sqrt{1 - \rho^2_{1,d;3,\ldots,d-1}}\sqrt{1 - \rho^2_{2,d;3,\ldots,d-1}}}, \tag{5.23}$$

cf. Yule and Kendall (1965, p. 290). Using this formula, the bivariate copulas' parameters can be calculated.

Example 5.8 (Three-Dimensional Gaussian Distribution)
Let us consider a three-dimensional Gaussian distribution with zero means, unit variances, and correlations $\rho_{12} = \rho_{23} = \rho_{13} = 0.4$. Applying (5.23) we calculate

$$\rho_{1,3;2} = \frac{\rho_{13} - \rho_{12}\rho_{23}}{\sqrt{1 - \rho^2_{12}}\sqrt{1 - \rho^2_{23}}} = \frac{0.4 - 0.16}{0.84} = \frac{2}{7}.$$

This distribution is an R-vine distribution with standard normal marginal distributions, the unique R-vine tree sequence $\mathcal{V}$ in three dimensions with $E_1 = \{\{1,2\},\{2,3\}\}$ (Figure 5.10) and

$$B = \Big\{B_{\{1,2\}} = C^{Gauss}_{P(0.4)}, B_{\{2,3\}} = C^{Gauss}_{P(0.4)}, B_{\{\{1,2\},\{2,3\}\}} = C^{Gauss}_{P(\frac{2}{7})}\Big\},$$

where $P(\rho) := \begin{pmatrix} 1 & \rho \\ \rho & 1 \end{pmatrix}$.

Knowing the bivariate conditional copulas of a multivariate Student's t-distribution, we can decompose it in a similar way.

Theorem 5.3.
The multivariate Student's t-copula with parameters ν and correlation matrix R can be constructed as a regular vine copula where all bivariate copulas corresponding to edges in T_i are bivariate Student's t-copulas with $\nu + i - 1$ degrees of freedom, and some appropriate 2×2 correlation matrices.

Proof. Using the properties summarized in Table 5.1, the proof is completely analogous to the Gaussian case. Since conditional copula parameters and partial correlations coincide for the multivariate Student's t-distribution (see Cambanis et al. (1981)) the inverse direction can also be proven in the same way as for the Gaussian copula. □

We have seen that by using only Gaussian or Student's t-pair copulas we regain the multivariate Gaussian or Student's t-distribution. But as all types of bivariate copulas can be used in a PCC, the construction is much more flexible and types of dependence like asymmetric tail dependence (which the Gaussian/Student's t-copulas do not exhibit) can be modeled.

Tail Dependence

As we have seen in Chapter 4, Lemmas 4.7 and 4.8, the bivariate Gaussian copula is upper- and lower-tail independent, whereas the Student's t-copula has both upper- and lower-tail dependence. Passing over from tail dependence coefficients to tail dependence functions

$$b(\mathbf{w}) = \lim_{u \to 0} \frac{C(u\,\mathbf{w})}{u}, \quad \text{for } \mathbf{w} \in [0,1]^d,$$

this stays true for their multivariate counterparts.

Vine copulas can be used to construct multivariate copulas with asymmetric tail dependence. While we refer the reader to the original work of Joe et al. (2010) for a complete treatment of the subject, we want to highlight their main results. In Theorem 4.1, Joe et al. (2010) show under some regularity conditions that if all bivariate copulas in T_1 of a D-vine are upper- (lower-) tail dependent, then also the whole vine copula exhibits upper- (lower-) tail dependence. While this shows that the R-vine copula model is a reasonable approach for describing data which exhibits flexible multivariate tail dependence, it only gives a vague idea about which dependencies can be modeled. Joe et al. (2010) clarify this in Section 5, comparing the possible range of bivariate tail dependence for three-dimensional vine copula models and the three-dimensional Student's t-copula. They point out that while the overall range of dependence is similar, the upper- and lower-tail dependence can differ for each margin in the vine model.

Flexibility vs. Parametric Margins

The possibility to choose from $d!/2 \cdot 2^{\binom{d-2}{2}}$ R-vine tree sequences in d dimensions (see Morales-Nápoles (2010)) and to use arbitrary bivariate copulas for each edge of the vine is of course not only reflected in the tails of the vine copula but over the whole range of the distribution. Taking all possible choices into consideration, the PCC offers great flexibility and understanding how the properties of the multivariate copula arise from the two-dimensional building blocks is still an area of ongoing research. As Berg and Aas (2009) note, a downside of the flexibility of vine copulas is that neither their unspecified bivariate margins nor the resulting multivariate distribution belong to a known parametric family. They can be obtained only through (numerical) integration and simulation, which is, on the one hand, undesirable as it makes it difficult to understand how changes in one edge of the vine copula affect the overall distribution, or specific margins, and thus complicates model selection. On the other hand, this is the reason why pair copula constructions can model marginal dependencies which cannot be covered by common parametric families. When the $(1,2)$ and $(2,3)$ margins are specified in three dimensions together with the conditional copula of $(1,3|2)$, the $(1,3)$ margin is left unspecified. Simulating it for different copula families and parameter values, we see that the level curves can be very asymmetric. Examples are displayed in Figure 5.13 using standard normal margins. They have been obtained from two-dimensional kernel density estimates using 1 000 000 data points simulated from the three-dimensional copula beforehand. The types and parameters of the bivariate building blocks are listed in Table 5.2. The plots demonstrate the flexibility of the vine distributions even in three dimensions.

Table 5.2 Model specifications for the plots in Figure 5.13. The abbreviations correspond to (C)layton, G(umbel), (S)tudent, (J)oe, and (F)rank.

Scenario	(1,2) C	(1,2) Par.	(1,2) τ	(2,3) C	(2,3) Par.	(2,3) τ	(1,3|2) C	(1,3|2) Par.	(1,3|2) τ
1	G	5	.80	C	$-.7$	$-.54$	C	.7	.26
2	S	(.8, 1.2)	.59	G	1.75	.43	S	$(-.95, 2.5)$	$-.80$
3	180J	2	.35	J	24	.92	180C	20	.91
4	F	-34	$-.89$	C	20	.91	F	34	.89

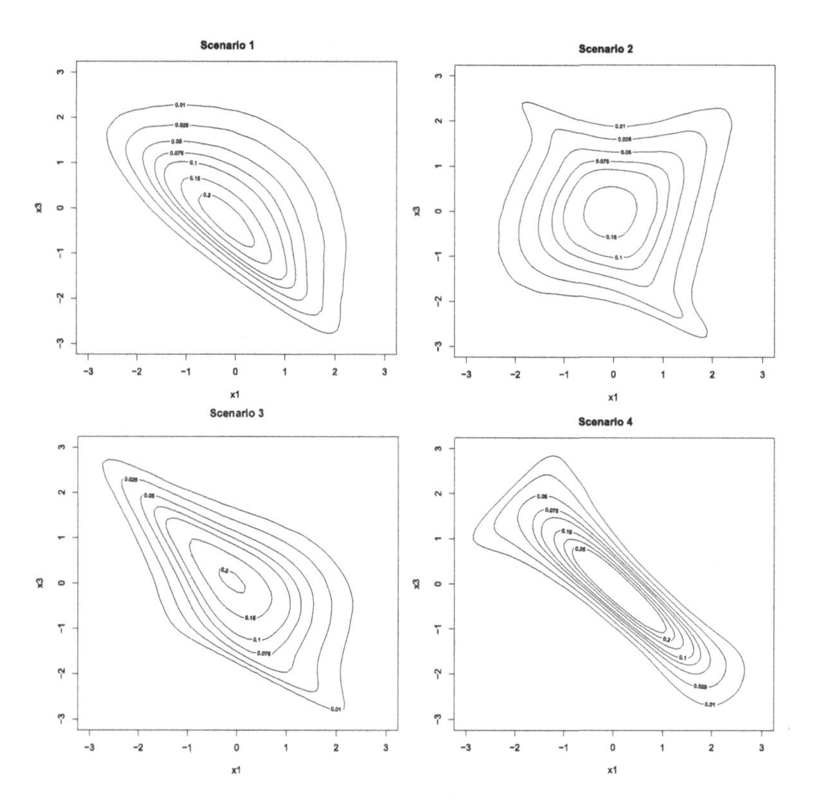

Fig. 5.13 Estimated level curves of the bivariate (1, 3) margin of D-vine distributions with standard normal margins corresponding to the four scenarios given in Table 5.2.

5.5 Application

To demonstrate the applicability of PCCs in practice, this section presents their use in a simple multivariate ARMA-GARCH model for VaR forecasting. The data set we consider consists of daily returns of ten selected stocks in the German stock index (DAX), namely: Allianz, BASF, Bayer, Daimler, Deutsche Bank, Deutsche Telekom, E.ON, RWE, SAP, and Siemens. This section is structured as follows: Section 5.5.1 introduces the marginal time series model which fits the margins of the given data set well. Section 5.5.2 gives references to the literature on how parameter estimation can be performed in the context of PCCs. Section 5.5.3 continues with more details on the selected portfolio and on how the VaR forecasts are calculated. Section 5.5.4 concludes with a statistical evaluation of the quality of our forecasts.

5.5.1 *Time Series Model for Each Margin*

First we describe the utilized marginal time series structure. For this let $\mathbf{X}_t := (X_{1t}, \ldots, X_{dt})$ for $t = 1, \ldots, T$ denote d dependent time series. For the ith marginal time series $\{X_{it}, t = 1, \ldots, T\}$ a univariate ARMA$(1,1)$–GARCH$(1,1)$ with Student's t-innovations is assumed, i.e.

$$X_{it} = \Phi_i X_{i(t-1)} + h_{it} \cdot Z_{it} + \theta_i h_{i(t-1)} \cdot Z_{i(t-1)}$$
$$Z_{it} \sim t(\nu_i) \text{ i.i.d. for } i = 1, \ldots, d,$$
$$h_{it} = \omega_i + \alpha_i \left(h_{i(t-1)} \cdot Z_{i(t-1)}\right)^2 + \beta_i h_{i(t-1)}^2,$$

where for $i = 1, \ldots, d$, $\alpha_i, \beta_i \in [0, 1]$, $\Phi_i \in [-1, 1]$, and $\theta_i \in (-\infty, \infty)$. We assume that the time series is stationary and further require $\alpha_i + \beta_i < 1$ for $i = 1, \ldots, d$. For the dependence structure between the d marginal time series, we impose a d-dimensional R-vine structure on the i.i.d. random innovation vectors $\mathbf{Z}_t = (Z_{1t}, \ldots, Z_{dt})$ for $t = 1, \ldots, T$, i.e. $\mathbf{Z}_t$ has the cdf given by

$$F(z_{1t}, \ldots, z_{dt}) = C\left(F_{\nu_1}^t(z_{1t}), \ldots, F_{\nu_d}^t(z_{dt})\right),$$

where C is a multivariate copula defined by an R-vine tree sequence $\mathcal{V}$ with specified bivariate copulas B. To summarize, we consider a multivariate ARMA–GARCH model, where the joint distribution of the innovations is not modeled by a multivariate normal distribution or t-distribution but by an R-vine distribution with Student's t-margins.

To see whether the marginal models are appropriate for daily returns in the DAX, we fit this model to data from the time period from January 16, 2001, to November 4, 2010. We then use the Kolmogorov–Smirnov test to check whether the standardized residuals are observations from a t-distribution as assumed, and the Ljung–Box test up to lags 5 and 25 to test the null hypothesis

$$H_0 : \text{"The observations are uncorrelated"}$$

against its alternative. For none of these tests can we reject the null hypothesis on the 5% level for any of the marginal time series of our dataset, indicating a reasonable marginal fit.

5.5.2 *Parameter Estimation*

Before the model of Section 5.5.1 can be used to forecast VaR by simulation using the introduced algorithms, we need to determine suitable marginal and copula model parameters for our dataset. Since estimation of copula models is not the main subject of this book we will only give references to the literature on how this problem can be treated. For the R-vine model there are two problems which need to be solved: (1) An R-vine tree structure and the bivariate copula for each edge need to be selected. (2) The parameters of the bivariate copulas and the marginal models have to be estimated. While there are different approaches for parameter estimation, model selection is usually performed in a two-step procedure: at first a model for each one-dimensional margin is fitted and then the residuals of the marginal models are used to select a suitable copula model. Dißmann (2010) uses a maximum spanning tree to select an R-vine tree sequence for a given dataset and determines the bivariate copulas with a goodness-of-fit test. Brechmann (2010) extends the procedure, discussing different criteria for the selection of bivariate copulas and for simplification of the vine structure (e.g. using only independence copulas on higher levels of the vine). We follow their approach to select an R-vine model for our dataset.

Since multivariate copula models often have large numbers of parameters, maximum likelihood (ML) estimation is also usually done in a two-step procedure. There are three different approaches: First, a parametric method (see Joe and Xu (1996)) in which, after fitting marginal models, the probability integral transform is used to obtain uniformly distributed data for the estimation of copula parameters. Then a non-parametric approach as in Genest et al. (1995), where the dataset is transformed by the empirical cdf instead of the cdf of a parametric model. These do both provide consistent and asymptotically normal estimators (see Genest et al. (1995) and Patton (2006)). However, they are not as efficient as joint ML, which is why Liu and Luger (2009) apply an iterative procedure to increase the efficiency. Kim et al. (2007) show that in case of severe misspecifications in the margins, the non-parametric approach outperforms all the others; a careful analysis of the data should however reveal such severe misspecifications. In our application with the challenge to model the time series structure of the dataset we will use a parametric procedure and fit an ARMA$(1,1)$–GARCH$(1,1)$ model to each margin using ML. Then, we fit the vine copula to the standardized residuals which are transformed to uniformly distributed data using the cdfs of the residual distributions in the marginal models.

While ML methods are more common in the literature on copula-GARCH models, Bayesian joint estimation for marginal and vine copula parameters is also possible as demonstrated in Hofmann and Czado (2010).

It is however computationally more intensive, which is why we restrict our attention to ML estimation for this application.

5.5.3 *Forecasting Value at Risk*

The VaR which we try to forecast using our copula-GARCH model is the (one-day and ten-day) VaR of a portfolio consisting of the ten stocks we have listed before; these are the stocks in the DAX with the highest market capitalization on October 25, 2010. We consider two portfolios: for the first the *value of positions* in the portfolio corresponds to the respective market capitalization; for the second the *number of stocks* for each position in the portfolio corresponds to the market capitalization.

We choose the one-day and ten-day horizons, as the ten-day VaR is what is required from financial institutions for assessing the market risk of their positions within the Basel framework (see Basel Committee (2006)) and the one-day horizon is needed to apply statistical procedures for the backtesting of our model, which we will explain in Section 5.5.4.

The testing period for which VaR forecasts are calculated is from September 11, 2009, to September 3, 2010, and consists of 250 trading days. The model parameters which are used for estimating it are obtained from a rolling window of 600 observations, requiring 250 model fits. For comparison purposes we fit four different multivariate copula models and include the independence copula as a benchmark:

(1) An R-vine PCC (45–90 copula parameters).
(2) A C-vine PCC (45–90 copula parameters).
(3) A multivariate Student's t-copula (46 copula parameters).
(4) A multivariate Gaussian copula (45 copula parameters).

The bivariate copula families which are utilized in the PCCs are the independence copula, Gaussian, Student's t, Gumbel, Clayton, Frank, Joe, BB1, BB7, and their rotated versions (see Section 5.3.1 for definitions and references).

While the R-vine model is the most general, simulating from a C-vine is computationally much faster, which is why we consider the C-vine as a second model. The independence copula ignores all forms of dependence between the marginal time series and we include it to be able to observe the benefits from using a multivariate model compared to univariate ARMA-GARCH models for each margin. Gaussian and Student's t are added for comparison purposes because they are the most commonly used multivariate copulas.

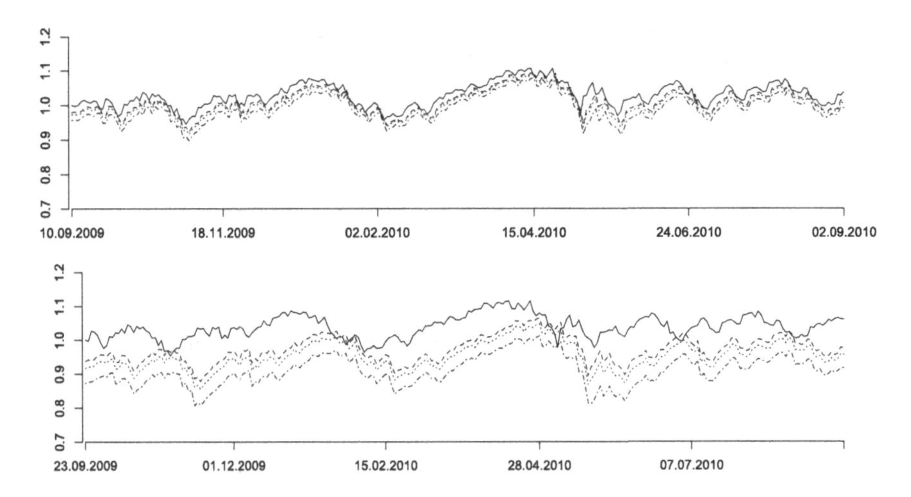

Fig. 5.14 The observed relative portfolio values of the first portfolio over 250 trading days (solid), together with the 90% VaR (dashed), the 95% VaR (dotted), and the 99% VaR (dashed and dotted) calculated from the R-vine model. (Top panel: one-day VaR, bottom panel: ten-day VaR.)

From each fitted model we simulate 100 000 observations of portfolio returns for the next ten days and use these to calculate a one-day and a ten-day VaR forecast from the quantiles of the simulated data. The VaR forecasts for the first portfolio together with the observed values are plotted in Figure 5.14 for the one-day forecast; it also shows the ten-day forecast with corresponding observations. The portfolio value at each point of time is normalized using the first value of the testing period.

As we can see, the values for the ten-day forecasts are much more spread out compared to the one-day forecasts, which reflects the greater uncertainty due to the longer time period.

5.5.4 *Backtesting Value at Risk*

Models for VaR values are only useful if they predict future risks accurately, and their quality should always be evaluated by backtesting them with appropriate methods. Backtesting a model means fitting it to historical data (which is exactly what we did with our simulations) and comparing model-based predictions to observed data afterwards. Considering for example forecasts of the 95% VaR we would expect that an exception occurs on average every 20th day and in a backtest it is checked whether this is the case. Backtesting of VaR models is also required in the Basel framework, where financial institutions have to backtest one-day-ahead forecasts of the 99% VaR over the last 250 trading days. Although we calculated both one-day and ten-day forecasts in our simulations, backtesting is always done for the one-day horizon. The Basel committee has chosen this forecasting period because the portfolio composition of larger trading entities changes very much over a ten-day horizon, which is why comparing the ten-day forecasts to actual trading returns is not meaningful (Basel Committee (1996)). While this kind of reasoning is not applicable in our situation with a fixed portfolio composition, there are also statistical arguments in favor of the one-day horizon, as we do not have to deal with overlapping observation periods in this case (Christoffersen and Pelletier (2004)).

As mentioned earlier, we do expect from a correct VaR model with coverage rate α that the number x of exceptions during T trading days equals on average $x = \alpha \cdot T$. From the calculated forecasts, we can statistically examine whether the frequency of exceptions lies within the range which is acceptable for a given confidence level and length of the testing period.

Statistical tests which rely on the frequency of exceptions are as follows:

(1) The traffic light approach of the Basel Committee (Basel Committee (1996)).
(2) The Kupiec proportion of failures (POF) test (Kupiec (1995)).

Tests based solely on the number of exceptions within the time T are called tests for *unconditional coverage* (UC).

A correct model should however not only show approximately the expected number of exceptions; the occurrences of exceptions should also be independent. Tests which incorporate this i.i.d. hypothesis are called tests for *conditional coverage* (CC). The tests we take into consideration are as follows

(1) The mixed Kupiec test proposed by Haas (2001).
(2) The GMM duration-based test of Candelon et al. (2011).

For details on the applied testing methods for the quality of VaR forecasts, we refer the reader to the cited literature.

5.5.5 *Backtest Results*

We employ the aforementioned tests to our one-day VaR forecasts at the 90%, 95%, and 99% levels. The resulting p-values of the test statistics are listed in Table 5.3 for the different copula models we are considering. Whereas we can not reject any of the tests for the C-vine or the R-vine on the 10% level, some of the tests are rejected for the multivariate Gaussian/Student's t-copula and all tests are heavily rejected for the independence copula. This effect is stronger in the second portfolio than in the first one where the composition is chosen to mimic the DAX composition.

This indicates that using a C-vine copula model for forecasting VaR might be the best choice for a DAX portfolio. While the results for the C-vine copula are similiar to the R-vine copula, simulating from a C-vine is computationally much faster, as we have seen in Section 5.3 (because we do not need to evaluate any h-functions).

While we can discriminate between the different models on the 90% and 95% levels, it is very difficult to do so on the 99% level due to the low number of expected value at risk exceptions over the testing period. This renders the employed tests very weak, and, e.g., the traffic light system of the Basel Committee would accept all models except for the independence copula as equally good risk models.

Table 5.3 P-values for the test statistics of the VaR backtest for the different models over a one-day horizon. The GMM test statistic for the 99% VaR obtained using the C-vine model could not be computed as only one VaR exception was observed; we denote these cases by "NA".

Portfolio 1 Value ~ Market Cap.		Kupiec POF	Mixed Kupiec	GMM of Order 2	GMM of Order 6	Traffic Light
	99%	0.74	0.95	0.84	0.99	
R-vine	95%	0.67	0.20	0.70	0.81	GREEN
	90%	0.67	0.10	0.58	0.66	
	99%	0.28	0.40	NA	NA	
C-vine	95%	0.67	0.20	0.70	0.81	GREEN
	90%	0.41	0.17	0.51	0.87	
	99%	0.38	0.31	0.62	0.99	
Student's t	95%	0.21	0.08	0.47	0.81	GREEN
	90%	0.41	0.18	0.51	0.87	
	99%	0.38	0.31	0.62	0.99	
Gaussian	95%	0.32	0.07	0.37	0.69	GREEN
	90%	0.41	0.18	0.51	0.87	
	99%	0.00	0.00	0.00	0.00	
Independence	95%	0.00	0.00	0.00	0.00	RED
	90%	0.00	0.00	0.00	0.00	

Portfolio 2 Value ~ Market Cap.		Kupiec POF	Mixed Kupiec	GMM of Order 2	GMM of Order 6	Traffic Light
	99%	0.74	0.95	0.84	0.99	
R-vine	95%	0.48	0.24	0.78	0.89	GREEN
	90%	0.31	0.10	0.33	0.51	
	99%	0.28	0.40	NA	NA	
C-vine	95%	0.48	0.24	0.79	0.88	GREEN
	90%	0.31	0.10	0.33	0.50	
	99%	0.76	0.90	0.92	0.99	
Student's t	95%	0.04	0.00	0.13	0.06	GREEN
	90%	0.07	0.02	0.10	0.30	
	99%	0.76	0.90	0.92	0.99	
Gaussian	95%	0.13	0.02	0.35	0.42	GREEN
	90%	0.11	0.01	0.12	0.19	
	99%	0.00	0.00	0.00	0.00	
Independence	95%	0.00	0.00	0.00	0.00	RED
	90%	0.00	0.00	0.00	0.00	

Chapter 6

Sampling Univariate Random Variables

This chapter was contributed by **Elke Korn** and **Ralf Korn**,[1] for which we would like to thank them very much. Much more on the subject matter is to be found in their recent monograph Korn et al. (2010).

6.1 General Aspects of Generating Random Variables

The basic ingredients for performing (stochastic) simulations are suitable random numbers (RNs). For simulation purposes we do not need true random numbers or cryptographically secure ones, but they should mimic a random behavior closely and should be (at least approximately) independent of each other. The main emphasis lies on producing them fast and being able to reproduce the produced sequence if we have to analyze the result. So, usually RNs for simulations are produced by deterministic algorithms. They only appear to be random, which is why we speak of *pseudorandom numbers*.

The task of generating general (pseudo) RNs consists of two distinct subtasks, which are even looked at by different researchers. The first part is generating uniformly distributed RNs $U \sim U[0,1]$, while the second part consists of transforming them to RNs with a specific distribution.

There are many ways to suitably transform the uniformly distributed RNs into a given distribution. They all have their specific applications, advantages, and disadvantages. The most important method is the *inversion method*, which maps exactly one uniformly distributed RN to one with the desired distribution. The monotonicity of the transformation can often be further exploited, no RN is lost, and usually every cycle of the algo-

[1]Fachbereich Mathematik, Technische Universität Kaiserslautern, Erwin Schrödinger Straße, 67663 Kaiserslautern, Germany, `korn@mathematik.uni-kl.de`.

rithm takes the same time, which is an advantage when the program is parallelized.

However, in some cases the inversion method is too complicated, not exact enough, or the calculations take too much time. Then, an *acception-rejection method* might be better. This method is usually fast and sometimes even quite simple. The value of a simple algorithm should not be underestimated. That way it is possible to write bug-free programs in little time, which is especially useful when you are hunting for bugs in other programs. But the acception-rejection method needs more than one uniformly distributed RN as input and the time for generating one number of the desired distribution can differ significantly.

Sometimes, there exist other transformation methods similar to the inversion method, which map two or more RNs to one or two new RNs. Then we no longer have a monotone transformation as in the inversion case, but the time cycle for producing a new RN is always of the same length. These formulas often involve trigonometric or log-type functions, and therefore they are usually slower and not suitable for hardware implementations which require simple functions. But some of them have the advantage of being very accurate, especially in the tails. This explains why we sometimes offer more than one algorithm for producing RNs with a desired distribution.

6.2 Generating Uniformly Distributed Random Variables

The basis for generating all kinds of random variables (RVs) is uniformly distributed RNs. Usually, a computer language will provide at least one routine for generating such RNs. These random number generators (RNGs) can be characterized by the following quintuple of parameters $(\mathcal{S}, \mu, f, \mathcal{U}, g)$ (see L'Ecuyer (1994)):

(1) $\mathcal{S}$ is the finite set of states, the so-called *state space*. Each $\boldsymbol{s}_n \in \mathcal{S}$ is a particular *state* at iteration step n, an important value which can be found as a variable in every RNG and changes after a new RN has been produced.
(2) μ is the probability measure for selecting $\boldsymbol{s}_0$, the initial state, from $\mathcal{S}$. $\boldsymbol{s}_0$ is called the *seed* of the RNG.
(3) The function f, also called the *transition function*, describes how the algorithm constructs a new state out of the previous one, i.e. $\boldsymbol{s}_{n+1} = f(\boldsymbol{s}_n)$.

(4) $\mathcal{U}$ is the *output space*. As we are considering uniformly distributed RNs, we will mainly look at discrete sets $\mathcal{U}$ from $[0,1]$, $[0,1)$, $(0,1]$, or $(0,1)$.
(5) The *output function* $g : \mathcal{S} \to \mathcal{U}$ maps the state $\boldsymbol{s}_n \in \mathcal{S}$ to a number $u_n \in \mathcal{U}$, the final random number we are interested in.

As the state space $\mathcal{S}$ is finite, the RNG will produce the same cycle of RNs again after a while. The smallest integer ρ for which $\boldsymbol{s}_{n+\rho} = \boldsymbol{s}_n$ is called the *period* of the RNG. In good RNGs the period is often of the same size as the state space, but sometimes the period is much smaller. In this case the RNG has several disjoint cycles, and then it is very important which seed to choose to start the generator.

6.2.1 *Quality Criteria for RNG*

To judge the quality of an RNG, we state various criteria:

(1) The period should be very long (nowadays a period of 2^{32} is considered too short). Otherwise it might happen that in a simulation the same cycle of RNs is used again. This means that the events in the simulation are no longer independent, and the result of the simulation could be complete nonsense.
(2) The RNG should produce RNs very fast.
(3) The RNG should have a possibility of being used in parallelized programs, e.g. it should provide several disjoint and very long substreams.
(4) The uniform distribution should be very well approximated. The finite set $\Phi_d := \{(u_0, \ldots, u_{d-1}) | s_0 \in \mathcal{S}\}$ of all vectors of d successive outputs from all possible initial states should cover the d-dimensional hypercube $[0,1]^d$ evenly. The size of this set equals the number of states $|\mathcal{S}| = |\Phi_d|$ as this set is seen as a multiset.
(5) The RNG should pass several statistical and theoretical tests. There are well-known test suites which can be used to check the distribution of the RNs. Examples are TestU01 by L'Ecuyer and Simard (2002), DieHard by Marsaglia (1996), and NIST by NIST USA (2011).

The research in the field of uniformly distributed RNs is very active. So we only give a weak recommendation of some RNGs here. Please take care, as they might even be outdated already. At the moment of writing, good generators for producing uniformly distributed RNs are the Mersenne Twister MT19338 (see Matsumoto and Nishimura (1998)) or other Mersenne Twister types (see Matsumoto and Saito (2008)),[2] combined LFSRs or combined MRGs (see L'Ecuyer (1999)), several WELL-RNGs (see L'Ecuyer et al. (2006)), XOR-Shift-RNG (see Marsaglia (2003)), or special lagged Fibonacci RNGs. To explain the abbreviations just used is beyond the scope of this survey. They will be explained in the given references.

6.2.2 *Common Causes of Trouble*

When working with RNGs, the following issues might cause trouble:

(1) RNGs have to be initialized. The seed must be chosen carefully so that the RNG works properly, which is usually done by an initialization routine. But this routine has to be started by the user.
(2) The output space $\mathcal{U}$ has to be checked for whether it contains 0s or 1s. This is often a source of errors when transformed with $\log(x)$ or $1/x$.
(3) When interested in extreme values, the output space should be checked for whether it includes extremely small numbers $\neq 0$ or numbers very close to 1. This is often not symmetric. Often you find 1×2^{-31} or $1 - 1 \times 10^{-7}$ as the nearest neighbors to 0 or 1, which might not be sufficient for simulation purposes.

6.3 The Inversion Method

The direct method for transforming uniformly distributed RNs into RNs of a given distribution is the inverse transformation method, also called the inversion method. To understand it, recall the definition of the generalized inverse (note that we can use "min" as a distribution function is right-continuous):

$$F^{-1}(u) := \min\{x \in \mathbb{R} : F(x) \geq u\}, \quad u \in (0,1).$$

Then, we recall from Lemma 1.4(3) the relation

$$\mathbb{P}\big(F^{-1}(U) \leq x\big) = \mathbb{P}\big(U \leq F(x)\big) = F(x), \quad U \sim U[0,1]. \tag{6.1}$$

This means that $F^{-1}(U)$ is distributed according to F and we are able to formulate the inversion method as Algorithm 6.1. Note in particular the monotone relationship between the uniformly distributed random variable U and the transformed variable.

[2] C++ code for the Mersenne twister is available from Rick Wagner `http://www-personal.umich.edu/~wagnerr/MersenneTwister.html`.

Algorithm 6.1 (Inversion Method)

FUNCTION inversion
  Generate $U \sim U[0,1]$
  RETURN $X := F^{-1}(U) \sim F$

Although various methods exist to generate random variables of a desired distribution from univariate RNs, the inversion method is in general the best choice for transforming RNs. The main reason for this is that it preserves structures. More precisely, if the distribution structure of the uniformly distributed pseudorandom numbers is good, so will be the structure of the transformed RNs. Another advantage is that it can easily be combined with variance reduction techniques such as antithetic variates.

If it is not possible to invert F analytically, it can be inverted numerically (by, e.g., the Newton–Raphson method) or with an explicit approximation formula. Explicit approximation formulas can then be improved by a Newton–Raphson, a regula falsi, or an interpolation step.

6.4 Generating Exponentially Distributed Random Numbers

The textbook example for the inversion method is the generation of exponentially distributed random variables. An $Exp(\lambda)$-distributed random variable X has the distribution function $F(x) = \big(1-\exp(-\lambda\, x)\big)\mathbb{1}_{\{x\geq 0\}}$, for $x \in \mathbb{R}$, and the inverse of F is given by $F^{-1}(u) = -\log(1-u)/\lambda$, $0 \leq u < 1$. Thus, $Y := -\log(1-U)/\lambda$ with $U \sim U(0,1)$ is $Exp(\lambda)$-distributed by the inversion method. As $(1-U)$ and U have the same distribution, one saves time by using $Y := -\log(U)/\lambda$. In software packages often other algorithms are implemented, as the evaluation of the logarithm function is rather time-consuming.

6.5 Acceptance-Rejection Method

Besides the inversion method, the acceptance-rejection method is the second major general method for sampling random variables from a given distribution. It was proposed by von Neumann (1951). It is typically used if the inversion of the distribution function F is very difficult, if evaluating it is very time-consuming, or if only approximations of the inverse exist.

The theoretical foundation of the acceptance-rejection method (AR method) is given by Theorem 6.1; see Devroye (1986, p. 40) for a simple proof. To formulate it, we need the following definition.

Definition 6.1 (Body of a Function)
Let $f : \mathbb{R}^d \to \mathbb{R}$ be a non-negative, integrable function. Then

$$B_f := \{(\boldsymbol{x}, z) \in \mathbb{R}^d \times \mathbb{R} \,|\, 0 \leq z \leq f(\boldsymbol{x})\}$$

is called the body of f.

Theorem 6.1.
Let $\boldsymbol{X}$ be an $\mathbb{R}^d$-valued random variable with density f, let $U \sim U[0,1)$ be independent of $\boldsymbol{X}$, and $c > 0$ a constant. Then, $(\boldsymbol{X}, U\, c\, f(\boldsymbol{X}))$ is uniformly distributed on $B_{c \cdot f}$. Vice versa, if the multivariate random variable $(\boldsymbol{X}, Z) \in \mathbb{R}^{d+1}$ is uniformly distributed on B_f, then $\boldsymbol{X}$ has density f on $\mathbb{R}^d$.

Hence, if we are able to pick points randomly (i.e. with uniform distribution) from the area B_f, then the first d components of every such point have the distribution with the desired density f. Thus, the task boils down to simulating uniformly distributed RNs on B_f. For this, we need the help of a random variable $\boldsymbol{Y}$ with density g, which we are able to sample easily, and which satisfies

$$f(\boldsymbol{x}) \leq c\, g(\boldsymbol{x}), \quad \forall\, \boldsymbol{x} \in \mathbb{R}^d, \tag{6.2}$$

with a constant $1 \leq c < \infty$. g is then called a *comparison density* for f.

As we can simulate RNs $\boldsymbol{Y}$ with a density g, by the first part of Theorem 6.1, we obtain uniformly distributed points $(\boldsymbol{Y}, U\, c\, g(\boldsymbol{Y}))$ on $B_{c \cdot g}$ via generating a RN $\boldsymbol{Y}$ with density g and another independent RN $U \sim U[0,1)$. If $U\, c\, g(\boldsymbol{Y}) \leq f(\boldsymbol{Y})$, then we have also found a point in B_f. Otherwise, we reject the point and have to perform a new trial. The requirement $c\, g(\boldsymbol{x}) \geq f(\boldsymbol{x})$ ensures that we do not cut out points from B_f. This indeed guarantees that accepted points are also uniformly distributed on B_f. Hence, we obtain Algorithm 6.2.

Algorithm 6.2 (Acceptance-Rejection Method)

FUNCTION rejection

REPEAT

Generate $U \sim U[0,1)$
Generate $\boldsymbol{Y}$ *with density* g
UNTIL $\left(U \leq f(\boldsymbol{Y}) / (c\,g(\boldsymbol{Y}))\right)$
RETURN $\boldsymbol{Y}$

By construction of the AR method, there is a positive probability that one needs more than one trial to produce one random number with a distribution according to density f. Indeed, we have

$$\begin{aligned}\mathbb{P}\Big(U \leq \frac{f(\boldsymbol{Y})}{c\,g(\boldsymbol{Y})}\Big) &= \mathbb{E}\Big[\mathbb{P}\Big(U \leq \frac{f(\boldsymbol{Y})}{c\,g(\boldsymbol{Y})} \;\Big|\; \boldsymbol{Y}\Big)\Big] \\ &= \mathbb{E}\Big[\frac{f(\boldsymbol{Y})}{c\,g(\boldsymbol{Y})}\Big] \\ &= \int \frac{f(\boldsymbol{y})}{c\,g(\boldsymbol{y})}\, g(\boldsymbol{y})\, d\boldsymbol{y} = \frac{1}{c}.\end{aligned}$$

Hence, the smaller the constant $c \geq 1$ can be chosen, the less likely the method produces points that will be rejected. So, a comparison density with a small constant c is desirable. However, the total speed of the method is also heavily dependent on the time needed to generate the sample $\boldsymbol{Y}$ and the time to calculate $f(\boldsymbol{Y})$.

Note that in the case of a bounded density f with compact support, the uniform density on the support of f can be used as a comparison density (although the constant c might be large).

One way to speed up the acceptance-rejection method is the use of so-called *squeeze functions* q_1, q_2 with

$$q_1(\boldsymbol{x}) \leq f(\boldsymbol{x}) \leq q_2(\boldsymbol{x}) \leq c\,g(\boldsymbol{x}), \quad \forall\, \boldsymbol{x} \in \mathbb{R}^d \tag{6.3}$$

that should be computable much faster than the function f. As a consequence of relation (6.3), a simulated number $\boldsymbol{Y}$ from the comparison density can directly be accepted if we have $U \leq q_1(\boldsymbol{Y})/(c\,g(\boldsymbol{Y}))$ or rejected in the case of $U > q_2(\boldsymbol{Y})/(c\,g(\boldsymbol{Y}))$. The function f only has to be evaluated if both cases do not apply.

For discrete probabilities the acceptance-rejection method has the very same form when we substitute the density with the probability mass function.

6.6 Generating Normally Distributed Random Numbers

When working with normally distributed RVs, we can always restrict ourselves to the case $X \sim \mathcal{N}(0,1)$, as we have the well-known relation $\sigma X + \mu \sim \mathcal{N}(\mu, \sigma^2)$.

6.6.1 *Calculating the Cumulative Normal*

First, we need a fast way to compute the distribution function $\Phi(x)$ of the normal distribution, as it has no simple analytical representation. There exist several approximations (see, e.g., Abramowitz and Stegun (1972) or Marsaglia (2004)), where we can choose between precise, fast, or simple (e.g. without the use of an exp function) variants. A good basis for approximations is the power series

$$\Phi(x) = \frac{1}{2} + \phi(x) \sum_{k=0}^{\infty} \frac{x^{2k+1}}{1 \cdot 3 \cdot 5 \cdots (2\,k+1)}, \; x \geq 0, \; \phi(x) := \frac{1}{\sqrt{2\pi}}\, e^{-x^2/2}.$$

Here we offer a fast-to-type approximation (see also Hastings (1955)) with just four constants:

$$\Phi(x) \approx 1 - \phi(x) \left(a_1\, t + a_2\, t^2 + a_3\, t^3 \right), \quad t = \frac{1}{1 + p\,x}, \quad x \geq 0,$$

where

$$p = 0.33267, \; a_1 = 0.4361836, \; a_2 = -0.1201676, \; a_3 = 0.937298.$$

The absolute error is uniformly less than 1×10^{-5}. Due to the symmetry of ϕ, we have $\Phi(x) = 1 - \Phi(-x)$ and can thus concentrate on $x \geq 0$.

When working with rare events, more precise approximations should be chosen. In particular, when large absolute values of x appear, it is better to use a specialized approximation of the survival function $\bar{\Phi}(x) = 1 - \Phi(x)$.

6.6.2 *Generating Normally Distributed Random Numbers via Inversion*

Without an explicit analytical representation of the distribution function, we cannot hope for an explicit formula of its inverse either. Of course, one can obtain the inverse via solving $\Phi(x) = u$ numerically with the Newton–Raphson algorithm

$$x_{n+1} = x_n - \frac{\Phi(x_n) - u}{\phi(x_n)}, \tag{6.4}$$

which works quite well here. To find a good starting point, we can use one of the many numerical approximations of the inverse.

Popular approximations are, e.g., the Moro inverse by Moro (1995), the Acklam inverse,[3] or the Beasley–Springer approximation by Beasley and Springer (1977). For hardware implementation, a piecewise polynomial interpolation together with a table of exact values is a further possibility. As an example, we present the Beasley–Springer algorithm, modified by Moro to achieve more accuracy in the tails (see Glasserman (2004, p. 67)):

$$\Phi^{-1}(u) \approx \frac{\sum_{n=0}^{3} a_n \left(u - \frac{1}{2}\right)^{2n+1}}{1 + \sum_{n=0}^{3} b_n \left(u - \frac{1}{2}\right)^{2n+2}}, \quad u \in [0.5, 0.92],$$

$$\Phi^{-1}(u) \approx \sum_{n=0}^{8} c_n t^n, \quad t = \log(-\log(1-u)), \quad u \in (0.92, 1),$$

$$\begin{aligned}
[a_k]_{k=0}^{3} &:= (2.50662823884, -18.61500062529,\\
&\quad 41.39119773534, -25.44106049637)\\
[b_k]_{k=0}^{3} &:= (-8.47351093090, 23.08336743743,\\
&\quad -21.06224101826, 3.13082909833)\\
[c_k]_{k=0}^{8} &:= (0.3374754822726147, 0.9761690190917186, 0.1607979714918209,\\
&\quad 0.0276438810333863, 0.0038405729373609, 0.0003951896511919,\\
&\quad 0.0000321767881768, 0.0000002888167364, 0.0000003960315187).
\end{aligned}$$

Again, we benefit from the symmetry of the normal distribution due to $\Phi^{-1}(1-u) = -\Phi^{-1}(u)$, which implies that it is sufficient to have approximations only on $[0.5, 1)$. This algorithm achieves a maximum absolute error of 3×10^{-9} for $u \in (1 \times 10^{-11}, 1 - 1 \times 10^{-11})$. If more precision in the tails is needed, we can add one or two Newton–Raphson steps (6.4) with a good approximation of Φ.

6.6.3 *Generating Normal Random Numbers with Polar Methods*

If we do not need the monotonicity of the inversion method, we can use the classical method for generating two independent $\mathcal{N}(0,1)$-numbers, the Box–Muller method. It transforms two independent $U(0,1]$-numbers into two independent standard normal RNs. As the functions log, cos, and sin have to be evaluated, this algorithm is rather slow.

[3] See `http://home.online.no/~pjacklam/notes/invnorm/`.

Algorithm 6.3 (The Box–Muller Method)

FUNCTION BoxMuller
 Generate independent $U_1, U_2 \sim U\,(0,1]$
 $Y_1 := \sin(2\,\pi\,U_2)\,\sqrt{-2\,\log(U_1)}$,
 $Y_2 := \cos(2\,\pi\,U_2)\,\sqrt{-2\,\log(U_1)}$
 RETURN independent $Y_1, Y_2 \sim \mathcal{N}(0,1)$

To make this method faster, Marsaglia and Bray added an acceptance-rejection step to get rid of the time-consuming functions sin and cos.

Algorithm 6.4 (The Marsaglia–Bray Algorithm)

FUNCTION MarsagliaBray
 REPEAT
  Generate independent $V_1, V_2 \sim U\,[-1,1]$
  $X = V_1^2 + V_2^2$
 UNTIL $(X \le 1)$
 $Y_1 := V_1\,\sqrt{-2\,\log(X)/X}$; $Y_2 := V_2\,\sqrt{-2\,\log(X)/X}$
 RETURN independent $Y_1, Y_2 \sim \mathcal{N}(0,1)$

Here, you have to take care of $X \neq 0$. Note that RNs $V \sim U[-1,1]$ can be generated via $V = 2\,U - 1$ with $U \sim U[0,1]$.

6.7 Generating Lognormal Random Numbers

As by its definition $X \sim \mathcal{LN}(\mu, \sigma^2)$ if and only if we have $\log(X) \sim \mathcal{N}(\mu, \sigma^2)$, we can simply generate $Y \sim \mathcal{N}(\mu, \sigma^2)$ by an appropriate algorithm and then set $X := \exp(Y)$.

6.8 Generating Gamma-Distributed Random Numbers

We first recall some properties of the Gamma distribution. Let $X \sim \Gamma(\beta, \eta)$ be a Gamma-distributed RV with shape parameter β and scale parameter η.

(1) **Moments:** We have $\mathbb{E}[X] = \beta/\eta$, $\text{Var}(X) = \beta/\eta^2$, and $\text{mode}(X) = (\beta - 1)/\eta$ for $\beta \geq 1$.
(2) **Scaling:** $X \sim \Gamma(\beta, \eta) \Rightarrow cX \sim \Gamma(\beta, \eta/c)$ for any $c > 0$.
(3) **Summation:** Given independent $X_i \sim \Gamma(\beta_i, \eta)$, $i = 1, \ldots, N$, one has

$$\sum_{i=1}^{N} X_i \sim \Gamma(\sum_{i=1}^{N} \beta_i, \eta).$$

(4) **Exponential distribution:** $\Gamma(1, \eta) = Exp(\eta)$.

As a consequence of the scaling property, we only need to be able to generate $\Gamma(\beta, 1)$-distributed RNs. By the last two properties we can generate $\Gamma(\beta, 1)$ RNs for positive integers β as the sum of β independent $Exp(1)$-distributed RNs. Because of the summation property, we can further split up the Gamma distribution into an integer part and a part with shape parameter $\beta < 1$,

$$\Gamma(\beta, 1) \sim \Gamma(\lfloor\beta\rfloor, 1) * \Gamma(\beta - \lfloor\beta\rfloor, 1),$$

where "*" denotes the convolution. We can thus concentrate on simulating the two ingredients, $\Gamma(\alpha, 1)$ for $\alpha < 1$ and $\Gamma(n, 1)$ for $n \in \mathbb{N}$, independently. However, this method of simulating $\Gamma(n, 1)$ is only suitable for small values of n. For large n, it is inefficient and also unstable.

6.8.1 *Generating Gamma-Distributed RNs with* $\beta > 1$

Well-known algorithms for $\beta > 1$ are the AR methods by Ahrens and Dieter, Best, and Cheng (for details see Devroye (1986, p. 407–413)). The rejection constant of the latter algorithm is

$$c = \frac{4\,\beta^{\beta}\,e^{-\beta}}{\lambda\,\Gamma(\beta)}, \quad \Gamma(x) := \int_0^{\infty} t^{x-1} e^{-t} dt, \tag{6.5}$$

which asymptotically tends to 1.13 for large β. Cheng's algorithm is based on the Burr XII density $g(x)$ with distribution function $G(x)$,

$$g(x) = \frac{\lambda\,\mu\,x^{\lambda-1}}{(\mu + x^{\lambda})^2}, \quad G(x) = \frac{x^{\lambda}}{\mu + x^{\lambda}}, \quad x \geq 0,$$

where λ and μ are chosen in dependence of β (see Algorithm 6.5). RNs with this distribution can be generated with the inversion method, since $G^{-1}(u) = \left((\mu u)/(1-u)\right)^{1/\lambda}$. Cheng chooses $\lambda = \sqrt{2\beta - 1}$ and $\mu = \beta^\lambda$. To speed up calculations a squeeze step (*) is added.

Algorithm 6.5 (Cheng's Algorithm for $\Gamma(\beta, 1)$ with $\beta > 1$)

FUNCTION Chengs_gamma (β)

REPEAT

Generate independent $U_1, U_2 \sim U(0,1)$

$$Y := \frac{1}{\sqrt{2\beta - 1}} \log\left(\frac{U_1}{1 - U_1}\right)$$

$$X := \beta\, e^Y; \quad Z := U_1^2\, U_2$$

$$R := \beta - \log(4) + \left(\beta + \sqrt{2\beta - 1}\right) Y - X$$

IF $\left(R \geq 4.5\, Z - 1 - \log(4.5)\right)$ *THEN RETURN* X (*)

UNTIL $\left(R \geq \log(Z)\right)$

RETURN $X \sim \Gamma(\beta, 1)$

6.8.2 *Generating Gamma-Distributed RNs with $\beta < 1$*

The case $\beta < 1$ can be reduced to the former case, because if the RNs $U \sim U(0,1)$ and $V \sim \Gamma(\beta+1, 1)$ are independent, then $Z = U^{1/\beta} V \sim \Gamma(\beta, 1)$ (see Stuart's theorem, e.g. in Devroye (1986, p. 182)). Note in particular that we have $U^{1/\beta} \sim Beta(\beta, 1)$.

But we still prefer a dedicated algorithm for $0 < \beta < 1$. The example we present here is constructed with the help of the Weibull distribution (see Devroye (1986, p. 415)), and was chosen because of its simple form. For further algorithms by Ahrens, Ahrens/Dieter, and Ahrens/Best (see Knuth (1998, p. 134, 586) or Devroye (1986, p. 419)). In the following algorithm, we have the rejection constant

$$c = \frac{\exp\left((1-\beta)\left(\beta^{\beta/(1-\beta)}\right)\right)}{\Gamma(\beta+1)} \leq 3.07, \quad \forall \beta \in (0,1).$$

The rejection constant tends to 1 as $\beta \nearrow 1$ or $\beta \searrow 0$.

Algorithm 6.6 (Algorithm for $\Gamma(\beta,1)$ with $0<\beta<1$)

FUNCTION Weibull_gamma (β)
Set $a := (1-\beta)\,\beta^{\beta/(1-\beta)}$
REPEAT
Generate independent $E_1, E_2 \sim Exp(1)$
Generate Weibull RN $X := E_1^{1/\beta}$
UNTIL $\left(X \le E_1 + E_2 - a\right)$
RETURN $X \sim \Gamma(\beta,1)$

6.8.3 *Relations to Other Distributions*

The following relations can be found in Devroye (1986, p. 403):

(1) If $X \sim \Gamma(\beta,\eta)$, then $1/X$ is inverse Gamma distributed with parameters β and η.
(2) If $X \sim \Gamma(a,\eta)$ and $Y \sim \Gamma(b,\eta)$, independent, then $X/(X+Y)$ is Beta distributed with parameters a, b.
(3) If $X \sim \Gamma(\beta, 1/2)$, then $X \sim \chi^2(2\beta)$.
(4) *Stuart's theorem*: If $Y \sim \Gamma(b,1)$, $Z \sim Beta(a, b-a)$ with $b > a > 0$, independent, then $X_1 := YZ \sim \Gamma(a,1)$ and $X_2 := Y(1-Z) \sim \Gamma(b-a,1)$, also independent. The case $b=1,\ 0<a<1$ leads to a method to generate $\Gamma(\beta,1)$-distributed RNs with $\beta<1$ on the basis of exponentially distributed RNs and Beta-distributed RNs.

6.9 Generating Chi-Square-Distributed RNs

The Chi-square distribution is a special case of the Gamma distribution as indicated in the properties of the Gamma distribution. So, Chi-square-distributed RNs can be generated with the help of Gamma-distributed RNs.

Algorithm 6.7 (Chi-Square-Distributed RNs)

FUNCTION chi_gamma(ν)
Generate $Y \sim \Gamma(\nu/2, 1)$
RETURN $X := 2Y \sim \chi^2(\nu)$

Chi-square-distributed RVs with degree $k \in \mathbb{N}$ can also be described with the help of normally distributed RVs: for $X_1, \ldots, X_k \sim \mathcal{N}(0,1)$ i.i.d. one has $Z := \sum_{i=1}^{k} X_i^2 \sim \chi^2(k)$. If $X_i \sim \mathcal{N}(\mu_i, 1)$, $i = 1, \ldots, k$, we obtain the non-central Chi-square distribution $\chi^2(\delta^2, k)$ with non-centrality parameter $\delta^2 = \sum_{i=1}^{k} \mu_i^2$ as the distribution of $Z := \sum_{i=1}^{k} X_i^2$. This leads to Algorithm 6.8 for sampling RNs with non-central Chi-square distribution.

Algorithm 6.8 **(χ^2-Distributed RNs via $\mathcal{N}$-Distributed RNs)**

FUNCTION chi_normal(δ^2, k)

Generate independent $N_i \sim \mathcal{N}(0,1),\ i = 1, \ldots, k$

Set $\hat{N}_1 := N_1 + \sqrt{\delta^2} \sim \mathcal{N}(\sqrt{\delta^2}, 1)$

RETURN $X := \hat{N}_1^2 + N_2^2 + \ldots + N_k^2 \sim \chi^2(\delta^2, k)$

If k is large, this method might be rather slow. As an alternative, we can use the decomposition of a non-central Chi-square RV into a non-central Chi-square part with one degree of freedom and a standard Chi-square part with one degree of freedom less:

$$\chi^2(\delta^2, \nu) = \chi^2(\delta^2, 1) + \chi^2(\nu - 1), \quad \nu > 0. \tag{6.6}$$

We thus only have to sample a normally distributed RV and a standard Chi-square RV.

6.10 Generating t-Distributed Random Numbers

The definition of the Student's t-distribution with $\nu \in \mathbb{N}$ degrees of freedom, $t(\nu)$, as a quotient of a standard normally distributed random variable $Z \sim \mathcal{N}(0,1)$ and a certain function of a Chi-square-distributed random variable V with ν degrees of freedom (see the sampling scheme below), independent of Z, gives an easy way to sample from this distribution:

Algorithm 6.9 **($t(\nu)$-Distributed RNs)**

FUNCTION Student_t(ν)

Generate $Z \sim \mathcal{N}(0,1)$

Generate $V \sim \chi^2(\nu)$

RETURN $X := \dfrac{Z}{\sqrt{V/\nu}} \sim t(\nu)$

This method can be easily extended to cover the case of a non-central t-distribution $t(\mu, \nu)$: simply replace Z in the last step of the algorithm by $Z + \mu$. The main disadvantage of this method could be speed in the cases where it is very costly to sample from the $\chi^2(\nu)$-distribution. To avoid the AR method, Bailey's formula (see Bailey (1994)) might be an alternative, although time-consuming functions are involved. Generate independent $U_1, U_2 \sim U(0,1]$. Then

$$X := \sqrt{\nu\,(U_1^{-2/\nu} - 1)}\,\cos(2\,\pi\,U_2) \sim t(\nu). \qquad (6.7)$$

6.11 Generating Pareto-Distributed Random Numbers

In the case of Pareto-distributed RNs we generate $U \sim U(0,1]$ and take $X := x_0/(U^{1/\alpha}) \sim Pareto(\alpha, x_0)$, which is a simple example for the inversion method.

6.12 Generating Inverse Gaussian-Distributed Random Numbers

An inverse-Gaussian-distributed RV $X \sim IG(\beta, \eta)$ is related to the normal distribution via

$$Y := \frac{\eta^2\,(X - \beta/\eta)^2}{X} \sim \chi^2(1),$$

which means that Y is the square of a normally distributed RV (see Devroye (1986, p. 148)). This property can be exploited to construct a method for generating inverse-Gaussian-distributed RNs and was introduced by Michael et al. (1976).

Algorithm 6.10 (Inverse Gaussian-Distributed RNs)

FUNCTION inverseGaussian (β, η)

Generate $Z \sim \mathcal{N}(0,1), U \sim U[0,1]$

Set $Y := Z^2$

Set $X := \dfrac{\beta}{\eta} + \dfrac{Y}{2\eta^2} - \dfrac{\sqrt{4Y\beta/\eta + Y^2/\eta^2}}{2\eta}$

IF $\left(U \leq \dfrac{\beta}{\eta \cdot X + \beta}\right)$ *THEN RETURN* $X \sim IG(\beta, \eta)$

ELSE RETURN $\dfrac{\beta^2}{\eta^2\,X} \sim IG(\beta, \eta)$

6.13 Generating Stable-Distributed Random Numbers

RNs with a stable distribution can be generated with a transformation algorithm, first described by Chambers et al. (1976) (see also Weron (1996)). Here in the special case with stable distributions having the Laplace transform

$$\psi(t) = \exp(-t^\alpha), \quad \alpha \in (0,1),$$

which belongs to the distribution $S_\alpha(\cos^{1/\alpha}(\pi\alpha/2), 1, 0)$ according to the parameterization of Samoronitska and Taqqu (1994) or Weron (1996), this method simplifies to Algorithm 6.11.

Algorithm 6.11 (Stable-Distributed RNs)

$$
\begin{aligned}
&\textit{FUNCTION stable}\,(\alpha) \\
&\qquad \textit{Generate } U \sim U(0,1),\ E \sim \textit{Exp}(1) \textit{ independent} \\
&\qquad \textit{Set } U := \pi(U - \tfrac{1}{2}),\ \textit{then } U \sim U(-\tfrac{\pi}{2}, \tfrac{\pi}{2}) \\
&\qquad \textit{Set } X := \frac{\sin(\alpha\,(\pi/2 + U))}{\cos(U)^{1/\alpha}} \left(\frac{\cos(U - \alpha\,(\pi/2 + U)}{E} \right)^{(1-\alpha)/\alpha} \\
&\qquad \textit{RETURN } X \sim \mathcal{S}(\alpha, 0)
\end{aligned}
$$

These distributions are often tilted with a parameter $h \geq 0$, then have the Laplace transform $\tilde{\psi}(t) = \exp(-(t+h)^\alpha + h^\alpha)$. The relation between the corresponding densities

$$\tilde{f}(x) = \frac{\exp(-hx)}{\psi(h)} f(x), \quad x \in [0, \infty),$$

leads to the following AR algorithm (see Hofert (2010, p. 100) and Hofert (2012) for a discussion of alternative and faster algorithms).

Algorithm 6.12 (Tilted-Stable-Distributed RNs)

$$
\begin{aligned}
&\textit{FUNCTION tilted\_stable}\,(\alpha, h) \\
&\qquad \textit{REPEAT} \\
&\qquad\qquad \textit{Sample } U \sim U[0,1] \\
&\qquad\qquad \textit{Generate } V := \textit{stable}(\alpha) \\
&\qquad \textit{UNTIL } (U \leq \exp(-h\,V)) \\
&\qquad \textit{RETURN } V \sim \mathcal{S}(\alpha, h)
\end{aligned}
$$

The expected number of iterations is $c = 1/\psi(h)$. If $\psi(h)$ is close to 0 the algorithm may be too slow.

6.14 Generating Discretely Distributed Random Numbers

The use of the inversion method to simulate discretely distributed random variables is straightforward. For the integer-valued random variable X with distribution given by $\mathbb{P}(X = i) = p_i$, $i \in \mathbb{N}$,

$$X := \min\left\{k \in \mathbb{N} \,\middle|\, \sum_{i=1}^{k} p_i \geq U\right\}, \quad U \sim U\left[0,1\right],$$

yields the desired random number. If this is coded in a straightforward way, the expected number of steps in search of the minimum is $\mathbb{E}[X+1]$. The search can be accelerated by performing a binary search or by using tables (see Devroye (1986, p. 89, 96)).

Algorithm 6.13 (Binary Search for Discrete RNs)

FUNCTION discrete (function CDF, Max*)*

 Generate $U \sim U[0,1]$

 Set $L := 1;\ R := Max$

 WHILE $(L < R - 1)$

  $m := \lfloor (L+R)/2 \rfloor$

  IF $\left(CDF(m) = \sum_{i=1}^{m} p_i < U\right)$ *THEN Set* $L := m$

  ELSE Set $R := m$

 END WHILE

 RETURN R

In Algorithm 6.13 we have to take care that the number Max is greater than or equal to the largest number that might appear in the calculations. The algorithm only makes sense if $Max > 2$, otherwise a binary search is superfluous.

The inversion method in the case of uniformly distributed integer-valued RVs is particularly easy. If we want $X \sim U\{0,1,\ldots,n-1\}$, i.e. U is uniformly distributed on $\{0,1,\ldots,n-1\}$, we generate $U \sim U[0,1)$ and return $X := \lfloor nU \rfloor$.

6.14.1 *Generating Random Numbers with Geometric and Binomial Distribution*

In the the case of RNs with a geometric distribution, i.e. $X \sim Geo(p)$, we generate $U \sim U[0,1)$ and return $X := \lceil \log(1-U)/\log(1-p) \rceil$. For the binomially distributed RNs we can use RNs with a geometric distribution.

Algorithm 6.14 (Binomially Distributed RNs)

$$
\begin{aligned}
&\textit{FUNCTION binomial}(n,p)\\
&\quad \textit{Set } X := -1;\ \textit{sum} := 0\\
&\quad \textit{REPEAT}\\
&\qquad \textit{Generate } V \sim Geo(p)\\
&\qquad \textit{sum} := \textit{sum} + V;\ X := X + 1\\
&\quad \textit{UNTIL } (\textit{sum} > n)\\
&\quad \textit{RETURN } X \sim Bin(n,p)
\end{aligned}
$$

For $p = 1/2$ the coin flip method is often much faster. Simply generate n i.i.d. random bits $\in \{0,1\}$ and add them up.

6.14.2 *Generating Poisson-Distributed Random Numbers*

Poisson-distributed RNs can be generated with the help of exponentially distributed RNs, as the waiting time between two Poisson events is exponentially distributed.

Algorithm 6.15 (Poisson-Distributed RNs)

$$
\begin{aligned}
&\textit{FUNCTION Poisson1}(\lambda)\\
&\quad \textit{Set } X := -1\\
&\quad \textit{product} := 1\\
&\quad \textit{REPEAT}\\
&\qquad \textit{Generate } U \sim U[0,1]\\
&\qquad \textit{product} := \textit{product} \cdot U\\
&\qquad X := X + 1\\
&\quad \textit{UNTIL } (\textit{product} < \exp(-\lambda))\\
&\quad \textit{RETURN } \ X \sim Poi(\lambda)
\end{aligned}
$$

If λ is large, Algorithm 6.15 might be rather slow. Instead, one could use an inversion algorithm which has been sped up with the help of the mode of the Poisson distribution, which is $\lfloor \lambda \rfloor$.

Algorithm 6.16 (Poisson-Distributed RNs)

CALCULATE $b := \mathbb{P}(X \leq \lfloor \lambda \rfloor)$; $p := \mathbb{P}(X = \lfloor \lambda \rfloor)$

FUNCTION Poisson2 (λ, b, p)
 Set $X := \lfloor \lambda \rfloor$; $sum := b$; $product := p$
 Generate $U \sim U(0, 1]$
 IF $(U > b)$
  REPEAT
   $X := X + 1$; $product := product \cdot \lambda / X$;
   $sum := sum + product$
  UNTIL $(U \leq sum)$
 END IF
 IF $(U \leq b - p)$
  $sum := sum - p$
  REPEAT
   $product := product \cdot X / \lambda$; $X := X - 1$;
   $sum := sum - product$
  UNTIL $(U > sum)$
 END IF
 RETURN $X \sim Poi(\lambda)$

Chapter 7

The Monte Carlo Method

This chapter was contributed by **Elke Korn** and **Ralf Korn**,[1] for which we would like to thank them very much. Much more on the subject matter is to be found in their recent monograph Korn et al. (2010).

7.1 First Aspects of the Monte Carlo Method

The technical term *Monte Carlo method* is used for a great variety of subjects, methods, and applications in many areas. They range from high-dimensional numerical integration via the calculation of success probabilities in games of chance to the simulation of complicated phenomena in nature. However, they are all based on the approximation of an expectation of a random variable X by the arithmetic mean of i.i.d. realizations of X, i.e. the relation

$$\mathbb{E}[X] \approx \frac{1}{n} \sum_{i=1}^{n} X_i =: \bar{X}_n. \tag{7.1}$$

Here, X is a real-valued random variable with finite expectation, n a positive (sufficiently large) integer, and the X_i are independent realizations of random variables with the same distribution as X. We call this type of approximation the *crude Monte Carlo method* or simply the *Monte Carlo method*, $\bar{X}_n$ the *crude Monte Carlo estimator* (for short: CMC).

It has been widely agreed to name J. von Neumann and S. Ulam as the inventors of the Monte Carlo method. The method was secretly developed and used during World War II. The first publication presenting it to an (academic) audience was Metropolis and Ulam (1949). The name "Monte

[1]Fachbereich Mathematik, Technische Universität Kaiserslautern, Erwin Schrödinger Straße, 67663 Kaiserslautern, Germany, korn@mathematik.uni-kl.de.

Carlo method" should indicate that one uses a sort of gambling to obtain an approximation procedure. Moreover, prior to the invention of pseudo-random numbers, one sometimes used reported tables of roulette outcomes as a source of i.i.d. random numbers.

The properties of CMC, yielding an unbiased and strongly consistent estimator for the expectation, follow from the linearity of the expectation operator and the strong law of large numbers for i.i.d. random variables as stated in Theorem 7.1.

Theorem 7.1 (CMC is Strongly Consistent)
Let X be a real-valued random variable with finite expectation. Then the crude Monte Carlo estimator is an unbiased and strongly consistent estimator for $\mathbb{E}[X]$, i.e. we have

$$\mathbb{E}\left[\bar{X}_n\right] = \mathbb{E}\left[X\right], \quad \bar{X}_n \stackrel{n\to\infty}{\longrightarrow} \mathbb{E}\left[X\right] \text{ almost surely.} \tag{7.2}$$

While these two properties are nice from a statistical point of view, calculating the variance of $\bar{X}_n$ reveals the weakness of CMC:

$$\mathrm{Var}(\bar{X}_n) = \frac{1}{n}\mathrm{Var}(X), \tag{7.3}$$

i.e. the standard deviation of CMC decreases as $1/\sqrt{n}$ with increasing numbers of samples n, given that $\mathrm{Var}(X)$ is finite. In this situation, the central limit theorem yields an asymptotic 95% confidence interval for $\mathbb{E}[X]$, namely

$$\left[\bar{X}_n - 1.96\,\frac{\sigma}{\sqrt{n}},\ \bar{X}_n + 1.96\,\frac{\sigma}{\sqrt{n}}\right] \tag{7.4}$$

with $\sigma^2 := \mathrm{Var}(X)$ and where 1.96 equals the 97.5% quantile of the standard Gaussian distribution. There are two particular things to be learnt from this fact:

(1) A Monte Carlo estimate should always be given together with a confidence interval to be able to judge its accuracy.
(2) To increase the accuracy of CMC by one order (i.e. to reduce the length of the confidence interval by a factor of 0.1), one needs a new run of CMC which then has to be based on $100\,n$ simulated random variables.

The second point indicates the slow convergence of CMC, its main disadvantage.

Remark 7.1 (On Applications of the Monte Carlo Method)

(1) The special case of $X = \mathbb{1}_A$, *where* $\mathbb{1}_A$ *is the indicator function of a certain event* A, *yields the CMC estimate for* $\mathbb{P}(A)$ *by the relative frequency of the occurrence of* A *in a sequence of independent trials.*

(2) The idea behind Monte Carlo integration is that an integral can often be rewritten as an expectation of a random variable that has a probability distribution with a density. In such a case, the integral can then be estimated as an expectation by CMC.

(3) The confidence interval (7.4) *still contains the typically unknown standard deviation* σ *of* X. *However, for large values of* n *we can simply replace it with*

$$\bar{X}_n \pm 1.96 \frac{\sigma}{\sqrt{n}} \approx \bar{X}_n \pm 1.96 \frac{S_n}{\sqrt{n}}, \tag{7.5}$$

where S_n^2 *the unbiased estimator of the (unknown) variance of* X, *i.e.*

$$S_n^2 = \frac{1}{n-1} \sum_{i=1}^{n} (X_i - \bar{X}_n)^2.$$

(4) CMC can also be used to estimate $\mathbb{E}[f(\boldsymbol{X})]$ *for an* $\mathbb{R}^d$*-valued random vector* $\boldsymbol{X}$ *and a real-valued function* f *via*

$$\mathbb{E}[f(\boldsymbol{X})] \approx \frac{1}{n} \sum_{i=1}^{n} f(\boldsymbol{X}_i),$$

given this expected value is finite. Note that all properties, i.e. strong consistency and unbiasedness, of CMC established for the univariate situation carry over. In the case of $\sigma_f^2 := Var\big(f(\boldsymbol{X})\big) < \infty$ *we can also use the formulas for the confidence intervals of CMC when we replace the estimator for* σ *with the one for* σ_f. *This fact is often stated as the independence of the Monte Carlo method from the dimension of the underlying random variable.*

(5) So far, copulas have not entered the scene in our presentation of the Monte Carlo method. However, if we consider the situation of $X = f(Y_1, \ldots, Y_d)$, *where* f *is a real-valued function and the random variables* Y_i *are related via a copula* C, *then the copula enters CMC via the sampling distribution of the vector* $(Y_1, \ldots, Y_d)$.

Remark 7.2 (Variance Reduction)
The slow convergence indicated earlier is hard to improve. A popular way to overcome it (at least partially), is by using the so-called variance reduction methods. *Their main idea often is* not *to sample a set of random variables* $X_1, \ldots, X_n$ *with the same distribution as* X. *Instead, one looks for a sample* $Y_1, \ldots, Y_n$ *such that we have*

$$\mathbb{E}\left[\bar{Y}_n\right] = \mathbb{E}\left[X\right], \quad Var(\bar{Y}_n) \leq Var(\bar{X}_n). \tag{7.6}$$

While the first relation ensures that the new estimator $\bar{Y}_n$ *has the correct mean, the second relation indicates that it has a smaller variance. This is what we call a variance reduction method. The variance-reducing property roughly has two implications: either (1) for a given sample size* n *the estimator* $\bar{Y}_n$ *is more accurate than the CMC* $\bar{X}_n$, *or (2) for a given accuracy, expressed in the length of the corresponding confidence intervals, the estimator* $\bar{Y}_n$ *needs a smaller sample size* n. *If the effort of generating samples from the distribution of* Y_i *is equal to that of sampling from the distribution of* X, *then a variance reduction method is more efficient for estimating* $\mathbb{E}\left[X\right]$ *than CMC.*

Algorithm 7.1 (Crude Monte Carlo Method)
Let f *be a given function and let* $sampleDISTR_X$ *be a procedure that returns an independent random sample value from the distribution of* X.

FUNCTION CMC(integer: n)

$FOR\ i = 1, \ldots, n$

$Set\ X_i := sampleDISTR_X$

END FOR

$$RETURN\ \ \bar{f}\left(X\right)_n := \frac{1}{n}\sum_{i=1}^{n} f\left(X_i\right)$$

7.2 Variance Reduction Methods

As already indicated in the previous section, the construction of variance reduction methods is the principal tool for speeding up calculations in the Monte Carlo method. It is beyond the scope of this book to give a complete survey of variance reduction techniques (we refer the interested reader to Korn et al. (2010, Chapter 3) for a recent monograph on the application of the Monte Carlo method in finance and insurance). We will concentrate on the following methods:

(1) antithetic variates (Section 7.2.1),
(2) control variates (Section 7.2.3), and
(3) importance sampling (Section 7.2.5),

and particularly highlight their use in connection with applications of copulas.[2]

7.2.1 *Antithetic Variates*

The method of antithetic variates aims at the introduction of a certain kind of symmetry that imitates properties of the underlying probability distribution into the sample. To explain it, we look at the CMC estimator

$$\bar{f}(U) = \frac{1}{n}\sum_{i=1}^{n} f(U_i),$$

for $\mathbb{E}[f(U)]$ with U uniformly distributed on $[0,1]$, where U_i are independent copies of U. To introduce symmetry, the *antithetic variates estimator* (CMCAV) additionally uses the numbers $1-U_1,\ldots,1-U_n$ in

$$\bar{f}_{anti}(U) = \frac{1}{2}\left(\frac{1}{n}\sum_{i=1}^{n} f(U_i) + \frac{1}{n}\sum_{i=1}^{n} f(1-U_i)\right). \tag{7.7}$$

As U and $1-U$ have the same distribution, the antithetic estimator remains unbiased. With $\sigma^2 = \mathrm{Var}\big(f(U)\big)$, the variance of the antithetic estimator is given by

$$\mathrm{Var}\big(\bar{f}_{anti}(U)\big) = \frac{\sigma^2}{2n} + \frac{1}{2n}\mathrm{Cov}\big(f(U), f(1-U)\big), \tag{7.8}$$

i.e. we have a smaller variance compared to the CMC estimator based on $2n$ random numbers if $f(U)$ and $f(1-U)$ are negatively correlated. On top of that, we also save computational effort as we only have to generate n random numbers instead of $2n$. By choosing $g(x) = -f(1-x)$ in Chebyshev's covariance inequality (see, e.g., Korn et al. (2010, Proposition 3.11)) and using Equation (7.8), we directly obtain Theorem 7.2.

Theorem 7.2 (Variance Reduction by Antithetic Variates)
Let f be a non-decreasing or a non-increasing function, and let U be uniformly distributed on $[0,1]$ with $Cov\big(f(U), f(1-U)\big)$ being finite. Then we have

$$Cov\big(f(U),\ f(1-U)\big) \le 0.$$

[2]The ideas for applying certain variance reduction techniques in the copula context have originated from discussions with Jan-Frederik Mai and Matthias Scherer.

In particular, the antithetic Monte Carlo estimator based on n random numbers has a smaller variance than the crude Monte Carlo estimator based on $2n$ random numbers.

Remark 7.3 (Variance Reduction by Antithetic Variates)
Theorem 7.2 justifies the remark already made in connection with the inversion method for uniformly distributed random variables U. As the distribution function F of a random variable Y is a non-decreasing function, so is F^{-1}. Hence, to estimate $\mathbb{E}\big[h(Y)\big]$ with a non-decreasing (a non-increasing function) h, application of the theorem to the non-decreasing (non-increasing) function $f(x) := h\big(F^{-1}(x)\big)$ shows the variance reduction property of antithetic variates.

Algorithm 7.2 (MC with Antithetic Variates: Uniform Case)

FUNCTION CMCAV(integer: n)
  FOR $i = 1, \ldots, n$
    Set $U_i :=$ *sample_U*$[0,1]$
  END FOR
  RETURN $\bar{f}_{anti}(U) = \frac{1}{2}\left(\frac{1}{n}\sum_{i=1}^{n} f(U_i) + \frac{1}{n}\sum_{i=1}^{n} f(1-U_i)\right)$

Antithetic variates can also be used for other symmetric distributions such as the normal distribution: for $X_i \sim \mathcal{N}(\mu, \sigma^2)$, the suitable antithetic variate is

$$\tilde{X}_i = 2\mu - X_i.$$

For $\mu = 0$, the antithetic variate is thus $-X_i$. With the use of Chebyshev's covariance inequality, a variance reduction theorem similar to Theorem 7.2 is valid. If we consider expectations of functions h of *independent* uniformly distributed random variables U_i,

$$\mathbb{E}[Z] = \mathbb{E}\big[h(U_1, \ldots, U_d)\big],$$

then the antithetic variates method can be applied component wise (note that $\Pi = \hat{\Pi}$, i.e. the independence copula is radially symmetric). In addition to the d-dimensional vector $\boldsymbol{U}^j = (U_1^j, \ldots, U_d^j)$ one can also use

$$\tilde{\boldsymbol{U}}^j = (1 - U_1^j, \ldots, 1 - U_d^j)$$

for constructing an antithetic variate Monte Carlo estimator. It can be shown that if h is non-decreasing in each component, then this method yields a variance reduction. This is also the case if we only use the antithetic variate $1 - U_m^j$ in some component m.

Remark 7.4 (Antithetic Variates: Confidence Intervals)
As the antithetic variate Monte Carlo estimator is a function of the n independent samples

$$h(U_i) = \frac{1}{2}\left(f(U_i) + f(\tilde{U}_i)\right),$$

with $\tilde{U}_i = 1 - U_i$ the antithetic variate to U_i, we obtain a confidence interval as in the case of the CMC estimator but have to use the variance estimator

$$\bar{\sigma}_{anti}^2 = \frac{1}{n-1}\sum_{i=1}^{n}\left(\frac{1}{2}\left(f(U_i) + f(\tilde{U}_i)\right) - \bar{f}_{anti}(U)\right)^2,$$

leading to the approximate 95% confidence interval for $\mathbb{E}[f(U)]$ of

$$\left[\bar{f}_{anti}(U) - 1.96\,\frac{\bar{\sigma}_{anti}}{\sqrt{n}},\, \bar{f}_{anti}(U) + 1.96\,\frac{\bar{\sigma}_{anti}}{\sqrt{n}}\right].$$

There is a natural application of the antithetic variates approach when considering radially symmetric copulas.

7.2.2 *Antithetic Variates for Radially Symmetric Copulas*

We first recall from the introduction that if C is a copula with $(U_1, \ldots, U_d) \sim C$, the random vector $(U_1, \ldots, U_d)$ with uniform marginals is radially symmetric about $(1/2, \ldots, 1/2)$, if we have $(U_1, \ldots, U_d) \stackrel{d}{=} (1 - U_1, \ldots, 1 - U_d)$. This is exactly the necessary situation for the application of the antithetic variates approach as described earlier: when simulating $(U_1, \ldots, U_d)$, we can also use $(1 - U_1, \ldots, 1 - U_d)$ without additional simulation effort and then use the antithetic MC estimator

$$\bar{f}_{anti}(U_1, \ldots, U_d) :=$$
$$\frac{1}{2}\left(\frac{1}{n}\sum_{i=1}^{n} f\left(U_1^{(i)}, \ldots, U_d^{(i)}\right) + \frac{1}{n}\sum_{i=1}^{n} f\left(1 - U_1^{(i)}, \ldots, 1 - U_d^{(i)}\right)\right)$$

as an approximation for $\mathbb{E}[f(U_1, \ldots, U_d)]$. We illustrate its use with the following simple two-dimensional example with

(1) univariate marginal laws $U_1, U_2 \sim U[0,1]$, and
(2) dependence structure $C_\rho^{Gauss}(u_1, u_2) := F_\rho\left(\Phi^{-1}(u_1), \Phi^{-1}(u_2)\right)$ for $u_i \in [0,1]$, i.e. the two-dimensional Gaussian copula[3] with correlation $\rho \in [-1,1]$.

Our aim is to compute $\mathbb{E}[U_1 \cdot U_2]$.

As the Gaussian copula is radially symmetric about the origin, we can directly apply the antithetic variates approach. Numerical results for $n = 10\,000$ simulations in Table 7.1 show that, compared to the crude Monte Carlo approach, we can reduce the length of the 95% confidence interval by factors between 1 and 4. Note in particular that for perfectly negatively correlated margins, we obtain no shortening of the interval while for perfect positively correlated ones we obtain a factor of 4. The reason for this is that in the perfectly negatively correlated case, we have $U_1 = 1 - U_2$, so the products $U_1 U_2$ and $(1-U_1)(1-U_2)$ agree. Thus, in this case, using antithetic variates does not help.

Table 7.1 Estimating $\mathbb{E}[U_1 U_2]$ without and with antithetic variates.

ρ	CMC	CMCAV	Length Reduction
-1	$[0.1641, 0.1671]$	$[0.1641, 0.1671]$	1.00
-0.5	$[0.2064, 0.2130]$	$[0.2083, 0.2114]$	2.13
0	$[0.2458, 0.2544]$	$[0.2484, 0.2517]$	2.61
0.5	$[0.2861, 0.2964]$	$[0.2888, 0.2920]$	3.22
1	$[0.3295, 0.3412]$	$[0.3322, 0.3351]$	4.03

The table also underlines that while the use of antithetic variates helps to reduce variance, it typically does not lead to a dramatic variance reduction. Its main advantage lies in the simplicity of its application. Thus, it is still worth using antithetic variates for radially symmetric copulas.

7.2.3 *Control Variates*

The main idea of control variates consists in finding a random variable Y for which we know $\mathbb{E}[Y]$ and which is perfectly positively correlated with

[3] F_ρ denotes the distribution function of the $\mathcal{N}_2(\mathbf{0}, \Sigma)$-distribution, where Σ is a 2×2 correlation matrix with off-diagonal entry ρ, and Φ^{-1} denote the quantile functions of the univariate standard normal distribution.

X, i.e. $\text{Corr}(X,Y) \approx 1$. Then, using the relations

$$\mathbb{E}[X] = \mathbb{E}[X - Y] + \mathbb{E}[Y], \tag{7.9}$$

$$\text{Var}(X) - \text{Var}(X - Y) = 2\,\text{Cov}(X,Y) - \text{Var}(Y) \tag{7.10}$$

motivates the introduction of the *control variate Monte Carlo estimator* (CVMC)

$$\mathbb{E}[X] \approx \frac{1}{n}\sum_{i=1}^{n}(X_i - Y_i) + \mathbb{E}[Y] =: \bar{X}_n^{(Y)}.$$

In particular, Equation (7.10) shows that CVMC has a lower variance than CMC if the correlation between X and Y is sufficiently high. Given a candidate Y for a control variate, one can even construct an optimal control variate $c^* Y$ out of it.

Theorem 7.3 (Optimal Control Variate Multiplier)
For given real-valued random variables X and Y, the CVMC $\bar{X}_n^{c^* Y}$, with

$$c^* = \frac{\sigma_{XY}}{\sigma_Y^2}$$

and $\sigma_{XY} := Cov(X,Y)$, has the smallest variance for all CVMC of the form $\bar{X}_n^{cY}$ with $c \in \mathbb{R}$. The use of $\bar{X}_n^{c^* Y}$ leads to a variance reduction of the estimator of

$$\frac{Var(\bar{X}_n) - Var(\bar{X}_n^{c^* Y})}{Var(\bar{X}_n)} = \rho_{X,c^*Y}^2 \tag{7.11}$$

with $\rho_{X,Y} = Corr(X,Y)$.

Remark 7.5 (Control Variates)

(1) The proof of Theorem 7.3 simply consists of the explicit calculation of $Var(\bar{X}_n^{cY})$, followed by minimizing the resulting concave, quadratic function in c.

(2) Equation (7.11) clearly highlights that we can obtain a significant relative variance reduction by CVMC only for high values of ρ_{X,c^*Y}^2.

(3) As in general the (exact) value of σ_{XY} is unknown, it has to be estimated and iteratively updated during the Monte Carlo simulation procedure.

(4) Finding a good control variate typically requires good intuition. There is no general algorithm for constructing it.

(5) *One can combine the control variate method with antithetic variates, which often results in a higher variance reduction. One can also use a multiple control variate approach, i.e. one can use more than one control variate. An application in a multidimensional setting is the unconditional mean control variate approach as described in Korn et al. (2010, p. 75).*

Algorithm 7.3 (Monte Carlo Method with Control Variate)
Let sampleDISTR_X *return a random number from the distribution of* X. *Let* Y *be the control variate.*

FUNCTION CVMC(integer: n)

FOR $i = 1, \ldots, n$

Set $X_i :=$ *sampleDISTR_X*; $Y_i :=$ *sampleDISTR_Y*

END FOR

$$RETURN\ \ \bar{X}_n^{(Y)} = \frac{1}{n}\sum_{i=1}^{n}(X_i - Y_i) + \mathbb{E}[Y]$$

We highlight the use and the performance of the control variate method in a copula framework via the following application.

7.2.4 *Approximation via a Simpler Dependence Structure*

Let us consider a situation where we want to calculate the expectation of a function of random variables $\mathbb{E}[f(X_1, \ldots, X_d)]$, where the dependence structure is given by a copula C, i.e. we have as the distribution function of $(X_1, \ldots, X_d)$

$$F(x_1, \ldots, x_d) = C\big(F_1(x_1), \ldots, F_d(x_d)\big).$$

There are now various possibilities for a control variate approach. An obvious candidate is to use random variables $Y_1, \ldots, Y_d$ with the same marginal distributions as the original X-variables that are related by a simpler copula function $\tilde{C}$, in the sense that we can calculate the expectation $\mathbb{E}[f(Y_1, \ldots, Y_d)]$ explicitly.

Of course it depends on all the input factors, distributions, and the function f how the copula used for the control variate is chosen. A first try can always be the independence copula, but in some applications one might have even better candidates. We illustrate the approach with a toy example where we consider

(1) univariate marginal laws $X_1 \sim Bin(1, p_1)$, $X_2 \sim Bin(1, p_2)$, and $p_i \in [0, 1]$, and
(2) dependence structure $C_\rho^{Gauss}(u_1, u_2) = F_\rho\left(\Phi^{-1}(u_1), \Phi^{-1}(u_2)\right)$, for $u_i \in [0, 1]$, i.e. the two-dimensional Gaussian copula with correlation $\rho \in [-1, 1]$.

Our aim is to compute

$$\mathbb{E}[X_1 \cdot X_2] = \mathbb{P}(X_1 = 1,\ X_2 = 1).$$

This example can be motivated as the calculation of the probability of a joint default of two bonds of two countries which are politically and/or economically related. It can thus be an ingredient in the price calculation of a suitably constructed credit derivative. As control variate we use (Y_1, Y_2) with independent components, in particular

(1) $Y_1 = X_1$ which can thus be reused for the control variate,
(2) $\tilde{C}(u_1, u_2) = u_1 \cdot u_2$ for $u_i \in [0, 1]$.

The numerical results for $n = 10\,000$ in Table 7.2 show that we can achieve the highest variance reduction, measured in terms of the reduction of length of the confidence interval for $\mathbb{E}[X_1 \cdot X_2]$, when we have $p_1 = p_2 = 0.5$ and $\rho \approx 0$. Indeed, for $\rho = 0$ we can obviously reduce the variance completely. In contrast, for $\rho \approx 1$ we achieve nearly no variance reduction by our control variate approach. Furthermore, for $\rho \approx -1$, the use of the independence copula as control variate even leads to a wider confidence interval, however, still containing the true value. The corresponding optimal multiplier $c^* = 0.075$ underlines the fact that the chosen control variate is not really suitable for variance reduction purposes. Further, we see in Table 7.2 that for smaller values of the (marginal) success probabilities, the independence copula as the control variate dependence structure becomes less and less attractive, which is also in line with our intuition as then the value 1 is very unlikely to be observed jointly. In all examples, the different MC estimators are comparable with regard to their accuracy.

For $p_1 = p_2 = 0.25$, $\rho = -0.99$, the exact value of the joint probability is so small, approximately 0.00000032, that CMC typically leads to a confidence interval that only contains 0, as no joint value of 1 is observed in the sample. Although CVMC then still provides a confidence interval of $[-0.0036, 0.0058]$, which can easily be improved by replacing the left border with 0, we cannot use CVMC_{opt} as the covariance estimator then typically also suggests an optimal value of 0.

Table 7.2 Probability of joint success of dependent bivariate Bernoulli variables.

p_1, p_2	ρ	CMC	CVMC	CVMC$_{opt}$	c^*
0.50, 0.50	0.01	[0.2465, 0.2635]	[0.2509, 0.2525]	[0.2509, 0.2525]	0.999
	0.30	[0.2938, 0.3118]	[0.2952, 0.3038]	[0.2955, 0.3039]	0.940
	0.90	[0.4267, 0.4461]	[0.4255, 0.4407]	[0.4265, 0.4411]	0.778
	−0.99	[0.0204, 0.0264]	[0.0119, 0.0283]	[0.0203, 0.0260]	0.075
0.25, 0.25	0.01	[0.0579, 0.0675]	[0.0631, 0.0645]	[0.0631, 0.0645]	0.999
	0.30	[0.0851, 0.0963]	[0.0885, 0.0951]	[0.0884, 0.0951]	0.954
	0.90	[0.1864, 0.2020]	[0.1886, 0.2020]	[0.1885, 0.2018]	0.852

7.2.5 *Importance Sampling*

Importance sampling is usually considered to be the most effective variance reduction method when it can be applied. However, it requires more technical work than the application of the two previous methods. The main idea can be demonstrated via the following example, where we assume that the bivariate random vector (X, Y) has a joint density $g(x, y)$. We want to estimate $\mathbb{E}[f(X, Y)]$. A typical situation where importance sampling is needed occurs when $f(x, y)$ is very big only for values of (x, y) that are very unlikely under the distribution with density $g(x, y)$. As an example, think of joint large values of a bivariate normal distribution. A related univariate example is given in Korn et al. (2010, Example 3.23, p. 91ff). Thus, sampling under the original distribution might not yield a single value of (x, y) with a large value of $f(x, y)$. Hence, the resulting CMC estimator significantly underestimates $\mathbb{E}[f(X, Y)]$.

The importance sampling solution to this problem is to sample under a new density $\tilde{g}(x, y)$, which puts more probability mass on the *important values* (x, y), i.e. those where $f(x, y)$ is big. This, of course, has to be coupled with a suitable transformation of the criterion function $f(x, y)$. For this, look at the following relation where we assume that $\tilde{g}(x, y)$ is strictly positive:

$$
\begin{aligned}
\mathbb{E}[f(X, Y)] &= \int f(x, y)\, g(x, y)\, dxdy \\
&= \int \frac{f(x, y)\, g(x, y)}{\tilde{g}(x, y)} \tilde{g}(x, y)\, dxdy \\
&= \tilde{\mathbb{E}}\left[\frac{f(X, Y)\, g(X, Y)}{\tilde{g}(X, Y)}\right].
\end{aligned}
$$

Here, $\tilde{\mathbb{E}}[.]$ denotes the expectation with respect to the density $\tilde{g}(x, y)$. A similar transformation applies for discrete distributions. In both situations the importance sampling distribution must have a support that contains that of the original distribution to avoid a division by 0, i.e. we need an absolutely continuous change of measure.

For a general multivariate random vector $\boldsymbol{X}$ with density $g(\boldsymbol{x})$, the importance sampling Monte Carlo estimator (ISMC) based on the sample $\boldsymbol{X}_1, \ldots, \boldsymbol{X}_n$ from $\tilde{g}$ is given by

$$\mathbb{E}\left[f\left(\boldsymbol{X}\right)\right] \approx \frac{1}{n}\sum_{i=1}^{n}\frac{f\left(\boldsymbol{X}_i\right)g\left(\boldsymbol{X}_i\right)}{\tilde{g}\left(\boldsymbol{X}_i\right)} =: \bar{f}_{imp}^{(\tilde{g})}\left(\boldsymbol{X}\right).$$

The main problem with importance sampling is finding an appropriate importance sampling density $\tilde{g}(x,y)$. In the situation of the (multivariate) normal distribution, this is often achieved by either a shift of the mean to the appropriate region or by spreading out the distribution via an appropriate scaling of the covariance matrix.

Algorithm 7.4 (MCM with Importance Sampling)
Let sampleDISTR$_g$ return a random number from the distribution with density g. Let $\tilde{g}$ be a density function with supp$(g) \subset$ supp$(\tilde{g})$.

FUNCTION CVMC(integer: n)

FOR $i = 1, \ldots, n$

Set $X_i :=$ *sampleDISTR*$_{\tilde{g}}$

END FOR

RETURN $\bar{f}_{imp}^{(\tilde{g})}\left(X\right) = \frac{1}{n}\sum_{i=1}^{n}\frac{f\left(X_i\right)g\left(X_i\right)}{\tilde{g}\left(X_i\right)}$

As already stated, the algorithm works in exactly the same way for a discrete distribution if $g(x)$ denotes the probability function and $\tilde{g}(x)$ the importance sampling probability function. As a special example we consider the situation of the dependent bivariate Bernoulli variables from the control variate section.

7.2.6 *Importance Sampling via Increasing the Dependence*

Let us again assume

(1) marginal laws $X \sim Bin\left(1, p_1\right)$, $Y \sim Bin\left(1, p_2\right)$, and $p_i \in [0,1]$, and
(2) dependence structure $C_\rho^{Gauss}(u_1, u_2) = F_\rho\left(\Phi^{-1}\left(u_1\right), \Phi^{-1}\left(u_2\right)\right)$ for $u_i \in [0,1]$, i.e. the dependence structure is given by the two-dimensional Gaussian copula with correlation $\rho \in [-1,1]$.

Our aim is to compute

$$\mathbb{E}\big[f(X,Y)\big] := 10\,000 \cdot \mathbb{E}[X \cdot Y] = 10\,000 \cdot \mathbb{P}(X=1,\ Y=1).$$

As in the control variate example, we can imagine the situation of a specially designed credit derivative. Here, one only receives a payment if the two credits jointly default or are both paid back in full.

In line with the motivating remarks, we should switch to a distribution which puts more emphasis on the diagonal, i.e. which increases the probabilities of X and Y attaining the same values. We will do this by imitating the scaling approach in the univariate normal distribution setting. More precisely, we are going to increase the dependency between the two random variables via choosing the correlation factor ρ in the Gaussian copula to be as big as possible. As we have to restrict ourselves to an absolutely continuous change of measure, we always choose $\rho = 0.999$ and take the corresponding joint distribution of the two Bernoulli variables as the importance sampling distribution. The effect of this choice on the joint distribution and also the improvement of the Monte Carlo estimation using importance sampling are reported in Table 7.3. To demonstrate the power of importance sampling we have chosen to use only $n = 1\,000$ samples.

In the case of symmetric marginal success probabilities (i.e. $p_1 = p_2 = 0.5$), note that the biggest variance reduction is obtained for the case of $\rho = -0.99$, i.e. the negatively correlated situation which is least favorable for the expected payoff. However, even in the situation of $\rho = 0.9$, importance sampling reduces the length of the confidence interval by a factor bigger than 3.

In the case of highly asymmetric success probabilities (i.e. $1 - p_1 = p_2 = 0.9$), the effect of importance sampling is mostly negligible concerning the length of the confidence intervals. However, the importance sampling estimator (given by the mean of the confidence interval) is typically closer to the true value than the CMC estimator. Again, the biggest gain is realized for the negatively correlated situation.

In Table 7.4, we illustrate the effect of importance sampling in this situation also by comparing the original (true) distribution with the importance sampling distribution which tries to put as much probability mass on the important values as possible. As for given marginal success probabilities p_1, p_2, the importance sampling probabilities p_{ii}^{imp}, $i = 0, 1$, are the same, independent of the correlation parameter ρ of the Gaussian copula. We always state the importance sampling probabilities only once per example. It can clearly be seen that the variance reduction, and thus the advantage,

Table 7.3 Confidence interval for the expected loss for two dependent Bernoulli variables (rounded to integers).

p_1, p_2	ρ	Exact Value	CMC	MC_{imp}
0.50, 0.50	0.01	5032	$[4970, 5590]$	$[4989, 5066]$
	0.30	5970	$[5980, 6580]$	$[5920, 6011]$
	0.90	8564	$[8523, 8937]$	$[8492, 8623]$
	−0.99	451	$[400, 680]$	$[447, 454]$
0.10, 0.90	0.01	181	$[170, 219]$	$[163, 209]$
	0.30	194	$[183, 233]$	$[173, 224]$
	0.90	200	$[182, 232]$	$[181, 231]$
	−0.99	198	$[113, 287]$	$[179, 229]$

importance sampling method is bigger, the more the importance sampling distribution differs from the true distribution.

Table 7.4 Joint probability and importance sampling probability distributions for two dependent Bernoulli variables (rounded to four digits).

p_1, p_2 last line	ρ	p_{00} p_{00}^{imp}	p_{01} p_{01}^{imp}	p_{10} p_{10}^{imp}	p_{11} p_{11}^{imp}
0.50, 0.50	0.01	0.2516	0.2484	0.2484	0.2516
	0.30	0.2985	0.2015	0.2015	0.2985
	0.90	0.4282	0.0718	0.0718	0.4282
	−0.99	0.0225	0.4775	0.4775	0.0225
imp. prob.	0.999	0.4929	0.0071	0.0071	0.4929
0.10, 0.90	0.01	0.0903	0.8097	0.0097	0.0903
	−0.99	0.0099	0.8901	0.0901	0.0099
imp. prob.	0.999	0.1000	0.8000	0.0000	0.1000

One might also think of situations where it is more suitable to keep the dependence structure between the two Bernoulli variables and to change the individual success probabilities p_i, $i = 1, 2$, instead.

7.2.7 *Further Comments on Variance Reduction Methods*

For further variance reduction methods (such as conditional sampling, stratified sampling, or weighted Monte Carlo), we refer the reader to Korn et al. (2010, Chapter 3) or to Glasserman (2004). Often, these methods are very close to (or even specialized versions of) the three main variance reduction methods we have presented.

As the Monte Carlo method tends to be very slow but is often the only numerical method that is easily usable in a multidimensional setting, one should always try to speed it up (or increase its accuracy) via variance reduction. However, if one is only interested in one particular expected value $\mathbb{E}[X]$, then one should also keep in mind that the time needed to create a particularly smart variance reduction method can easily be much longer than the time needed to run a crude Monte Carlo simulation with a sufficiently high number of samples n.

Appendix A

Supplemental Material

A.1 Validating a Sampling Algorithm

After one has implemented some sampling algorithm for $\boldsymbol{U} = (U_1, \ldots, U_d) \sim C$, one is well advised to carefully review its accuracy. Assume we have simulated a sequence $\boldsymbol{U}^{(1)}, \ldots, \boldsymbol{U}^{(n)}$ of independent random vectors with a distribution function given by the copula C. A check of the validity of the sampling scheme might include the following steps:

(1) **Univariate marginals**: C being a copula, each univariate marginal law must be uniform on $[0,1]$. Considering this, one starts by checking the range of $U_k^{(l)}$, $k = 1, \ldots, d$, $l = 1, \ldots, n$; all entries must be within $[0,1]$. The case $U_k^{(l)} \in \{0,1\}$ is already suspicious and indicates numerical problems. Next, we check if the univariate marginals $k = 1, \ldots, d$ are truly uniform on $[0,1]$. For this, we might use standard tests such as the Kolmogorov–Smirnov test, which is provided in most statistical software packages. Alternatively, a simple box plot of $U_k^{(l)}$, $l = 1, \ldots, n$, applied to all univariate marginals $k = 1, \ldots, d$, also helps to identify misspecified univariate marginals.

(2) **Stylized facts of** C: To visualize if the sample corresponds to the copula C, one might draw scatterplots of various two- or three-dimensional marginals. These scatterplots allow one to visualize the stylized properties of the simulated copula, so we can compare these with the properties of the copula C. Properties that are sometimes easy to detect are (a) the location of mass as a subset of $[0,1]^d$, (b) a singular component, (c) positive upper- and lower-tail dependence, (d) exchangeability, and (e) radial symmetry.

(3) **Measures of dependence**: Often, one knows the theoretical value of measures of dependence of C such as Kendall's tau or Spearman's rho.

One can then compute the corresponding empirical versions of these measures of dependence from the sample; the values should be "about the same". Often, these measures of dependence even have a known (asymptotic) distribution, which allows us to construct tests to clarify what "about the same" means. The same holds for empirical moments or in general functions of the dependence structure.

(4) **Parameter sensitivity**: It is often the case that the strength of dependence of a copula is monotone in a parameter θ. Simulating the copula with different parameters should provide evidence of this fact. Moreover, we sometimes know limiting cases for the parameters, where the copula might converge, e.g., to the independence copula. This should also be reflected in empirical measures of dependence and scatterplots when the parameter is moved to a known limit.

(5) **The diagonal**: A test based on the diagonal section of the copula can be constructed from the observation

$$\mathbb{P}(U_1 \leq u, \ldots, U_d \leq u) = C(u, \ldots, u) = \mathbb{P}\big(\max_{k=1,\ldots,d}\{U_k\} \leq u\big).$$

Note that $u \mapsto C(u, \ldots, u)$, $u \in [0,1]$, is the (univariate) distribution function of $\max\{U_1, \ldots, U_d\}$, so tests for the respective univariate distribution can be applied.

(6) **Parameter range**: Some algorithms are vulnerable to numerical errors when the parameter becomes extreme (in the sense that the parameter tends to some limit of its range). Also, it is possible that the speed of the algorithm drops significantly for extreme parameters. In this regard, one is well advised to carefully examine such behavior (and to suitably restrict the parameter range) before one uses the sampling scheme as a black box within some more complex algorithm.

A.2 Introduction to Lévy Subordinators

Lévy processes are continuous-time equivalents of discrete-time random walks. The increments of a Lévy process are stationary and do not depend on the past, not even on the current value of the process. Stochastic processes of this type are used as building blocks for probabilistic models in many applications, e.g. in financial engineering. Prominent examples of Lévy processes are Brownian motion and the Poisson process. If a Lévy process has almost surely non-decreasing paths, it is called a Lévy subordinator. For background on Lévy processes beyond this introduction we refer the reader to the textbooks by Bertoin (1996, 1999), Sato (1999), Schoutens (2003), Applebaum (2004), Cont and Tankov (2004), and Schilling et al. (2010).

Definition A.1 (Classical Lévy Subordinator)
A $[0,\infty)$-valued stochastic process $\Lambda = \{\Lambda_t\}_{t\geq 0}$ on a probability space $(\Omega, \mathcal{F}, \mathbb{P})$ is a classical Lévy subordinator *if it is a non-decreasing Lévy process, i.e. $\Lambda_0 = 0$ holds $\mathbb{P}$-almost surely, Λ has càdlàg paths,*[4] *and the following conditions are satisfied:*

(1) Λ is stochastically continuous, i.e.

$$\forall t \geq 0, \forall \epsilon > 0, \quad \text{it holds that } \lim_{h\downarrow 0} \mathbb{P}\big(|\Lambda_{t+h} - \Lambda_t| \geq \epsilon\big) = 0.$$

(2) Λ has independent increments, i.e. for all $0 \leq t_0 \leq t_1 \leq \ldots \leq t_n$ the random variables $\Lambda_{t_0} - \Lambda_0, \Lambda_{t_1} - \Lambda_{t_0}, \ldots, \Lambda_{t_n} - \Lambda_{t_{n-1}}$ are stochastically independent.
(3) Λ has stationary increments, i.e. the law of $\Lambda_{t+h} - \Lambda_t$ is independent of $t \geq 0$ for each $h \geq 0$, i.e. $\Lambda_{t+h} - \Lambda_t \stackrel{d}{=} \Lambda_h$.
(4) $t \mapsto \Lambda_t$ is almost surely non-decreasing.

It is important to note that condition (1) in the above definition does not imply that Λ has continuous paths. It basically means that the jump times of paths of Λ are not allowed to be deterministic, i.e. not known in advance. The simplest example of a classical Lévy subordinator is a *(homogeneous) Poisson process*, denoted $N = \{N_t\}_{t\geq 0}$. It can be constructed as follows. On a probability space $(\Omega, \mathcal{F}, \mathbb{P})$ let $\{E_i\}_{i\in\mathbb{N}}$ be a sequence of i.i.d. random variables (interpreted as waiting times) with $E_1 \sim Exp(\beta)$ for a parameter $\beta > 0$. Then, N is defined via

$$N_t := \sum_{n=1}^{\infty} \mathbb{1}_{\{E_1+\ldots+E_n \leq t\}}, \quad t \geq 0. \tag{A.1}$$

Thus, a path of N starts at $N_0 = 0$, remains there until time E_1, and then jumps to 1. It remains in state 1 until time $E_1 + E_2$ and then jumps to 2, and so on. Thus, the state space of N is $\mathbb{N}_0$. The fact that N has i.i.d. increments originates from the lack of memory property of the exponential distribution of the random variables $\{E_i\}_{i\in\mathbb{N}}$. The parameter β is called the *intensity* of the Poisson process N. The name is justified by the fact that N_t

[4] Càdlàg is the abbreviation of "continue à droite, limite à gauche", which is French for "right-continuous with left limits". This means that $\mathbb{P}$-almost surely $\lim_{s\uparrow t} \Lambda_s$ exists for $t > 0$ and $\lim_{s\downarrow t} \Lambda_s = \Lambda_t$ for $t \geq 0$.

is $Poi(\beta\, t)$-distributed for all $t > 0$ (see Cont and Tankov (2004, Proposition 2.12, p. 48)). For our purpose it is convenient to extend Definition A.1 to include the (absorbing) state infinity as a possible value for Λ_t, $t > 0$.

Definition A.2 (Lévy Subordinator)

A $[0,\infty) \cup \{\infty\}$*-valued stochastic process* $\Lambda = \{\Lambda_t\}_{t\geq 0}$ *is called a* Lévy subordinator *if it is defined for* $t \geq 0$ *by* $\Lambda_t := \tilde{\Lambda}_t + \infty \cdot \mathbb{1}_{\{N_t \geq 1\}}$*, where* $\tilde{\Lambda} = \{\tilde{\Lambda}_t\}_{t\geq 0}$ *is a classical* ($[0,\infty)$*-valued) Lévy subordinator and* $N = \{N_t\}_{t\geq 0}$ *is an independent Poisson process. The intensity of the Poisson process* N *is called the* killing rate *of* Λ*. As a convention, it is allowed to be 0, in which case we mean that* $\Lambda = \tilde{\Lambda}$.

In the literature, e.g. in Applebaum (2004) and Bertoin (1996), a process Λ according to Definition A.2 is sometimes called a *killed subordinator.*[5] This emphasizes the intuitive interpretation that the process is "killed" when it jumps to infinity, since it is absorbed by this state due to its non-decreasing paths. Adding the state infinity, sometimes called the *cemetery state*, leads to a *compactification* of the state space $[0,\infty)$. Such an analytical technique is useful for including "marginal cases" in derivations. Due to such technical reasons, by the term "Lévy subordinator" we always refer to the extended Definition A.2, which includes the case of a classical Lévy subordinator.

Lévy processes are most easily treated by means of their characteristic function. Due to positiveness in the case of a Lévy subordinator, i.e. a non-decreasing Lévy process, it is even more convenient to consider the (existing) Laplace transform. The analytical form of the Laplace transform is nowadays known as the *Lévy–Khinchin representation* (see Theorem A.4). It relies on the fact that for each $t > 0$ the distribution of Λ_t, Λ being a classical Lévy subordinator, is *infinitely divisible*. More precisely, Definition A.1 implies for all $n \in \mathbb{N}$ and $t > 0$ that

$$\begin{aligned}\Lambda_t &= \left(\Lambda_{\frac{t}{n}} - \Lambda_{\frac{0}{n}}\right) + \left(\Lambda_{\frac{2t}{n}} - \Lambda_{\frac{t}{n}}\right) + \ldots + \left(\Lambda_{\frac{nt}{n}} - \Lambda_{\frac{(n-1)\,t}{n}}\right)\\ &\stackrel{d}{=} \Lambda^{(1)}_{\frac{t}{n}} + \Lambda^{(2)}_{\frac{t}{n}} + \ldots + \Lambda^{(n)}_{\frac{t}{n}},\end{aligned}$$

where $\Lambda^{(i)}$, for $i \in \mathbb{N}$, are i.i.d. copies of Λ, i.e. the random variable Λ_t can be represented in distribution as the sum of n i.i.d. random variables for each $n \in \mathbb{N}$, a distributional property called "infinite divisibility" (see Sato (1999) and Applebaum (2004) for further details). Conversely, there exists a classical Lévy subordinator Λ such that Λ_1 is distributed according to

[5]This name is not used by all authors; we use the terminology of Bertoin (1999) who omits the term "killed".

any given infinitely divisible distribution on $[0,\infty)$. This correspondence allows us to transfer results about infinitely divisible distributions to Lévy processes. For instance, Theorem A.4 characterizes a Lévy subordinator by means of a constant $\mu \geq 0$ and a measure ν on $(0,\infty]$.

Theorem A.4 (Lévy (1934), Khinchin (1937))
The Laplace transforms of a Lévy subordinator Λ on a probability space $(\Omega,\mathcal{F},\mathbb{P})$ admit the functional form

$$\mathbb{E}[e^{-x\,\Lambda_t}] = e^{-t\,\Psi(x)}, \quad x \geq 0, \quad t \geq 0, \tag{A.2}$$

where the function $\Psi : [0,\infty) \to [0,\infty)$ is called the Laplace exponent *of Λ. Moreover, there is a unique non-negative drift $\mu \geq 0$ and a unique positive measure ν on $(0,\infty]$, called the* Lévy measure *of Λ, such that*

$$\Psi(x) = \mu\,x + \int_{(0,\infty]} (1 - e^{-t\,x})\,\nu(dt), \quad x \geq 0. \tag{A.3}$$

The Lévy measure ν satisfies the conditions

$$\int_{(0,1]} t\,\nu(dt) < \infty, \quad \nu\big((\epsilon,\infty]\big) < \infty, \quad \text{for all } \epsilon > 0. \tag{A.4}$$

Conversely, given a drift $\mu \geq 0$ and a measure ν on $(0,\infty]$ satisfying (A.4), there exists a Lévy subordinator with drift μ and Lévy measure ν. Thus, the distributional properties of a Lévy subordinator are completely characterized by its so-called characteristics (μ,ν).

Proof. Originally due to Lévy (1934), Khinchin (1937), and Khinchin (1938). A sketch of the proof can be found in Bertoin (1999). □

Remark A.6 (Mass at Infinity)
The right-hand side of Equation (A.3) is a short-hand notation for

$$\Psi(x) = \mu\,x + \int_{(0,\infty)} (1 - e^{-t\,x})\,\nu(dt) + \mathbb{1}_{\{x>0\}}\,\nu(\{\infty\}), \quad x \geq 0. \tag{A.5}$$

It is justified by using the conventions $0 \cdot \infty = 0$ and $\exp(-\infty) = 0$, which imply that $\Psi(0) = 0$ even though one might have $\nu(\{\infty\}) > 0$. Positive mass of ν at ∞ introduces a discontinuity of Ψ at 0. However, Ψ is smooth on $(0,\infty)$. It is a so-called Bernstein function, *i.e. Ψ is infinitely often differentiable on $(0,\infty)$ and the first derivative $\Psi^{(1)}$ is completely monotone, i.e. $(-1)^k\,\Psi^{(k)} \geq 0$ for all $k \in \mathbb{N}$ (see, e.g., Applebaum (2004, Theorem 1.3.23(2), p. 52)). The number $\nu(\{\infty\}) \in [0,\infty)$ is precisely the killing rate. Hence, in the case of a classical Lévy subordinator it holds that $\nu(\{\infty\}) = 0$ and thus the last term $\mathbb{1}_{\{x>0\}}\,\nu(\{\infty\})$ in (A.5) vanishes.*

Example A.1 (Characteristics of a Poisson Process)
On a probability space $(\Omega, \mathcal{F}, \mathbb{P})$, consider a homogeneous Poisson process $N = \{N_t\}_{t \geq 0}$ with intensity $\beta > 0$, as defined in (A.1). Then for each $t > 0$ the random variable N_t is $Poi(\beta\, t)$-distributed, i.e. the Laplace transform of N_t is computed as

$$\mathbb{E}\big[e^{-x\, N_t}\big] = \sum_{k=0}^{\infty} e^{-x\, k}\, \frac{(\beta\, t)^k}{k!}\, e^{-\beta\, t} = e^{-t\, \beta\, (1-e^{-x})}, \quad x \geq 0.$$

Thus, the Laplace exponent of N is $\Psi(x) = \beta\,\big(1 - \exp(-x)\big)$, $x \geq 0$. Obviously, N has zero drift $\mu = 0$ and the Lévy measure ν is a one-point mass concentrated at 1. More precisely, $\nu(B) = \beta\, \mathbb{1}_{\{1 \in B\}}$, for $B \in \mathcal{B}\big((0,\infty]\big)$.

Intuitively, for any Borel set $B \in \mathcal{B}\big((0,\infty]\big)$ the value $\nu(B)$ corresponds to the expected number of jumps of Λ within one unit of time whose size is in B, i.e.

$$\nu(B) = \mathbb{E}\Big[\Big|\Big\{t \in (0,1]\,:\, \Delta\Lambda_t := \Lambda_t - \lim_{s \uparrow t} \Lambda_s \in B\Big\}\Big|\Big] \tag{A.6}$$

(see, e.g., Cont and Tankov (2004, Definition 3.4, p. 76)). Hence, ν bears information about the size and frequency of the jumps of Λ. Finally, new Lévy subordinators can be constructed from known ones by the idea of subordination: given two independent Lévy subordinators $\Lambda^{(1)}$ and $\Lambda^{(2)}$, the process

$$\Lambda_t := \Lambda^{(2)}_{\Lambda^{(1)}_t}, \quad t \geq 0,$$

is again a Lévy subordinator (see, e.g., Bertoin (1999, Proposition 8.6)). One can even compute its characteristics, a result which is due to Huff (1969). The original idea of subordination of a process is due to Bochner (1955). Moreover, it is an easy exercise to check that the sum of two independent Lévy subordinators is again a Lévy subordinator. Thus, it is theoretically possible to construct a huge repertoire of parametric families of Lévy subordinators from known ones. This fact is used to construct hierarchical Archimedean copulas in Section 2.4 and Lévy-frailty copulas in Section 3.3.

Four of the most popular Lévy subordinators are introduced in the sequel.

A.2.1 *Compound Poisson Subordinator*

Consider a probability space $(\Omega, \mathcal{F}, \mathbb{P})$ on which $\{J_i\}_{i \in \mathbb{N}}$ are i.i.d. non-negative random variables and $N = \{N_t\}_{t \geq 0}$ is an independent Poisson

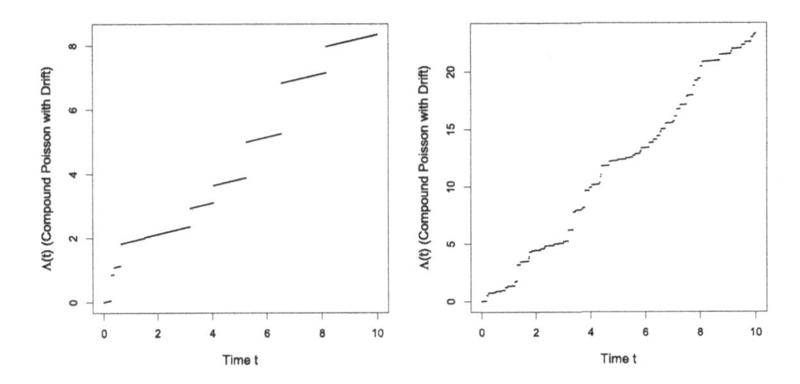

Fig. A.1 Simulated paths of a compound Poisson process with drift and $Exp(\eta)$-distributed jump sizes. In the graph on the left the jump intensity is $\beta = 1$, the drift $\mu = 0.2$, and $\eta = 2$. The parameters in the graph on the right are $(\mu, \eta, \beta) = (0.2, 4, 8)$. Both paths are simulated up to time $T = 10$.

process with intensity $\beta > 0$. With a non-negative drift $\mu \geq 0$ defining

$$\Lambda_t := \mu\, t + \sum_{i=1}^{N_t} J_i, \quad t \geq 0,$$

it follows from Cont and Tankov (2004, Proposition 3.3, p. 71) that $\Lambda = \{\Lambda_t\}_{t\geq 0}$ is a Lévy subordinator, called a *compound Poisson subordinator (with drift if $\mu \neq 0$)*. Furthermore it follows from Cont and Tankov (2004, Proposition 3.5, p. 75) that the Lévy measure ν of Λ has the special form

$$\nu(B) = \beta\, \mathbb{P}(J_1 \in B), \quad B \in \mathcal{B}\big((0,\infty)\big).$$

This implies that the Laplace exponent Ψ of Λ is given by

$$\Psi(x) = \mu\, x + \beta\left(1 - \mathbb{E}\left[e^{-x\, J_1}\right]\right), \quad x \geq 0. \tag{A.7}$$

Intuitively, Λ grows linearly with constant drift μ, it jumps whenever the Poisson process N jumps, and the ith jump has random jump size J_i. Thus, in a bounded time interval $[s,t]$, for $0 \leq s < t < \infty$, compound Poisson subordinators (with drift) almost surely exhibit only finitely many jumps. More precisely, the number of jumps of Λ in $[s,t]$ is $Poi\big(\beta\,(t-s)\big)$-distributed. A typical path of such a stochastic process is illustrated in Figure A.1. It is worth mentioning that a Lévy subordinator is of the compound Poisson type (possibly with drift) if and only if it has almost surely finitely many jumps on any bounded time interval. Compound Poisson subordinators are thus said to be *finite activity processes*. Analytically, a Lévy subordinator with Lévy measure ν is of the compound Poisson type if and only if there is an $\epsilon > 0$ such that $\nu\big((0,\epsilon)\big) < \infty$. This follows more or less from (A.4) and (A.6).

Remark A.7 (Compound Poisson Approximation)
In some sense, an arbitrary Lévy subordinator can be approximated arbitrarily close in distribution by a compound Poisson subordinator. Since compound Poisson subordinators can in principle be sampled accurately, such an approximation implies approximate sampling schemes for arbitrary Lévy subordinators (see Damien et al. (1995) for details).

A.2.2 *Gamma Subordinator*

Following Schoutens (2003, p. 52), a Lévy subordinator Λ is called *Gamma subordinator*, if it has zero drift and its Lévy measure ν, parameterized by $(\beta,\eta) \in (0,\infty)^2$, is absolutely continuous with respect to the Lebesgue measure on $(0,\infty)$ and has the special form

$$\nu(dt) = \beta\, e^{-\eta\, t}\, \frac{1}{t}\, \mathbb{1}_{\{t>0\}}\, dt.$$

It is easy to check that the measure ν defined in this way satisfies $(A.4)$ and hence defines a Lévy subordinator. It is possible to compute the Laplace exponent Ψ in closed form. Following Tricomi (1951), for continuous functions $f : (0,\infty) \to \mathbb{R}$ with existing limits $\lim_{t\downarrow 0} f(t) \in \mathbb{R}$ and $\lim_{t\to\infty} f(t) \in \mathbb{R}$, the so-called *Frullani theorem* states that

$$\int_{(0,\infty)} \big(f(a\,t) - f(b\,t)\big)\, \frac{1}{t}\, dt = \big(\lim_{t\downarrow 0} f(t) - \lim_{t\to\infty} f(t)\big)\, \log\Big(\frac{b}{a}\Big).$$

In particular, the function $f(t) := \exp(-t)$ is admissible in this formula and with $a := \eta$ and $b := \eta + x$ one may deduce

$$\Psi(x) = \int_{(0,\infty)} \big(1 - e^{-x\,t}\big)\, \beta\, e^{-\eta\, t}\, \frac{1}{t}\, dt = \beta\, \log\Big(1 + \frac{x}{\eta}\Big).$$

Therefore, the Lévy–Khinchin representation implies that the random variable Λ_t has Laplace transform

$$\mathbb{E}[e^{-x\,\Lambda_t}] = e^{-t\,\beta\, \log\left(1+\frac{x}{\eta}\right)} = \Big(1 + \frac{x}{\eta}\Big)^{-\beta\, t}, \quad x > 0,\, t > 0.$$

This is known to be the Laplace transform of a Gamma distribution, which explains the name of the process Λ. More precisely, for $t > 0$ we have $\Lambda_t \sim \Gamma(\beta\, t, \eta)$. Note that for each $\epsilon > 0$ it holds that $\nu\big((0,\epsilon)\big) = \infty$, meaning that a Gamma subordinator jumps almost surely infinitely often

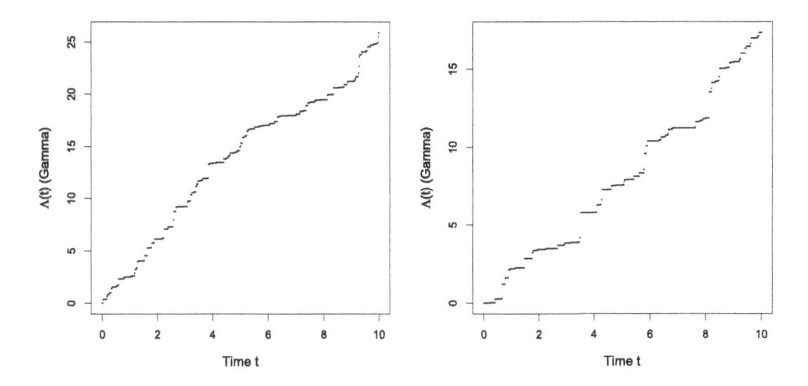

Fig. A.2 Simulated paths of a Gamma subordinator. Parameters used are $(\eta, \beta) = (2, 4.725)$ (left) and $(\eta, \beta) = (1, 2)$ (right). Both paths are simulated up to time $T = 10$ using $n = 1\,000$ grid points.

in a finite time interval. Lévy subordinators with this property are thus said to exhibit *infinite activity*. A typical path of a Gamma subordinator is illustrated in Figure A.2. Due to the infinite activity, simulating a path of $\{\Lambda_t\}_{t\in[0,T]}$ is impossible without discretization bias. The simulation is accomplished by defining a grid $0 < T/n < T\,2/n < \ldots < T\,(n-1)/n < T$ and accumulating i.i.d. random variables which are $\Gamma(\beta\,T/n, \eta)$-distributed.

A.2.3 *Inverse Gaussian Subordinator*

A Lévy subordinator Λ is called an *inverse Gaussian subordinator*, if it has zero drift and its Lévy measure ν, parameterized by $(\beta, \eta) \in (0, \infty)^2$, is absolutely continuous with respect to the Lebesgue measure on $(0, \infty)$ and has the form

$$\nu(dt) = \frac{1}{\sqrt{2\,\pi}}\,\frac{\beta}{t^{\frac{3}{2}}}\,e^{-\frac{1}{2}\,\eta^2\,t}\,\mathbb{1}_{\{t>0\}}\,dt.$$

For each $t > 0$ it is well known that $\Lambda_t \sim IG(\beta\,t, \eta)$ (see Seshadri (1993) for a proof and further results). The Laplace exponent of Λ is given by

$$\Psi(x) = \beta\,\big(\sqrt{2\,x + \eta^2} - \eta\big), \quad x \geq 0. \tag{A.8}$$

The name "inverse Gaussian" stems from the fact that Λ may be constructed as

$$\Lambda_t := \inf\big\{s > 0 \,:\, \eta\,s + X_s = \beta\,t\big\},$$

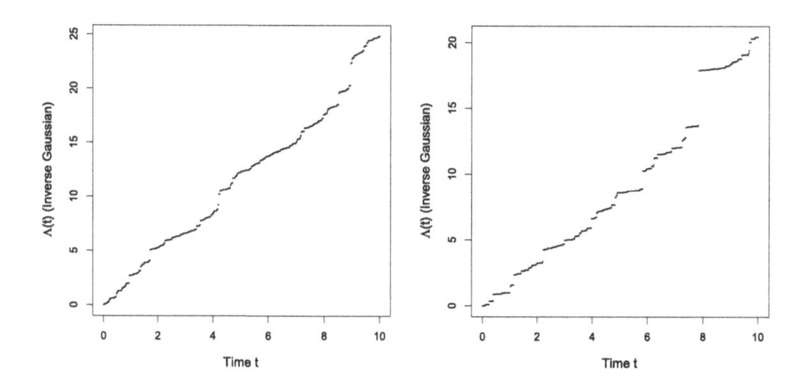

Fig. A.3 Simulated paths of an inverse Gaussian subordinator. The parameters are $(\eta,\beta)=(2,4.725)$ (left) and $(\eta,\beta)=(1,2)$ (right). Both paths are simulated up to time $T=10$ using $n=1\,000$ grid points.

where $X=\{X_t\}_{t\geq 0}$ is a standard Brownian motion on a probability space $(\Omega,\mathcal{F},\mathbb{P})$. Thus, Λ_t can be interpreted as the first hitting-time of the level $\beta\,t$ of a Brownian motion with drift. The resulting relation to the normal distribution justifies the wording. Moreover, Shuster (1968) shows how to express the distribution function of Λ_t in terms of the standard normal distribution function Φ: for all $t>0$, $x\geq 0$ it holds that

$$\mathbb{P}(\Lambda_t\leq x)=\Phi\Big(\eta\sqrt{x}-\frac{\beta\,t}{\sqrt{x}}\Big)+e^{2\,\beta\,t\,\eta}\,\Phi\Big(-\eta\sqrt{x}-\frac{\beta\,t}{\sqrt{x}}\Big),$$

$$\Phi(x)=\int_{-\infty}^{x}\frac{1}{\sqrt{2\,\pi}}\,e^{-\frac{s^2}{2}}\,ds.$$

Like the Gamma subordinator, an inverse Gaussian subordinator exhibits infinite activity, since $\nu\big((0,\epsilon)\big)=\infty$ for each $\epsilon>0$. Figure A.3 illustrates typical paths of such a process. The sampling is done similarly as in the case of a Gamma subordinator.

A.2.4 *Stable Subordinator*

A Lévy subordinator Λ is called an *α-stable subordinator* with parameter $\alpha\in(0,1)$, if it has zero drift $\mu=0$ and its Lévy measure ν is absolutely continuous with respect to the Lebesgue measure and is defined via

$$\nu(dt)=\frac{\alpha}{\Gamma(1-\alpha)}\,\frac{1}{t^{1+\alpha}}\,\mathbb{1}_{\{t>0\}}\,dt.$$

One immediately checks that $\nu\big((0,\epsilon)\big)=\infty$ for $\epsilon>0$. Hence, Λ is another example of an infinite activity process. It can be verified by an application of Fubini's theorem (see, e.g., Applebaum (2004, p. 69)) that the Laplace exponent of Λ is given by $\Psi(x) = x^{\alpha}$, $x \geq 0$. Therefore, it holds for each $t > 0$ that $t^{-1/\alpha}\,\Lambda_t \sim \mathcal{S}(\alpha)$. The density f_{Λ_t} of Λ_t is not known in closed form,[6] but Nolan (1997) uses Fourier inversion techniques to compute a numerically convenient form, which is given by

$$f_{\Lambda_t}(x) = t^{-\frac{1}{\alpha}}\, f_{\Lambda_1}\big(t^{-\frac{1}{\alpha}}\, x\big),$$

where

$$\begin{aligned}
f_{\Lambda_1}(x) &= \mathbb{1}_{\{x>0\}}\, \frac{\alpha\left(\frac{x}{\gamma}\right)^{\frac{1}{\alpha-1}}}{\gamma\,\pi\,(1-\alpha)} \int_{-\frac{\pi}{2}}^{\frac{\pi}{2}} g_\alpha(u)\, e^{-\left(\frac{x}{\gamma}\right)^{\frac{\alpha}{\alpha-1}} g_\alpha(u)}\, du,\\
g_\alpha(u) &= \left(\cos\left(\frac{\pi\,\alpha}{2}\right)\right)^{\frac{1}{\alpha-1}} \left(\frac{\cos u}{\sin\left(\alpha(\frac{\pi}{2}+u)\right)}\right)^{\frac{\alpha}{\alpha-1}} \frac{\cos\left(\frac{\pi}{2}\,\alpha + (\alpha-1)\,u\right)}{\cos u},\\
\gamma &= \left(\cos\left(\frac{\pi\,\alpha}{2}\right)\right)^{\frac{1}{\alpha}}. \qquad \text{(A.9)}
\end{aligned}$$

An α-stable subordinator has a heavy-tailed distribution. For instance, it is shown in Wolfe (1975) that

$$\mathbb{E}[\Lambda_t^{\beta}] = \begin{cases} \frac{t^{\frac{\beta}{\alpha}}\,\Gamma\left(1-\frac{\beta}{\alpha}\right)}{\Gamma(1-\beta)} & ,\ \beta \in (0,\alpha) \\ \infty & ,\ \beta \geq \alpha \end{cases}, \quad t > 0.$$

In particular, such distributions are standard examples for random variables without an existing first moment. This property, together with the convenient functional form of the Laplace exponent, makes this process interesting in many applications. Typical paths of an α-stable subordinator are illustrated in Figure A.4. The sampling is again accomplished by accumulating n i.i.d. random variables whose distribution equals that of $\Lambda_{1/n} \stackrel{d}{=} n^{-1/\alpha}\,\mathcal{S}(\alpha)$.

A.3 Scale Mixtures of Marshall–Olkin Copulas

In this section we briefly introduce a stochastic model, originally due to Mai et al. (2011), that combines the representation of extendible Archimedean copulas with the one of Lévy-frailty copulas (see Sections 2.2 and 3.3). The result is a parametric subfamily of copulas known as scale mixtures of Marshall–Olkin copulas (for background on the latter, see Li (2009)). Recall from Section 2.2 that extendible Archimedean copulas are constructed

[6]Except for $\alpha = 1/2$ (see, e.g., Applebaum (2004, p. 50)).

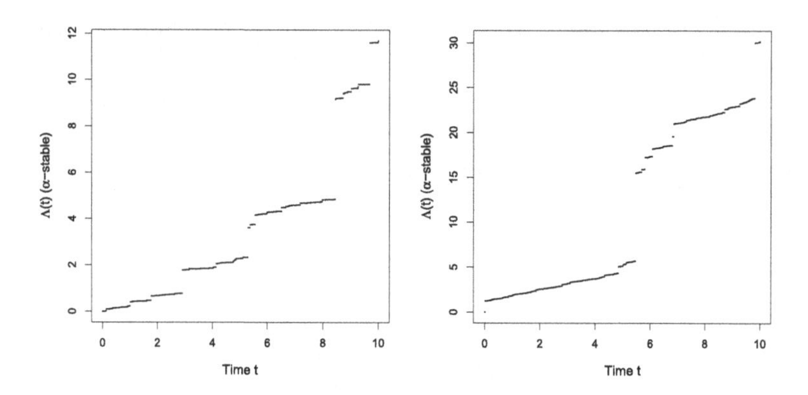

Fig. A.4 Simulated paths of an α-stable subordinator with $\alpha = 0.5$ (left) and $\alpha = 0.8$ (right). Both paths are simulated up to time $T = 10$ using $n = 1\,000$ grid points.

via a conditionally i.i.d. model with some positive random variable M as the common factor. Closely related, Lévy-frailty copulas are constructed in Section 3.3 from a random distribution function driven by a Lévy subordinator. Combining both, we consider some probability space $(\Omega, \mathcal{F}, \mathbb{P})$ on which we define

$$(U_1, \ldots, U_d) := \Big(\varphi\big(\inf\{t > 0 : \Lambda_{M\,t} > E_1\}\big), \ldots, \varphi\big(\inf\{t > 0 : \Lambda_{M\,t} > E_d\}\big)\Big),$$

where $\Lambda = \{\Lambda_t\}_{t \geq 0}$ is a Lévy subordinator with Laplace exponent Ψ satisfying $\Psi(1) = 1$, M is a positive random variable with Laplace transform φ, and $E_1, \ldots, E_d$ is a list of i.i.d. unit exponentials. These building blocks are stochastically independent. It is shown in Mai et al. (2011) that the distribution function of $(U_1, \ldots, U_d)$ is precisely the copula

$$C_{\varphi,\Psi}(u_1, \ldots, u_d) := \varphi\Big(\sum_{i=1}^{d} \varphi^{-1}(u_{(i)})\,\big(\Psi(i) - \Psi(i-1)\big)\Big), \tag{A.10}$$

where $u_{(1)} \leq \ldots \leq u_{(d)}$ is the ordered list of $u_1, \ldots, u_d \in [0,1]^d$. It is easy to observe that the model (and the copula) reduces to the respective building blocks/models in Algorithms 2.1 and 3.6 when $\Lambda_t = t$ is chosen as a degenerate subordinator with $\Psi(x) = x$ or when M is chosen to be a constant.

The random vector $(U_1, \ldots, U_d)$ inherits the statistical properties from its building blocks and is therefore quite flexible. For instance, the tail dependence coefficients are given by

$$UTD_{C_{\varphi,\Psi}} = \Psi(2) \lim_{x \uparrow \infty} \frac{\varphi'\big(\Psi(2)\,x\big)}{\varphi'(x)},$$

$$LTD_{C_{\varphi,\Psi}} = 2 - \Psi(2) \lim_{x \downarrow 0} \frac{\varphi'\big(\Psi(2)\,x\big)}{\varphi'(x)}$$

(see Bernhart et al. (2012)). The probabilities $\mathbb{P}(U_1 = \ldots = U_k)$, $k = 2, \ldots, d$, are precisely the same as in the case of a Lévy-frailty copula, i.e.

$$\mathbb{P}(U_1 = \ldots = U_k) = \frac{\sum_{i=0}^{k} \binom{k}{i} (-1)^{i+1} \Psi(i)}{\Psi(k)}, \quad k = 2, \ldots, d.$$

It is interesting to know that the random scaling in time by M overcomes the time-homogeneous increments of the Lévy subordinator and the resulting model no longer exhibits the lack of memory property. This might be desirable in real-world situations (e.g. multivariate default models) where the lack of memory property of the Lévy-frailty model is not realistic. An h-extendible variant of this dependence structure is proposed in Mai and Scherer (2011b) and analyzed in Bernhart et al. (2012). Sampling such random variables is possible by suitably including M into the sampling algorithm of the Lévy-frailty copula.

Example A.2 (An Archimedean Copula with an Armageddon Shock)

A parametric copula is obtained when an Archimedean model, constructed from a positive random variable M with Laplace transform φ and a list $E_1, \ldots, E_d$ of independent unit exponentials, is mixed with a Lévy subordinator that increases linearly with drift $\alpha \in [0,1)$ and suddenly jumps to ∞, i.e. its Lévy measure is $\nu(\{\infty\}) = (1-\alpha)$, $\nu((0,\infty)) = 0$. The subordinator is thus given by

$$\Lambda_t = \alpha\, t + \infty \cdot \mathbb{1}_{\{t \geq E\}}, \quad t \geq 0,$$

where E is an exponential random variable with intensity parameter $1-\alpha$. The resulting copula of

$$(U_1, \ldots, U_d) := \Big(\varphi\big(\inf\{t > 0 \,:\, \Lambda_{M\,t} > E_1\}\big), \ldots, \varphi\big(\inf\{t > 0 \,:\, \Lambda_{M\,t} > E_d\}\big)\Big),$$

interpolates between the comonotonicity copula and the chosen Archimedean copula and is given by

$$C_{\varphi,\Psi}(u_1, \ldots, u_d) = \varphi\Big((1-\alpha)\,\varphi^{-1}(u_{(1)}) + \alpha \sum_{i=1}^{d} \varphi^{-1}(u_i)\Big),$$

where $u_1, \ldots, u_d \in [0,1]$ and Ψ denotes the Laplace exponent of Λ.

Proof. We find for $x > 0$

$$\begin{aligned}\mathbb{E}\big[e^{-x\,\Lambda_t}\big] &= \mathbb{E}\big[e^{-x\,\alpha\,t - x\,\infty\,\mathbb{1}_{\{t \geq E\}}}\big] \\ &= e^{-x\,\alpha\,t}\big(0 \cdot \mathbb{P}(t \geq E) + 1 \cdot \mathbb{P}(t < E)\big) \\ &= e^{-t\,(\alpha\,x + (1-\alpha))} = e^{-t\,\Psi(x)},\end{aligned}$$

and $\Psi(0) = 0$. We observe that $\Psi(1) = 1$ and $\nu(\{\infty\}) = \lim_{x \searrow 0} \Psi(x) = 1 - \alpha \neq \Psi(0) = 0$ (the discontinuity of Ψ at 0 corresponds to the atom $\nu(\{\infty\}) = 1-\alpha$). Next, we note that $\Psi(1) - \Psi(0) = 1$ and $\Psi(i) - \Psi(i-1) = \alpha$ for $i \geq 2$. Hence, for $u_1, \ldots, u_d \in [0, 1]$

$$\begin{aligned}C_{\varphi,\Psi}(u_1, \ldots, u_d) &= \varphi\Big(\sum_{i=1}^{d} \varphi^{-1}(u_{(i)})\,\big(\Psi(i) - \Psi(i-1)\big)\Big) \\ &= \varphi\Big(\varphi^{-1}(u_{(1)}) + \alpha \sum_{i=2}^{d} \varphi^{-1}(u_{(i)})\Big) \\ &= \varphi\Big((1-\alpha)\,\varphi^{-1}(u_{(1)}) + \alpha \sum_{i=1}^{d} \varphi^{-1}(u_{(i)})\Big),\end{aligned}$$

the ordering can be dropped in the sum. □

Observe that (condition on Λ and M to obtain the last equation)

$$\begin{aligned}&\mathbb{P}(U_1 \leq u_1, \ldots, U_d \leq u_d) \\ &= \mathbb{P}\Big(\inf\big\{t \geq 0 \,:\, \Lambda_{M\,t} \geq E_k\big\} \geq \varphi^{-1}(u_k),\, k = 1, \ldots, d\Big) \\ &= \mathbb{P}\big(\Lambda_{M\,\varphi^{-1}(u_k)} \leq E_k,\, k = 1, \ldots, d\big) \\ &= \mathbb{E}\Big[\prod_{k=1}^{d} e^{-\Lambda_{M\,\varphi^{-1}(u_k)}}\Big].\end{aligned} \tag{A.11}$$

Concerning sampling, using the mixture model language of Algorithm 1.3 in Section 1.2.3, we obtain from Equation (A.11) that conditioned on $\sigma(\{\Lambda_t\}_{t\geq 0}, M) = \sigma(E, M)$, $U_1, \ldots, U_d$ are i.i.d. with random distribution function

$$\begin{aligned}F_x &= \exp(-\Lambda_{M\,\varphi^{-1}(x)}) = \exp\big(-\alpha\,M\,\varphi^{-1}(x) - \infty\mathbb{1}_{\{M\,\varphi^{-1}(x) \geq E\}}\big) \\ &= \exp\big(-\alpha\,M\,\varphi^{-1}(x)\big)\,\mathbb{1}_{\{x \geq \varphi(E/M)\}}, \quad x \in [0, 1].\end{aligned}$$

Given M and E, the quantile of this distribution function is given by

$$\begin{aligned}F_y^{-1} &= \inf\{x \in [0, 1] : F_x \geq y\} \\ &= \left\{\begin{array}{ll}\varphi(E/M) & : y \in (0, e^{-\alpha\,E}) \\ \varphi\big(-\log(y)/(\alpha\,M)\big) & : y \in [e^{-\alpha\,E}, 1)\end{array}\right\}, \quad y \in (0, 1).\end{aligned}$$

This suggests the following convenient sampling scheme:

Algorithm A.5 (Archimedean Copulas with Armageddon Shock)

(1) Sample independent $E \sim Exp(1-\alpha)$, M, *and* $V_1, \ldots, V_d \stackrel{i.i.d.}{\sim} U[0,1]$.
(2) Return $(U_1, \ldots, U_d)$, *where* $U_k := F_{V_k}^{-1}$, $k = 1, \ldots, d$.

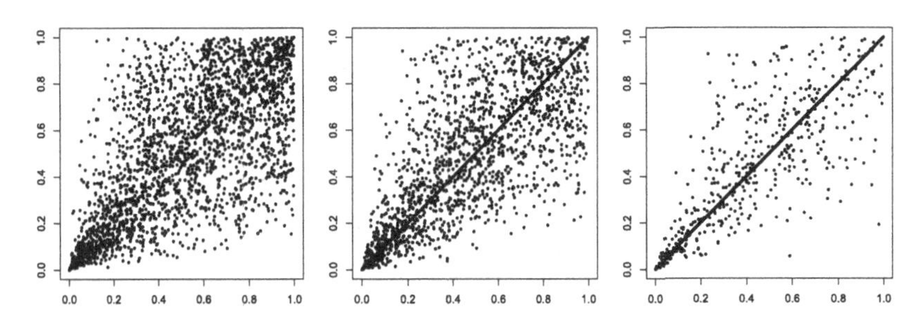

Fig. A.5 Scatterplot of 2 500 samples from a Clayton copula (see Equation (2.11)) with Armageddon shock with parameters $\vartheta = 2$ and $\alpha = 0.9$ (left), 0.5 (middle), and 0.1 (right). Note that the dependence interpolates between the classical Clayton copula and the comonotonicity copula.

A.4 Further Reading

With this book, we aimed at providing a probabilistic treatment of high-dimensional copulas with a specific focus on efficient sampling routines. Clearly, focusing on these aspects made it necessary to disregard other material that might also be interesting for the reader. Concerning the general theory of copulas, we have significantly been influenced by the monographs by Joe (1993) and Nelsen (2006). These were used as primary references for proofs that we had to omit to save space. Joe (1993) gives a very comprehensive treatment of various aspects of "dependence modeling"; with a strong focus on statistical issues and applications. Also, it contains a huge battery of parametric families of multivariate models/copulas. Nelsen (2006) mostly uses analytical techniques; his book is a very popular textbook on copulas. Another classical book is that by Schweizer and Sklar (1983). Readers particularly interested in elliptical (and related) distributions are referred to Fang et al. (1990). Our exposition of exchangeable Archimedean copulas is based on the work by McNeil and Nešlehová (2009).

More on the efficient sampling of Archimedean copulas is presented in the research papers of M. Hofert and A. McNeil. A recent book on pair copulas is by Kurowicka and Joe (2011). Classical papers on estimation are those by Genest et al. (1995) and Joe and Xu (1996). Dependence orderings are discussed in the work by Müller and Stoyan (2002). Finally, a classical treatment of exchangeable and extendible random vectors is given by Aldous (1985).

Concerning applications in the financial industry, a famous book on risk management with a large section on copulas is that by McNeil et al. (2005). Copula methods for financial applications are also presented in Cherubini et al. (2004) and Cherubini et al. (2012). The popularity of copulas in credit risk applications originates to a large extent from the respective chapter in the book by Schönbucher (2003). More on Monte Carlo techniques (with applications in finance and insurance) can be found in the books by Glasserman (2004) and Korn et al. (2010). A very popular treatment of univariate sampling techniques is given by Devroye (1986).

Bibliography

K. Aas, C. Czado, A. Frigessi, H. Bakken, Pair-copula constructions of multiple dependence, *Insurance: Mathematics and Economics* **44**(2) (2009) pp. 182–198.

M. Abramowitz, I.A. Stegun, *Handbook of Mathematical Functions*, Dover, New York (1972).

D.J. Aldous, Exchangeability and related topics, *École d'Été de Probabilités de Saint-Flour XIII-1983. Lecture Notes in Mathematics* **1117**, Springer, Berlin (1985) pp. 1–198.

M. Ali, N. Mikhail, M. Haq, A class of bivariate distributions including the bivariate logistic, *Journal of Multivariate Analysis* **8** (1978) pp. 405–412.

C. Alsina, M.J. Frank, B. Schweizer, *Associative Functions: Triangular Norms and Copulas*, World Scientific, Hackensack (2006).

D. Applebaum, *Lévy Processes and Stochastic Calculus*, Cambridge University Press, Cambridge (2004).

E. Arjas, T. Lehtonen, Approximating many server queues by means of single server queues, *Mathematics of Operations Research* **3** (1978) pp. 205–223.

B.C. Arnold, A characterization of the exponential distribution by multivariate geometric compounding, *Sankhyā: The Indian Journal of Statistics* **37**(1) (1975) pp. 164–173.

R.W. Bailey, Polar generation of random variates with the t-distribution, *Mathematics of Computation* **62** (1994) pp. 779–781.

R.E. Barlow, F. Proschan, *Statistical Theory of Reliability and Life Testing*, Holt, Rinehart and Winston, New York (1975).

O.E. Barndorff-Nielsen, N. Shephard, Normal modified stable processes, *Theory of Probability and Mathematical Statistics* **65** (2001) pp. 1–19.

Basel Committee on Banking Supervision, Supervisory framework for the use of "backtesting" in conjunction with the internal models approach to market risk capital requirements, available at *www.bis.org* (1996).

Basel Committee on Banking Supervision, Internal coverage of capital measurement and capital standards: A revised framework, comprehensive version, available at *www.bis.org* (2006).

J.D. Beasley, S.G. Springer, Algorithm AS 111: The percentage points of the normal distribution, *Applied Statistics* **26**(1) (1977) pp. 118–121.

T. Bedford, R.M. Cooke, Probability density decomposition for conditionally dependent random variables modeled by vines, *Annals of Mathematics and Artificial Intelligence* **32** (2001a) pp. 245–268.

T. Bedford, R.M. Cooke, Monte Carlo simulation of vine dependent random variables for applications in uncertainty analysis, *Proceedings of ESREL2001, Turin, Italy* (2001b).

T. Bedford, R.M. Cooke, Vines: A new graphical model for dependent random variables, *Annals of Statistics* **30**(4) (2002) pp. 1031–1068.

J. Beirlant, Y. Goegebeur, J. Segers, J. Teugels, *Statistics of Extremes: Theory and Applications*, Wiley Series in Probability and Statistics, Wiley, Chichester (2004).

D. Berg, K. Aas, Models for construction of multivariate dependence, *European Journal of Finance* **15**(7/8) (2009) pp. 639–659.

G. Bernhart, M. Escobar, J.-F. Mai, M. Scherer, Default models based on scale mixtures of Marshall–Olkin copulas: properties and applications, forthcoming in *Metrika* (2012).

S. Bernstein, Sur les fonctions absolument monotones, *Acta Mathematica* **52**(1) (1929) pp. 1–66.

J. Bertoin, *Lévy Processes*, Cambridge University Press, Cambridge (1996).

J. Bertoin, Subordinators: Examples and Applications, *École d'Été de Probabilités de Saint-Flour XXVII-1997. Lecture Notes in Mathematics* **1717** Springer, Berlin (1999) pp. 1–91.

P. Billingsley, *Probability and Measure*, Wiley Series in Probability and Statistics, Wiley, New York (1995).

S. Bochner, *Harmonic Analysis and the Theory of Probability*, University of California Press, Berkeley (1955).

L. Bondesson, On simulation from infinitely divisible distributions, *Advances in Applied Probability* **14** (1982) pp. 855–869.

E. Brechmann, Truncated and simplified regular vines and their applications, Diploma thesis, Technische Universität München (2010).

S. Cambanis, S. Hung, G. Simons, On the theory of elliptically contoured distributions, *Journal of Multivariate Analysis* **11** (1981) pp. 368–385.

B. Candelon, G. Colletaz, C. Hurlin, S. Tokpavi, Backtesting Value at Risk: A GMM duration-based test, *Journal of Financial Econometrics* **9**(2) (2011) pp. 314–343.

J.M. Chambers, C.L. Mallows, B.W. Stuck, A method for simulating stable random variables, *Journal of the American Statistical Association* **71** (1976) pp. 340–344.

A. Charpentier, J. Segers, Tails of multivariate Archimedean copulas, *Journal of Multivariate Analysis* **100**(7) (2009) pp. 1521–1537.

U. Cherubini, E. Luciano, W. Vecchiato, *Copula Methods in Finance*, John Wiley and Sons, London (2004).

U. Cherubini, S. Romagnoli, Computing the volume of n-dimensional copulas, *Applied Mathematical Finance* **16**(4) (2009) pp. 307–314.

U. Cherubini, F. Gobbi, S. Mulinacci, S. Romagnoli, *Dynamic Copula Methods in Finance*, John Wiley and Sons, London (2012).

P. Christoffersen, D. Pelletier, Backtesting Value at Risk: A duration-based approach, *Journal of Financial Econometrics* **2**(1) (2004) pp. 84–108.
D.G. Clayton, A model for association in bivariate life tables and its application in epidemiological studies of family tendency in chronic disease incidence, *Biometrika* **65** (1978) pp. 141–151.
R. Cont, P. Tankov, *Financial Modelling with Jump Processes*, Chapman and Hall/CRC Financial Mathematics Series, Boca Raton, Florida (2004).
R.D. Cook, M.E. Johnson, A family of distributions for modelling non-elliptically symmetric multivariate data, *Journal of the Royal Statistical Society: Series B* **43** (1981) pp. 210–218.
R.M. Corless, G.H. Gonnet, D.E.G. Hare, D.J. Jeffrey, D.E. Knuth, On the Lambert W function, *Advances in Computational Mathematics* **5** (1996) pp. 329–359.
C. Czado, T. Schmidt, *Mathematische Statistik*, Springer, Berlin (2011).
P. Damien, P.M. Laud, A.F.M. Smith, Approximate random variate generation from infinitely divisible distributions with applications to Bayesian inference, *Journal of the Royal Statistical Society* **57**(3) (1995) pp. 547–563.
W.F. Darsow, B. Nguyen, E.T. Olsen, Copulas and Markov processes, *Illinois Journal of Mathematics* **36**(4) (1992) pp. 600–642.
S. Daul, E. De Giorgi, F. Lindskog, A.J. McNeil, The grouped t-copula with an application to credit risk, *Risk* **16**(11) (2003) pp. 73–76.
H.A. David, H.N. Nagaraja, *Order Statistics*, Wiley Series in Probability and Statistics, Wiley, New York (1970).
B. de Finetti, La prévision: ses lois logiques, ses sources subjectives, *Annales de l'Institut Henri Poincaré* **7** (1937) pp. 1–68.
M. DeGroot, *Optimal Statistical Decisions*, John Wiley and Sons, New York (2004).
L. de Haan, S.I. Resnick, Limit theory for multivariate sample extremes, *Zeitschrift für Wahrscheinlichkeitstheorie und verwandte Gebiete* **40**(4) (1977) pp. 317–337.
H. Dette, W.J. Studden, *The Theory of Canonical Moments with Applications in Statistics, Probability, and Analysis*, Wiley Series in Probability and Statistics, Wiley, New York (1997).
L. Devroye, *Non-uniform Random Variate Generation*, Springer, New York (1986).
J. Dißmann, Statistical inference for regular vines and application, Diploma thesis, Technische Universität München (2010).
F. Durante, Construction of non-exchangeable bivariate distribution functions, *Statistical Papers* **50**(2) (2009) pp. 383–391.
P. Embrechts, F. Lindskog, A.J. McNeil, Modelling dependence with copulas and applications to risk management, in *Handbook of Heavy Tailed Distributions in Finance*, ed. S. Rachev, Elsevier/North-Holland, Amsterdam (2003) pp. 329–384.
L. Fahrmeir, A. Hamerle, *Multivariate statistische Verfahren*, de Gruyter, Berlin (1984).
M. Falk, J. Hüsler, R.-D. Reiss, *Laws of Small Numbers: Extremes and Rare Events*, Birkhäuser, Basel-Boston-Berlin (2004).

K.-T. Fang, S. Kotz, K.-W. Ng, *Symmetric Multivariate and Related Distributions*, Chapman and Hall, London (1990).

W. Feller, *An Introduction to Probability Theory and its Applications, Volume II*, second edition, John Wiley and Sons, New York (1966).

G. Fishman, *Monte Carlo: Concepts, Algorithms and Applications*, Springer, New York (1996).

G. Frahm, On the extremal dependence coefficient of multivariate distributions, *Statistics and Probability Letters* **76**(14) (2006) pp. 1470–1481.

M.J. Frank, On the simultaneous association of $F(x, y)$ and $x + y - F(x, y)$, *Aequationes Math* **21** (1979) pp. 37–38.

M. Fréchet, Les tableaux de corrélation dont les marges et des bornes sont données, *Annales de l'Université de Lyon, Sciences Mathématiques et Astronomie* **20** (1957) pp. 13–31.

J. Galambos, S. Kotz, *Characterizations of Probability Distributions*, Lecture Notes in Mathematics (Volume 675), Springer, Berlin (1978).

C. Genest, K. Ghoudi, L.-P. Rivest, A semiparametric estimation procedure of dependence parameters in multivariate families of distributions, *Biometrika* **82**(3) (1995) pp. 543–552.

C. Genest, R.J. MacKay, Copules archimédiennes et familles de lois bidimensionnelles dont les marges sont données, *Canadian Journal of Statistics* **14** (1986) pp. 145–159.

C. Genest, J. Nešlehová, A primer on copulas for count data, *Astin Bulletin* **37**(2) (2007) pp. 475–515.

C. Genest, L.-P. Rivest, A characterization of Gumbel's family of extreme value distributions, *Statistics and Probability Letters* **8** (1989) pp. 207–211.

K. Giesecke, A simple exponential model for dependent defaults, *Journal of Fixed Income* **13**(3) (2003) pp. 74–83.

P. Glasserman, *Monte Carlo Methods in Financial Engineering*, Springer, New York (2004).

A. Gnedin, J. Pitman, Moments of convex distribution functions and completely alternating sequences, in *Probability and Statistics: Essays in Honor of David A. Freedman, Vol. 2*, ed., D. Nolan, T. Speed, Institute of Mathematical Statistics, Beachwood, Ohio (2008) pp. 30–41.

G.H. Golub, C.F. van Loan, *Matrix Computations*, John Hopkins University Press, Baltimore, Maryland (1989).

E.J. Gumbel, Distributions des valeurs extrêmes en plusiers dimensions, *Publications de l'Institut de Statistique de l'Université de Paris* **9** (1960a) pp. 171–173.

E.J. Gumbel, Bivariate exponential distributions, *Journal of the American Statistical Association* **55**(292) (1960b) pp. 698–707.

M. Haas, New methods in backtesting, *Financial Engineering Research Center Caesar*, Bonn (2001).

I.H. Haff, K. Aas, A. Frigessi, On the simplified pair-copula construction: Simply useful or too simplistic?, *Journal of Multivariate Analysis* **101** (2010) pp. 1296–1310.

C. Hastings, *Approximations for Digital Computers*, Princeton University Press, Princeton, New Jersey (1955).

F. Hausdorff, Summationsmethoden und Momentfolgen I, *Mathematische Zeitschrift* **9**(3–4) (1921) pp. 74–109.

F. Hausdorff, Momentenprobleme für ein endliches Intervall, *Mathematische Zeitschrift* **16** (1923) pp. 220–248.

C. Hering, M. Hofert, J.-F. Mai, M. Scherer, Constructing hierarchical Archimedean copulas with Lévy subordinators, *Journal of Multivariate Analysis* **101**(6) (2010) pp. 1428–1433.

C. Hering, J.-F. Mai, Moment-based estimation of extendible Marshall–Olkin distributions, forthcoming in *Metrika* (2012).

M. Hofert, Sampling Archimedean copulas, *Computational Statistics and Data Analysis* **52** (2008) pp. 5163–5174.

M. Hofert, Sampling nested Archimedean copulas with applications to CDO pricing, Dissertation Universität Ulm (2010).

M. Hofert, Efficiently sampling Archimedean copulas, *Computational Statistics & Data Analysis* **55** (2011) pp. 57–70.

M. Hofert, Sampling exponentially tilted stable distributions, forthcoming in *ACM Transactions on Modeling and Computer Simulation* (2012).

M. Hofert, M. Mächler, Nested Archimedean copulas meet R: The nacopula package, *Journal of Statistical Software* **39** (2011) pp. 1-20.

M. Hofert, M. Mächler, A.J. McNeil, Likelihood inference for Archimedean copulas, working paper (2011).

M. Hofmann, C. Czado, Assessing the VaR of a portfolio using D-vine copula based multivariate GARCH models, working paper (2010).

B.W. Huff, The strict subordination of a differential process, *Sankhya: The Indian Journal of Statistics, Series A* **31**(4) (1969) pp. 403–412.

N. Jacob, R.L. Schilling, Function spaces as Dirichlet spaces (about a paper by Maz'ya and Nagel), *Journal for Analysis and its Applications* **24**(1) (2005) pp. 3–28.

H. Joe, Parametric families of multivariate distributions with given margins, *Journal of Multivariate Analysis* **46** (1993) pp. 262–282.

H. Joe, Families of m-variate distributions with given margins and $m(m-1)/2$ bivariate dependence parameters, in *Distributions with Fixed Marginals and Related Topics*, ed. L. Rüschendorf, B. Schweizer, M.D. Taylor, IMS, Seattle, Washington (1996).

H. Joe, *Multivariate Models and Dependence Concepts*, Chapman and Hall/CRC, London (1997).

H. Joe, H. Li, A.K. Nikoloulopoulos, Tail dependence functions and vine copulas, *Journal of Multivariate Analysis* **101** (2010) pp. 252–270.

H. Joe, J.J. Xu, The estimation method of inference functions of margins for multivariate models, Technical Report 166, Dep. of Statistics, Univ. of British Columbia (1996).

S. Karlin, L.S. Shapley, *Geometry of Moment Spaces*, Memoirs of the American Mathematical Society, Vol. 12, American Mathematical Society, Providence, RI (1953).

A. Kemp, Efficient generation of logarithmically distributed pseudo-random variables, *Journal of the Royal Statistical Society: Series C (Applied Statistics)* **30**(3) (1981) pp. 249–253.

A. Khinchin, Zur Theorie der unbeschränkt teilbaren Verteilungsgesetze, *Matematicheskii Sbornik* **44**(1) (1937) pp. 79–119.

A. Khinchin, *Limit Laws for Sums of Independent Random Variables*, ONTI, Moscow-Leningrad (1938).

A. Khoudraji, Contributions à l'etude des copules et à la modélisation des valeurs extrêmes bivariées, PhD thesis, Université de Laval, Québec (1995).

G. Kim, M.J. Silvapulle, P. Silvapulle, Comparison of semiparametric and parametric methods for estimating copulas, *Computational Statistics and Data Analysis* **51**(6) (2007) pp. 2836–2850.

C.H. Kimberling, A probabilistic interpretation of complete monotonicity, *Aequationes Mathematicae* **10** (1974) pp. 152–164.

G. Kimeldorf, A.R. Sampson, Uniform representations of bivariate distributions, *Communications in Statistics* **4** (1975) pp. 617–627.

D.E. Knuth, *The Art of Computer Programming, Volume 2 (Seminumerical Algorithms)*, 3rd edition, Addison-Wesley, Reading, Massachusetts (1998).

R. Korn, E. Korn, G. Kroisandt, *Monte Carlo Methods and Models in Finance and Insurance*, Chapman and Hall/CRC, London (2010).

P. Kupiec, Techniques for verifying the accuracy of risk management models, *Journal of Derivatives* **3** (1995) pp. 73–84.

D. Kurowicka, Some results for different strategies to choose optimal vine truncation based on wind speed data, conference presentation, 3rd Vine Copula Workshop, Oslo (2009).

D. Kurowicka, R.M. Cooke, Sampling algorithms for generating joint uniform distributions using the vine-copula method, *Computational Statistics and Data Analysis* **51** (2007) pp. 2889–2906.

D. Kurowicka, R. Joe, *Dependence Modeling: Vine Copula Handbook*, World Scientific, Singapore (2011).

P. L'Ecuyer, Uniform random number generation, *Annals of Operations Research*, **53** (1994) pp. 77–120.

P. L'Ecuyer, Good parameters and implementations for combined multiple recursive random number generators, *Operations Research* **47**(1) (1999) pp. 159–164.

P. L'Ecuyer, R. Simard, TestU01: A software library in ANSI C for empirical testing of random number generators, available at *http://www.iro.umontreal.ca/~lecuyer* (2002).

P. L'Ecuyer, F. Panneton, M. Matsumoto, Improved long-period generators based on linear recurrences modulo 2, *ACM Transactions on Mathematical Software* **32**(1) (2006) pp. 1–16.

M. Larsson, J. Nešlehová, Extremal behaviour of Archimedean copulas, *Advances in Applied Probability* **43** (2011) pp. 195–216.

P. Lévy, *Théorie de l'addition des variables aléatoires*, second edition (first edition 1934), Gauthier-Villars, Paris (1954).

H. Li, Tail dependence comparison of survival Marshall–Olkin copulas, *Methodology and Computing in Applied Probability* **10**(1) (2008) pp. 39–54.

H. Li, Orthant tail dependence of multivariate extreme value distributions, *Journal of Multivariate Analysis* **100**(1) (2009) pp. 243–256.

X. Li, P. Mikusinski, H. Sherwood, M.D. Taylor, Some integration-by-parts formulas involving 2-copulas, in *Distributions with given Marginals and Statistical Modelling*, ed. C.M. Cuadras, J. Fortiana, J.A. Lallena Rodriguez, Kluwer, Dordrecht (2002) pp. 153–159.

E. Liebscher, Construction of asymmetric multivariate copulas, *Journal of Multivariate Analysis* **99**(10) (2008) pp. 2234–2250.

F. Lindskog, A.J. McNeil, Common Poisson shock models: Applications to insurance and credit risk modelling, *Astin Bulletin* **33**(2) (2003) pp. 209–238.

F. Lindskog, A. McNeil, U. Schmock, Kendall's tau for elliptical distributions, working paper (2002), available at *www.risklab.ch/Papers.html#KendallsTau*.

C.H. Ling, Representation of associative functions, *Publication Mathematicae Debrecen* **12** (1965) pp. 189–212.

Y. Liu, R. Luger, Efficient estimation of copula-GARCH models, *Computational Statistics and Data Analysis* **53**(6) (2009) pp. 2284–2297.

J.-F. Mai, Extendibility of Marshall–Olkin distributions via Lévy subordinators and an application to portfolio credit risk, Dissertation, Technische Universität München, available at *https://mediatum2.ub.tum.de/node?id=969547* (2010).

J.-F. Mai, M. Scherer, Lévy-frailty copulas, *Journal of Multivariate Analysis* **100**(7) (2009a) pp. 1567–1585.

J.-F. Mai, M. Scherer, Efficiently sampling exchangeable Cuadras-Augé copulas in high dimensions, *Information Sciences* **179** (2009b) pp. 2872–2877.

J.-F. Mai, M. Scherer, The Pickands representation of survival Marshall–Olkin copulas, *Statistics and Probability Letters* **80**(5–6) (2010) pp. 357–360.

J.-F. Mai, M. Scherer, Reparameterizing Marshall–Olkin copulas with applications to sampling, *Journal of Statistical Computation and Simulation* **81**(1) (2011a) pp. 59–78.

J.-F. Mai, M. Scherer, H-extendible copulas, working paper (2011b).

J.-F. Mai, M. Scherer, Extendibility of Marshall–Olkin distributions and inverse Pascal triangles, forthcoming in *Brazilian Journal of Probability and Statistics* (2011c).

J.-F. Mai, M. Scherer, Sampling exchangeable and hierarchical Marshall–Olkin distributions, forthcoming in *Communications in Statistics: Theory and Methods* (2011d).

J.-F. Mai, M. Scherer, Bivariate extreme-value copulas with discrete Pickands dependence measure, *Extremes* **14**(3) (2011e) pp. 311–324.

J.-F. Mai, M. Scherer, R. Zagst, CIID frailty models and implied copulas, working paper (2011).

K.V. Mardia, Multivariate Pareto distributions, *The Annals of Mathematical Statistics* **33** (1962) pp. 1008–1015.

G. Marsaglia, Choosing a point from the surface of a sphere, *The Annals of Mathematical Statistics* **43**(2) (1972) pp. 645–646.

G. Marsaglia, The Marsaglia random number CD-ROM including the Diehard battery of tests of randomness, available at *http://stat.fsu.edu/pub/diehard* (1996).

G. Marsaglia, Xorshift RNGs, *Journal of Statistical Software* **8**(14) (2003) pp. 1–6.

G. Marsaglia, Evaluating the normal distribution, *Journal of Statistical Software* **11**(4) (2004) pp. 1–11.

A.W. Marshall, I. Olkin, A multivariate exponential distribution, *Journal of the American Statistical Association* **62**(317) (1967) pp. 30–44.

A.W. Marshall, I. Olkin, Families of multivariate distributions, *Journal of the American Statistical Association* **83**(403) (1988) pp. 834–841.

M. Matsumoto, T. Nishimura, Mersenne twister: A 623-dimensionally equidistributed uniform pseudo-random number generator, *ACM Transactions on Modeling and Computer Simulation* **8**(1) (1998) pp. 3–30.

M. Matsumoto, M. Saito, SIMD-oriented fast Mersenne twister: A 128-bit pseudorandom number generator, in *Monte Carlo and Quasi-Monte Carlo Methods 2006*, ed. A. Keller, S. Heinrich, H. Niederreiter, Springer, Berlin (2008) pp. 607–622.

A.J. McNeil, Sampling nested Archimedean copulas, *Journal of Statistical Computation and Simulation* **78**(6) (2008) pp. 567–581.

A.J. McNeil, R. Frey, P. Embrechts, *Quantitative Risk Management*, Princeton University Press, Princeton, New Jersey (2005).

A.J. McNeil, J. Nešlehová, Multivariate Archimedean copulas, d-monotone functions and l_1-norm symmetric distributions, *Annals of Statistics* **37**(5B) (2009) pp. 3059–3097.

N. Metropolis, S. Ulam, The Monte Carlo method, *Journal of the American Statistical Association* **44** (1949) pp. 335–341.

J.R. Michael, W.R. Schucany, R.W. Haas, Generating random variates using transformations with multiple roots, *The American Statistician* **30**(2) (1976) pp. 88–90.

P. Mikusinski, H. Sherwood, M. Taylor, The Fréchet bounds revisited, *Real Analysis Exchange* **17** (1992) pp. 759–764.

D.S. Moore, M.C. Spruill, Unified large-sample theory of general chi-squared statistics for tests of fit, *Annals of Statistics* **3**(3) (1975) pp. 599–616.

O. Morales-Nápoles, Counting vines, in *Dependence modeling: Vine copula handbook*, ed. D. Kurowicka, H. Joe, World Scientific, Singapore (2010).

B. Moro, The full Monte, *Risk Magazine* **8**(2) (1995) pp. 57–58.

R. Moynihan, On τ_T semigroups of probability distribution functions II, *Aequationes Mathematicae* **17** (1978) pp. 19–40.

A. Müller, D. Stoyan, *Comparison Methods for Stochastic Models and Risks*, Wiley, Chichester (2002).

M.E. Muller, A note on a method for generating points uniformly on N-dimensional spheres, *Communications of the ACM* **2** (1959) pp. 19–20.

C.H. Müntz, Über den Approximationssatz von Weierstrass, *Festschrift H.A. Schwarz* (1914) pp. 303–312.

National Institute of Standards and Technology, USA, available at *http://csrc.nist.gov/rng*.

R.B. Nelsen, Dependence and order in families of Archimedean copulas, *Journal of Multivariate Analysis* **60**(1) (1997) pp. 111–122.

R.B. Nelsen, *An Introduction to Copulas*, second edition, Springer, New York (2006).

J. von Neumann, Various techniques used in connection with random digits, in *Monte Carlo Method*, ed. A.S. Householder, G.E. Forsythe, H.H. Germond, National Bureau of Standard Series, Vol. 12, US Government Printing Office, Washington, DC (1951) pp. 36–38.

J.P. Nolan, Numerical calculation of stable densities and distribution functions, *Communications in Statistics: Stochastic Models* **13**(4) (1997) pp. 759–774.

G.L. O'Brien, The comparison method for stochastic processes, *Annals of Probability* **3** (1975) pp. 80–88.

D. Oakes, Multivariate survival distributions, *Journal of Nonparametric Statistics* **3**(3-4) (1994) pp. 343–354.

F. Oberhettinger, L. Badii, *Tables of Laplace Transforms*, Springer, Berlin (1973).

A.J. Patton, Estimation of multivariate models for time series of possibly different lengths, *Journal of Applied Econometrics* **21**(2) (2006) pp. 147–173.

W.R. Pestman, *Mathematical Statistics*, de Gruyter, Berlin (1998).

J. Pickands, Multivariate extreme value distributions, *Proceedings of the 43rd Session ISI, Buenos Aires* (1981) pp. 859–878.

A.P. Prudnikov, Y.A. Brychkov, O.I. Marichev, *Integraly i Rjady (Integrals and Series)*, Nauka, Moskow (1981).

S.I. Resnick, *Extreme Values, Regular Variation and Point Processes*, Springer, Berlin (1987).

P. Ressel, Monotonicity properties of multivariate distribution and survival functions with an application to Lévy-frailty copulas, *Journal of Multivariate Analysis* **102**(3) (2011) pp. 393–404.

M.L. Rizzo, *Statistical Computing with R*, Chapman and Hall, London (2007).

C. Ruiz-Rivas, C.M. Cuadras, Inference properties of a one-parameter curved exponential family of distributions with given marginals, *Journal of Multivariate Analysis* **27**(2) (1988) pp. 447–456.

L. Rüschendorf, Stochastically ordered distributions and monotonicity of the OC-function of sequential probability ratio tests, *Mathematische Operationsforschung und Statistik Series Statistics* **12**(3) (1981) pp. 327–338.

L. Rüschendorf, On the distributional transform, Sklar's theorem, and the empirical copula process, *Journal of Statistical Planning and Inference* **139** (2009) pp. 3921–3927.

G. Samoronitska, M.S. Taqqu, *Stable non-Gaussian Random Processes: Stochastic Models with Infinite Variance*, CRC Press, Boca Raton, Florida (1994).

K.-I. Sato, *Lévy Processes and Infinitely Divisible Distributions*, Cambridge University Press, Cambridge (1999).

M. Scarsini, On measures of concordance, *Stochastica* **8** (1984) pp. 201–218.

U. Schepsmeier, Maximum likelihood estimation of C-vine pair-copula constructions based on bivariate copulas from different families, Diploma thesis, Technische Universität München (2010).

R.L. Schilling, R. Song, Z. Vondracek, *Bernstein Functions*, De Gruyter, Berlin (2010).

F. Schmid, R. Schmidt, Bootstrapping Spearman's multivariate rho, *Proceedings in Computational Statistics, ed. A. Rizzi, M. Vichi, Springer, Heidelberg* (2006) pp. 759–766.

F. Schmid, R. Schmidt, Multivariate conditional versions of Spearman's rho and related measures of tail dependence, *Journal of Multivariate Analysis* **98**(6) (2007a) pp. 1123–1140.

F. Schmid, R. Schmidt, Nonparametric inference on multivariate versions of Blomqvist's beta and related measures of tail dependence, *Metrika* **66**(3) (2007b) pp. 323–354.

R. Schmidt, Tail dependence for elliptically contoured distributions, *Mathematical Methods of Operations Research* **55** (2002) pp. 301–327.

V. Schmitz, Revealing the dependence structure between $X_{(1)}$ and $X_{(n)}$, *Journal of Statistical Planning and Inference* **123** (2004) pp. 41–47.

I.J. Schoenberg, Metric spaces and positive definite functions, *Transactions of the American Mathematical Society* **44** (1938) pp. 522–536.

P.J. Schönbucher, *Credit Derivatives Pricing Models*, Wiley, New York (2003).

U. Schöning, *Algorithmik*, Spektrum Akademischer Verlag GmbH, Heidelberg (2001).

W. Schoutens, *Lévy Processes in Finance: Pricing Financial Derivatives*, Wiley Series in Probability and Statistics, Wiley, Chichester (2003).

B. Schweizer, A. Sklar, Associative functions and abstract semigroups, *Publicationes Mathematicae Debrecen* **10** (1963) pp. 69–81.

B. Schweizer, A. Sklar, *Probabilistic Metric Spaces*, North-Holland/Elsevier, New York (1983).

V. Seshadri, *The Inverse Gaussian Distribution: A Case Study in Exponential Families*, Oxford University Press, New York (1993).

J. Shuster, On the Inverse Gaussian distribution, *Journal of the American Statistical Association* **63**(324) (1968) pp. 1514–1516.

A. Sklar, Fonctions de répartition à n dimensions et leurs marges, *Publications de l'Institut de Statistique de L'Université de Paris* **8** (1959) pp. 229–231.

O. Szász, Über die Approximation stetiger Funktionen durch lineare Aggregate von Potenzen, *Mathematische Annalen* **77**(4) (1916) pp. 482–496.

K. Takahasi, Note on the multivariate Burr's distribution, *Annals of the Institute of Statistical Mathematics* **17** (1965) pp. 257–260.

F.G. Tricomi, On the theorem of Frullani, *American Mathematical Monthly* **58**(3) (1951) pp. 158–164.

R. Weron, On the Chambers-Mallows-Stuck method for simulating skewed stable random variables, *Statistics and Probability Letters* **28** (1996) pp. 165–171.

D.V. Widder, *The Laplace Transform*, second edition, Princeton University Press, Princeton, New Jersey (1946).

R.E. Williamson, Multiply monotone functions and their Laplace transforms, *Duke Mathematical Journal* **23**(2) (1956) pp. 189–207.

V. Witkovský, Computing the distribution of a linear combination of inverted gamma variables, *Kybernetika* **37**(1) (2001) pp. 79–90.

S.J. Wolfe, On moments of probability distribution functions, *Lecture Notes in Mathematics, Vol. 457*, Springer, Berlin (1975) pp. 306–316.

E.F. Wolff, N-dimensional measures of dependence, *Stochastica* **4**(3) (1980) pp. 175–188.

G.U. Yule, M.G. Kendall, *An Introduction to the Theory of Statistics*, 14th edition, Charles Griffin and Company, London (1965).

Index